THE DETERMINATION
OF TRACE METALS
IN NATURAL WATERS

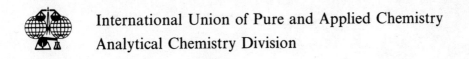

International Union of Pure and Applied Chemistry
Analytical Chemistry Division

THE DETERMINATION
OF TRACE METALS
IN NATURAL WATERS

EDITED BY

T. S. WEST
The Macaulay Institute
Aberdeen, UK

The Late H. W. NÜRNBERG
Institut für Chemie
der Kernforschungsanlage
Julich, GmbH
Federal Republic of Germany

BLACKWELL SCIENTIFIC PUBLICATIONS

OXFORD LONDON EDINBURGH

BOSTON PALO ALTO MELBOURNE

© 1988 International Union of Pure and
Applied Chemistry
and published for them by
Blackwell Scientific Publications
Editorial Offices:
Osney Mead, Oxford OX2 0EL
  (*Orders*: Tel. 0865 240201)
8 John Street, London WC1N 2ES
23 Ainslie Place, Edinburgh EH3 6AJ
3 Cambridge Center, Suite
  208, Cambridge, Massachusetts 02142, USA
667 Lytton Avenue, Palo Alto
  California 94301, USA
107 Barry Street, Carlton
  Victoria 3053, Australia

First published 1988

Set by Macmillan India Ltd
Printed and bound in Great Britain by
Butler & Tanner Ltd, Frome and London

DISTRIBUTORS

USA and Canada
  Blackwell Scientific Publications Inc
  P O Box 50009, Palo Alto
  California 94303
  (*Orders*: Tel. (415) 965–4081)

Australia
  Blackwell Scientific Publications
  (Australia) Pty Ltd
  107 Barry Street
  Carlton, Victoria 3053
  (*Orders*: Tel. (03) 347 0300)

British Library
Cataloguing in Publication Data
The Determination of trace metals in natural
  waters.
  1. Water—Analysis 2. Trace elements
  —Analysis 3. Chemistry, Analytic
  I. West, T. S.   II. Nurnberg, H. W.
  III. International union of Pure and
  Applied Chemistry, *Analytical Chemistry
  Division*
  551.4   QD142

  ISBN 0-632-02021-0

Library of Congress
Cataloging-in-Publication Data
The Determination of trace metals in
  natural waters.
  At head of title: International Union of
  Pure and Applied Chemistry,
  Analytical Chemistry Division.
  Includes index.
  1. Water—Analysis. 2. Trace elements in
  water.
  I. West, T. S. (Thomas Summers)
  II. Nürnberg, H. W. III. International
  Union of Pure and Applied Chemistry.
  Analytical Chemistry Division.
  QD142.D48 1988 546′.226 87-27640

  ISBN 0-632-02021-0

Contents

vi   CONTENTS

ix CONTENTS

8 Adsorption of Trade Elements by Suspended Particulate Matter in Aquatic Systems

A. C. M. BOURG

List of Contributors

G. Ackermann, *Lehrstuhl für Analytische Chemie, Section Chemie, Bergakademie Freiberg, Leipziger Strasse, DDR-9200 Freiberg, German Democratic Republic*

S. Ahrland, *Department of Inorganic Chemistry 1, Chemical Centre, University of Lund, S-22100 Lund, Sweden*

G. den Boef, *Laboratorium voor Analytische Schiekunde, J.H. van't Hoff Institut, Universiteit van Amsterdam, Nieuwe Achtergracht 166, 1018WV Amsterdam, Netherlands*

A. C. M. Bourg, *Service Géologique National, Dept. EAU, Bureau de Recherches Géologiques et Minières, F-45060, Orélans Cedex 2, France*

J. A. C. Broekaert, *Institut für Spektrochemie u, Angewandte Spektroskopie, Postfach 778, Bunsen-Kirchhoff-Strasse 11, D-4600 Dortmund, Federal Republic of Germany*

J. Buffle, *Départment de Chimie Minérale Analytique et Appliquée de l'Université, Section Chemie—Sciences II, 30 quai Ernest-Ansermet, CH-1211 Genève 4, Switzerland*

D. Thorburn Burns, *Department of Chemistry (Analytical), Queen's University of Belfast, Stranmillis Road, Belfast, BT9 5AG, N. Ireland, UK*

L. R. P. Butler, *CSIR, PO Box 395, Pretoria 001, South Africa*

A. K. Covington, *Electrochemistry Research Laboratories, Department of Physical Chemistry, University of Newcastle-upon-Tyne, NE1 7PU, UK*

W. Davidson, *Freshwater Biological Association, The Ferry House, Ambleside, Cumbria, LA 22, UK*

A. Hulanicki, *Wydzial Chemii, Uniwersytet Warszawski, ul. L Pasteura 1, PL-02 093, Warszawa, Poland*

R. Jenkins, *International Centre for Diffraction Data, 1601 Park Lane, Swarthmore, PA. 19081, USA*

K. Laqua, *Institut für Spektrochemie u. Angewandle, Spektroskopie, Postfach 778, Bunsen-Kirchhoff-Strasse 11, D-4600, Dortmund, Federal Republic of Germany*

L. Mart, *Institut für Chemie der Kernforschungsanlage Julich, GmbH, Institut 4: Angewandle Physikalische Chemie, D-5170, Julich 1, Federal Republic of Germany*

W. H. Melhuish, *Institute of Nuclear Sciences, DSIR, Private Bag, Lower Hutt, New Zealand*

J. W. Murray, *School of Oceanography, University of Washington, Seattle, WA 98195, USA*

The late H. W. Nürnberg, *Institut für Chemie Kernforschungsanlage Julich, GmbH, Institut 4: Angewandle Physikalische Chemie, D-5170, Julich 1, Federal Republic of Germany*

F. Pellerin, *Hopital Maison de Cure Medicale, Corentin-Celton, 37, Boulevard Gambetta, 92133 Issy-Les-Moulinex, France*

E. Plško, *Geologicky Ulstav, Universzity Komenského, Zadunajská 15, CD-851 01, Bratislava, Czechoslovakia*

J. Robin, *Laboratoire de Physicochemie Industrielle, INSA de Lyon, 20 Av Albert Einstein, F-69621, Villeurbanne Cédex, France*

I. Rubeska, *UNDP, Box 7285 ADC, Passay City, Metro Manila, Philippines*

S. B. Savvin, *V.I. Vernadskii Institute of Geochemistry and Analytical Chemistry, Academy of Sciences of USSR, Kosygin St., 19, SU-117975, SZD-1, Moscow, V-334, USSR*

L. Sommer, *Katedra Analytiché Chemie, Prirocovedecke Fakulty Univerzity J.E. Parkyne, Kotlarska 2, CS-611 37 Brno, Czechoslovakia*

E. Steinnes, *Department of Chemistry, University of Trondheim, 7055 Dragvoll, Norway*

W. I. Stephen, *Chemistry Department, University of Birmingham, Birmingham B15 2TT, UK*

A. Strasheim, *Chemistry Department, The University, Pretoria 0002, South Africa*

A. M. Ure, *Department of Spectrochemistry, Macaulay Institute, Craigiebuckler, Aberdeen, AB9 2QJ, UK*

E. Wänninen, *Laboratory of Analytical Chemistry, Åbo Akademi, Bishopsgatan 8, SF-20500, Abo 50, Finland*

T. S. West, *Macaulay Institute, Craigiebuckler, Aberdeen, AB9 2QJ, UK*

P. D. Whalley, *Freshwater Biological Association, The Ferry House, Ambleside, Cumbria, LA 22, UK*

M. Whitfield, *Marine Biological Association, The Laboratory, Citadel Hill, Plymouth PL1 2PB, UK*

Editor's Preface

The idea of this book on the determination of ten biosignificant trace metals in natural waters was evolved by the Analytical Chemistry Division of the International Union of Pure and Applied Chemistry several years ago. It rapidly became a 'whole Division' project involving chemists from many countries although the authors of the present 10 sections are necessarily limited in number. Professor Pellerin, President of the Division from 1981–85, has set out the *raison d'être* of the book in his introduction and no further comment is, therefore, required.

The work was planned to conclude and be presented at the 1985 General Assembly of IUPAC under the joint editorship of Professor H. W. Nürnberg, and the writer, with the objective of publication in early 1986, but the untimely death of Professor Nürnberg in spring 1985 and other circumstances subsequently made this target date impossible.

Sampling for analysis

Analysis depends critically on the acquisition of a sample that is truly representative of the material to be analysed. At first sight it might appear that few things would be easier to sample than water, but even this is fraught with difficulties. In addition, at the trace element level, there are serious problems that arise from significant contamination during sampling and storage due to the presence of extraneous materials, including the sampling device and the storage vessel and, equally, a serious risk of loss of analyte due to adsorption, vaporisation, biological, chemical and other physical changes. However, sampling is a problem that is common to all the trace element procedures whether they be electroanalytical, radiochemical, spectrochemical, etc. For this reason **sampling techniques are dealt with in detail in Section 3 only**, although the essential principles of sampling, storage and transportation are mentioned very briefly in the introduction to Section 2.

Section 1 on spectrophotometric methods is given at the level of practical detail as in a laboratory manual, since such methods are likely to be applicable in most laboratories and for this reason the section is arranged element by element and on an individual author basis. The material in this chapter was co-ordinated on behalf of the members of Commission V.1 by the Secretary of Division V, G. den Boef with the help of A. Hulanicki and D. T. Burns.

Following this, spectrochemical methods such as atomic absorption spectrometry, which is also generally available and widely used is discussed in Section 2 along with other more specialized techniques of spectrometry.

Section 3, apart from dealing with sampling techniques for trace metals in sea water, flowing streams, sedimented waters, etc., discusses electrochemical techniques in a fair degree of detail and partly on an individual element basis because electrochemical techniques are also widely available. Section 4 is on general aspects of trace metal analysis by neutron activation methods, but in much less detail and not on an individual element basis. Section 5 deals with pH measurement in natural waters – a considerable problem, but one of fundamental importance. Speciation in general, which is of profound importance for the chemical, biological and even physical reactivity of trace metals is discussed in considerable detail in the next five chapters, *viz.*: Electrochemical measurement of speciation (6), complexation by inorganic ligands in sea water (7), complexation by suspended particulate matter (8), speciation in oxygenated estuarine waters (9), chemical mechanisms in sea water (10).

The trace metals whose analyses are discussed in this book are those of principal biosignificance either from the viewpoint of their essentiality to various life-forms or of their toxicity, namely

Cd, Co, Cr, Cu, Hg, Mo, Ni, Pb, Se, Zn.

It is appreciated that there are, of course, many other biosignificant trace elements and indeed some of them are also referred to, e.g. Fe, Mn and V, in the texts of the various sections.

T. S. WEST

Introduction

Professor F. PELLERIN
President of the Analytical Chemistry Division 1981–85

Each field of chemistry is confronted with the problem of monitoring of water quality and research on trace chemical compounds. Obtaining perfectly pure water is a major problem; the chemistry of solutions and the laws and mechanisms of reactions were for a long time limited to aqueous solutions before non-aqueous solvents were developed. But water is also intimately linked to life – it is even the *sine qua non* of life – and has a considerable biological and socio-economic importance, which makes the protection and treatment and, therefore, the quality control, of natural waters a matter of great general interest on a global scale. Water analysis gives rise to a variety of tasks calling for the most modern means of production and the most elaborate analytical techniques which benefit daily and immediately from scientific and methodological advances. Amongst the many problems to be solved, that of research on trace metals arises every day and has many aspects, all of which call for the same types of analytical methods.

Underground waters that are well protected from pollution nevertheless include inorganic trace elements that are much sought after for their curative properties, but the consumption of such waters is prohibited because of the risk of poisoning. Waters circulating on the surface of the globe or in oceans polluted by biological and industrial wastes must undergo considerable purification, which implies the determination of trace concentrations at nanogram or picogram levels; such levels are, nevertheless, sometimes capable of modifying biological equilibria, ecosystems and our environment irreversibly. Trace elements intervene in biological processes and enzyme reactions, enhancing or inhibiting them. In everyday life, drinking water itself is subject to a particularly careful and continuous analysis, in view of the many risks of pollution. Trace metals dumped in metallic or plastic containers can contaminate or pollute foodstuffs; also, in the field of therapeutics, traces of aluminium in water used for haemodialysis can cause brain disorders.

Whatever the specialization of scientists working on the determination of trace elements in natural waters, the collaboration of the analytical chemist is indispensable to provide the best suited methods from the various branches of analytical chemistry, and it is for this reason that the Division of Analytical Chemistry of IUPAC makes a particularly important contribution. In effect, the various Commissions of the Division address all aspects of modern analytical chemistry. The Division promotes chemical and instrumental methods of analysis (spectral, electrochemical, chromatographic). Its Commissions on Microchemical Techniques and Trace analysis, and on

Radiochemistry and Nuclear Techniques, undertake the development of trace analysis.

The recording, analysis and interpretation of data are the object of the work of other specialized Commissions. The Division and its Commissions thus have a research and analysis potential that can be, and is, applied to the detection of trace metals in natural waters.

This project, initiated in 1979 by T. S. WEST, President of the Division, involves the participation of most of the Commissions of the Division and is intended to offer chemists and related scientists a survey of available methods, showing their respective performances, their areas of application and, in general, information that will enable them to apply the data to the resolution of specific problems. In the field of the determination of trace metals in natural waters, the work to be covered was enormous and it was necessary to organize the method of work.

This book is dedicated to Professor Wolfgang NÜRNBERG who died suddenly in May 1985. From the start of the project, Professor NÜRNBERG showed great enthusiasm for coordinating the Commissions' work, linking the reports, harmonizing the format. At the same time he brought us his internationally renowned competence in the field of the determination of traces of elements as well as the experience of his own research work.

The work was carried out in two stages. The first stage consisted, in Commissions making an inventory of methods, giving their performances, precisions, sensitivities and limits, the possibilities for development and the types of application; in other words, the objective of this inventory was to guide the user in his choice of method. The outputs of the Commissions were compiled by the Co-ordinator. It was clear from the outset that it would constitute a major work.

The specialists in each Commission, selected the methods best suited to the analysis of natural waters on the basis of their personal experience. The choice of methods was not limited to the most sophisticated or those of highest performance; in many countries or in particular circumstances it might be necessary to have a simple analytical set-up.

Finally, the work as a whole is based on the expertise of all the members of the Division of Analytical Chemistry who, since 1979, in the various Commissions, have taken part in its preparation and have made an effort for which they should be thanked.

We hope that this manual will meet the needs of the many potential users for reliable guidelines and methods for the analysis of trace elements in natural waters, e.g. analytical chemists, hydrologists, oceanographers, toxicologists, ecologists, etc. involved in the quality control of natural waters and the detection of traces.

The role of the Division of Analytical Chemistry of IUPAC is to provide a clear picture based on an objective well-informed choice of available methods. This book on trace analysis in water, required analytical chemists not only for the analysis, but also for guidance in the choice of methods. We hope that it will render a service to analytical chemists in the solution of their specific problems and stimulate them to undertake and develop new research.

Section 1 Spectrophotometric and Fluorimetric Methods

Section Editors

Commission V.1

IUPAC

G. DEN BOEF

Laboratorium voor Analytische Schiekunde
J. H. van't Hoff Instituut
Universiteit van Amsterdam
Nieuwe Achtergracht 166
1018 WV Amsterdam
Netherlands

A. HULANICKI

Wydzial Chemii
Uniwersytet Warszawski
ul. L. Pasteura 1
PL-02 093 Warszawa
Poland

D. T. BURNS

Department of Chemistry (Analytical)
Queen's University of Belfast
Stranmillis Rd
Belfast, BT9 5AG,
N. Ireland, UK

1.1 Selenium

Prepared by
D. THORBURN BURNS

Department of Chemistry (Analytical), Queens University of Belfast, Stranmillis Rd, Belfast BT9 5AG, N. Ireland, UK

1.1.1 Introduction

Selenium is both an essential micronutrient and a toxic species [1]. The tolerance range between beneficial and toxic concentrations is narrow for many biological systems [2,3]. The concentration in natural surface and marine waters is low except where these are in contact with seleniferous soils or where there is pollution. A number of industries (glass, ceramics, electronics and xerography) emit selenium bearing wastes and hence robust, interference-free methods over wide concentration ranges are required to meet the variable analytical requirements of environmental monitoring [4].

The formal oxidation states of selenium are $-II$, 0, $+IV$ and $+VI$. The predominant oxidation state in natural waters is not well established and may vary with the conditions. Sillén [5] suggested, on thermodynamic grounds, that Se(VI) would be the dominant species in sea water, a view supported by certain experimental findings [6]. However, Chau and Riley [7] concluded that Se(IV) would be the most probable stable state under normal aerobic conditions and confirmed this for the Irish Sea and for the English Channel. More recently Yoshii *et al.* [8] Sugimura and Suzuki [9] and Shimoishi and Tôei [10] found both Se(IV) and Se(VI) in Japanese coastal sea waters. The very low levels of Se(IV) in certain British rivers could be accounted for by scavenging of Se(IV) by particulate hydrous iron oxide [11]. Data on speciation and the concentration levels in various waters have been critically reviewed by Robberecht and van Grieken [12]. Levels of selenium to be expected in natural waters are:

potable waters [13]; less than	$10 \ \mu g \ dm^{-3}$
seepage from seleniferous soils, up to	$500 \ \mu g \ dm^{-3}$
river waters	$10–350 \ ng \ dm^{-3}$
estuarine waters [11]; up to	$400 \ ng \ dm^{-3}$
sea waters [7]; up to	$50 \ ng \ dm^{-3}$

The stability of aqueous samples of Se(IV) and Se(VI) varies with the pH, type of water and type of container. Shendrikar and West [14] reported 2% loss over 15 days for 1 ppm Se, preserved with 0.5% nitric acid, stored in polyethylene bottles. Nitrate, however, interferes with the hydride generation method for selenium; adjustment to pH 1.5 with sulfuric acid provides optimum conditions of preservation, up to 125 days for Se(IV) and Se(VI), and does not interfere in hydride generation analysis [15].

1.1.2 Spectrophotometric and fluorimetric determination of selenium

1.1.2.1 General aspects

Spectrophotometric methods are usually specific for Se(IV) and hence also for Se(VI) and total Se can be determined by difference after selective oxidation/reduction. The most important classes of reagents are the o-diamines and sulfur-containing ligands [16]. The o-diamines react to form piazselenols which may be measured spectrophotometrically [17], fluorimetrically [18] or by GLC which is very sensitive by electrocapture detection using, for example, dibromo-derivatives [10].

3,3'-diaminobenzidine introduced by Hoste and Gillis [17] is the most popular spectrophotometric reagent for selenium. The coloured product is the monopiazselenol [18] not the dipiazselenol as assumed by Hoste and Gillis [17] and later by Cheng [19], who improved the method by the addition of EDTA to mask interfering ions and by extraction of the complex piazselenol into toluene. Further selectivity may be achieved by the introduction of a distillation of selenium tetrabromide [20]. For the determination of selenium in fertilizers the S.A.C. (Society for Analytical Chemistry) recommends a further separation, extraction with zinc dithiol [21]. The procedure described below for total selenium is based on the well tried APHA (American Public Health Association) method [13]; should increased sensitivity be required the extracts may be examined by spectrofluorimetry.

Se(IV) may be determined in the presence of other forms of selenium using diaminobenzidines [10]. *N.B. These substances are carcinogens.* The difference between total dissolved selenium and available selenium(IV) represents an operationally defined fraction which could include several forms of selenium including organically bound forms [11]. Selenium(IV) may be coprecipitated with iron(III) hydroxide which provides speciation and concentration suitable for use in determination of selenium in sea water [7, 8]. Se(IV) is also selectively adsorbed on hydrous aluminium oxide [7,22]. Selective reduction of Se(IV) and Se(VI) to Se(o) followed by adsorption on activated charcoal prior to XRF analysis has been applied to a variety of water samples [23].

1.1.2.2 Procedures

1.1.2.2.1 Spectrophotometric determination of selenium using 3,3'-diaminobenzidine [13]

(i) *Principles of the method for total selenium*

Acid permanganate oxidation converts all inorganic selenium compounds to Se(VI) but many organic compounds are not completely oxidised. It is unlikely however that any selenium–carbon bonds will remain intact after this treatment. Inorganic selenium is oxidised by acid permanganate even in the presence of much greater concentrations of organic matter than would be anticipated in most water supplies.

Substantial losses of selenium occur when solutions of sodium selenate are evaporated to dryness, but in the presence of calcium all the selenium is recovered. An excess of calcium over sulfate is not necessary.

Selenium(VI) is reduced to selenium(IV) in warm 4 mol dm^{-3} hydrochloric acid solution. The temperature, time and acid concentrations are specified in order to obtain quantitative reduction without loss of selenium.

The optimum pH for the formation of piazselenol is approximately 1.5. Above pH 2 the rate of formation is critically dependent on pH. Careful pH adjustment is essential to achieve reproducible results; pH indicators are not satisfactory for this purpose. After formation of the piazselenol the pH is adjusted to above 6 when the distribution ratio of the piazselenol into organic extractants is almost independent of pH.

(ii) Interferences

No inorganic compounds give a positive interference. Coloured toluene-extractable organic compounds may exist, but it is unlikely that they would resist the initial acid permanganate oxidation. Negative interference results from the presence of compounds that oxidise the reagent. The addition of EDTA, under the conditions given, masks the effect of at least 2.5 mg iron(III). Manganese has no effect in reasonable concentrations. Iodide and, to a lesser extent, bromide cause low results as shown by the data in Table 1.1.

Table 1.1. Selenium recovery in the presence of bromide and iodide (25 μg Se)

Iodide (mg)	Selenium recovered (%)		
	Br (0 mg)	Br (1.25 mg)	Br (2.50 mg)
0	100	100	96
0.5	95	94	95
1.25	84	80	—
2.50	75	—	70

(iii) Limit of detection

The piazselenol is an intense yellow complex with absorption maxima at 340 and 420 nm, the molar absorptivities in toluene solution are 2.89×10^4 and 1.02×10^4 dm^3 mol^{-1} cm^{-1}, respectively. The limit of detection is 1 μg Se with a 4 cm cell at 420 nm.

(iv) Reagents

Stock selenium solution, 1.00 g dm^{-3}. Accurately weigh 1.000 g of analytical grade selenium into a small beaker. Add 5 cm^3 concentrated nitric acid and warm until reaction is complete; cautiously evaporate to dryness. Transfer quantitatively to a 1 dm^3 volumetric flask and dilute to volume with distilled water.

Standard selenium solution, 1.00 mg dm^{-3}. Dilute 1.00 cm^3 of stock solution to 1 dm^3 with distilled water. Prepare fresh daily.

Methyl orange indicator solution. Dissolve 500 mg in distilled water and dilute to 1 dm^3.

Hydrochloric acid solutions; 0.1 mol dm^{-3}, concentrated*, and $1:1$**.

Calcium chloride solution. Dissolve 30 g $CaCl_2.2H_2O$ in distilled water and dilute to 1 dm^3.

Potassium permanganate solution, 0.02 mol dm^{-3}. Dissolve 3.2 g $KMnO_4$ in 1 dm^3 of distilled water.

Sodium hydroxide, 0.1 mol dm^{-3}.

Ammonium chloride solution*. Dissolve 250 g NH_4Cl in 1 dm^3 of distilled water.

EDTA–sulfate reagent*. Dissolve 100 g disodium EDTA. $2H_2O$ plus 200 g sodium sulfate in 1 dm^3 distilled water. Add concentrated ammonia solution dropwise whilst stirring until dissolution is complete.

Ammonia solutions. 5 mol dm^{-3}, dilute concentrated ammonia solution $1:2$ with distilled water; concentrated**.

Diaminobenzidine solution: CAUTION HANDLE THIS REAGENT WITH EX-TREME CARE [24]. Dissolve 100 mg 3,3'-diaminobenzidine hydrochloride in 10 cm^3 distilled water. Prepare no more than 8 h before use as this solution is unstable.

Toluene.

Sodium sulfate. Na_2SO_4 anhydrous. Only required if centrifuge is not available.

Acid potassium bromide reagent**. Dissolve 10 g potassium bromide in 25 cm^3 distilled water. Cautiously add 25 cm^3 concentrated sulfuric acid, mixing and cooling under the tap after each $2–3 \text{ cm}^3$ increment of acid is added. This solution must be prepared immediately before use since potassium hydrogen sulfate will precipitate on cooling. Reheating to dissolve this salt will drive off some hydrogen bromide.

Hydrogen peroxide**. 30% w/v.

Phenol solution**. Dissolve 5 g phenol in 100 cm^3 distilled water.

(v) *Procedure A—Direct determination without distillation*

Prepare standards containing 0, 10.0, 25.0 and 50.0 μg Se in 500 cm^3 conical flasks. Dilute to approximately 250 cm^3 with distilled water, add 10 drops methyl orange indicator solution, 2 cm^3 0.1 mol dm^{-3} hydrochloric acid, 5 cm^3 calcium chloride solution, 3 drops 0.02 mol dm^{-3} potassium permanganate solution and 5 cm^3 of clean glass beads to prevent bumping. Boil vigorously for 5 minutes.

To 1 dm^3 water sample in a 2 dm^3 beaker add 10 drops of methyl orange indicator solution. Titrate to the methyl orange end-point with 0.1 mol dm^{-3} hydrochloric acid and add 2 cm^3 excess. Add 3 drops

* Reagents used in Procedure A only.
** Reagents used in Procedure B only.

$0.02 \, \text{mol dm}^{-3}$ potassium permanganate solution, $5 \, \text{cm}^3$ calcium chloride solution and $5 \, \text{cm}^3$ of clean glass beads. Heat to boiling, add further potassium permanganate, as required, to maintain a purple tint. Any precipitate of manganese(IV) oxide can be ignored as it will have no adverse effects. Reduce the volume to $250 \, \text{cm}^3$ and quantitatively transfer to a $500 \, \text{cm}^3$ conical flask.

Add $5 \, \text{cm}^3$ of $0.1 \, \text{mol dm}^{-3}$ sodium hydroxide solution to each flask and evaporate to dryness. Avoid prolonged heating of the residues.

Cool the flasks, add $5 \, \text{cm}^3$ concentrated hydrochloric acid and $10 \, \text{cm}^3$ ammonium chloride solution to each. Heat in a boiling water or steam bath for 10 ± 0.5 minutes.

Transfer the warm solutions and any ammonium chloride precipitate to $100 \, \text{cm}^3$ beakers and wash out the flasks with $5 \, \text{cm}^3$ EDTA–sulphate reagent and $5 \, \text{cm}^3$ $5 \, \text{mol dm}^{-3}$ ammonia solution. Adjust the pH to 1.5 ± 0.3 with ammonia solution using a pH meter. Any precipitate of EDTA will not interfere. Add $1 \, \text{cm}^3$ diaminobenzidine reagent by means of a safety pipette and heat in a boiling water or steam bath for about 5 minutes.

Before extraction of piazselenol, cool the beakers and contents, add concentrated ammonia solution and adjust the pH to 8 ± 1. Any precipitated EDTA will dissolve at this stage. Pour the sample and standard solutions into $50 \, \text{cm}^3$ graduated cylinders and adjust the volumes to $50 \pm 1 \, \text{cm}^3$ with washing from the beakers. Pour the contents of the cylinders into $250 \, \text{cm}^3$ separating funnels, add $10 \, \text{cm}^3$ toluene to each and shake for 30 ± 5 s. Allow to separate, discard the aqueous layers and transfer the organic phases to 12 or $13 \, \text{cm}^3$ centrifuge tubes. Centrifuge briefly to clear the toluene extracts of water droplets. If a centrifuge is not available filter the organic phases through dry filter papers to which 0.1 g anhydrous sodium sulfate has been added.

Determine the absorbances at 420 nm using a spectrophotometer or filter colorimeter with a violet filter (maximum transmittance near 420 nm) using 1 or 4 cm glass cells as appropriate. The piazselenol colour is stable, but any evaporation of toluene will concentrate the colour. Beer's law is obeyed up to 50 μg Se.

Precision and accuracy. A synthetic unknown sample containing 20 μg dm^{-3} Se, 40 μg dm^{-3} As, 250 μg dm^{-3} Be, 240 μg dm^{-3} B and 6 μg dm^{-3} V in distilled water was analysed in 35 laboratories by the diaminobenzidine method, with a relative standard deviation of 21.2% and a relative error of 5.0% [13].

The direct method is preferred in the absence of a large amount of iodide since it takes less time.

(vi) *Procedure B—Determination with distillation*

Selenium may be quantitatively separated from most other elements by distillation of the volatile selenium tetrabromide from an acid solution containing bromine. In the recommended method the bromine is generated by reaction of bromide with hydrogen peroxide. This procedure avoids the hazards of dealing and dispensing elemental bromine. Selenium tetrabromide and a small excess of bromine is adsorbed in water. The excess bromine is removed by precipitation as tribromophenol and selenium(IV) is determined with 3,3'-diaminobenzidine as before. No substances are known to interfere.

The procedure of determination follows the scheme given for Procedure A until addition of 5 cm^3 of 0.1 mol dm^{-3} sodium hydroxide and evaporation to dryness. To each cooled distillation flask (Fig. 1.1) add 50 cm^3 acid potassium bromide. Add 1 cm^3 30% w/v hydrogen peroxide and immediately fit the flask to the condenser through which cold water is running.

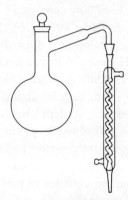

Fig. 1.1. Distillation apparatus.

Immerse the tip of the condenser in a 100 cm^3 beaker containing 20 cm^3 distilled water. Distil in a fume cupboard until the colour of bromine is gone from the flask. Wash out the distillate remaining in the condenser into the beaker with 5 cm^3 distilled water.

Add phenol solution to the distillate until the colour of bromine is discharged. A white precipitate of tribromophenol will form, any yellow tetrabromophenol formed does not interfere. Alternatively [25] the excess bromine may be removed by addition of hydroxylammonium chloride. Adjust the pH to 1.5±0.3 using concentrated ammonia or hydrochloric acid as appropriate. Add 1 cm^3 diaminobenzidine reagent by means of a safety pipette and heat in a boiling water or steam bath for about 5 minutes.

The extraction of piazselenol and measurement of absorbance is carried out as for direct determination. Absorbance readings must be made within 2 h after extraction because in the presence of phenol the solutions slowly acquire a green tint.

1.1.2.3 Spectrofluorimetric determination of selenium using 3,3'-diaminobenzidine

The toluene extract of the piazselenol may be examined fluorimetrically [18] (λ_{ex} 420 nm, λ_{em} 606 nm) with an increase in sensitivity compared to the absorbance measurement, the increase in sensitivity varies with the instrument; simple filter instruments may provide 40 fold reduction in the limit of detection [21]. 2,3-diaminonaphthalene is more sensitive but subject to greater interference [26]. It has been applied to the analysis of lake sediments [27], clear waters [28], effluent streams [29] and sea water [30].

1.1.3   References

1 W.C. Cooper and R.A. Zingara (Eds), *Selenium*, Van Nostrand, Reinhold, New York, (1974).
2 D.M. McKown and J. Morris, *J. Radioanal. Chem.*, **43**, 411 (1978).
3 J.E. Oldfield, W.A. Alloway, H.A. Laitinen, H.W. Lakin and O.H. Muth, *Geochemistry and the Environment*, Vol. I, National Academy of Sciences, Calif., pp. 57–63 (1972).
4 Committee on Medical and Biological Effects of Environmental Pollutants, *Selenium*, National Academy of Sciences Pub., Washington, pp. 28–50 (1976).
5 L.G. Sillén, Svensk Kem. Tidskr., **75**, 161 (1963).
6 Y. Sugimura, Y. Suzuki and Y. Miyake, *J. Oceanogr. Soc. Japan*, **32**, 235 (1976).
7 Y.K. Chau and J.P. Riley, *Anal. Chim. Acta*, **33**, 36 (1965).
8 O. Yoshii, K. Hiraki, Y. Nishikawa and T. Shigematsu, *Bunseki Kagaku*, **26**, 91 (1977).
9 Y. Sugimura and Y. Suzuki, *J. Oceanogr. Soc. Japan*, **33**, 23 (1977).
10 Y. Shimoishi and K. Tôei, *Anal. Chim. Acta*, **100**, 65 (1978).
11 C.I. Measures and J.D. Burton, *Nature*, **273**, 293 (1978).
12 H. Robberecht and R. van Grieken, *Talanta*, **29**, 823 (1982).
13 A.E. Greenberg, J.J. Connors, D. Jenkins and M.A. Franson (Eds), *Standard Methods for the Examination of Water and Waste-water*, 15th edn American Public Health Association, Washington, (1984).
14 A.D. Shendrikar and P.W. West, *Anal. Chim. Acta*, **74**, 189 (1975).
15 V. Cheam and H. Agemian, *Anal. Chim. Acta*, **113**, 237 (1980).
16 W.J. Williams, *Handbook of Anion Determination*, Butterworth, London, (1979).
17 J. Hoste and J. Gillis, *Anal. Chim. Acta*, **12**, 158 (1955).
18 C.A. Parker and L.G. Harvey, *Analyst*, **86**, 54 (1961).
19 K.L. Cheng, *Anal. Chem.*, **28**, 1738 (1956).
20 J.R. Rossum and P.A. Villarruz, *J. Am. Water Works Assn*, **54**, 746 (1962).
21 N.W. Hansen (Ed), *Official, Standardised and Recommended Methods of Analysis*, The Society for Analytical Chemistry, London, (1973).
22 D. Knab and E.S. Gladney, *Anal. Chem.*, **52**, 825 (1980).
23 H.J. Robberecht and R.E. van Grieken, *Anal. Chem.*, **52**, 449 (1980).
24 R.J. Lewis and R.L. Tatken (Eds), *Registry of Toxic Effects of Chemical Substances*, U.S. Dept. of Health and Human Services, Ohio (1971).
25 L. Bareza and L. Sommer, *Z. anal. Chem.*, **192**, 304 (1963).
26 C.A. Parker and L.G. Parker, *Analyst*, **87**, 558 (1962).
27 J.H. Wiersma and G.F. Lee, *Environ. Sci. Technol.*, **5**, 1203 (1971).
28 J.M. Rankin, *Environ. Sci. Technol.*, **7**, 823 (1973).
29 J.A. Raihle, *Environ. Sci. Technol.*, **6**, 621 (1972).
30 K. Hiraki, O. Yoshii, H. Hirayama, Y. Nishikawa and T. Shigematsu, *Bunseki Kagaku*, **22**, 712 (1973).

1.2   Chromium

Prepared by
E. WÄNNINEN
*Laboratory of Analytical Chemistry, Åbo Akademi, Bishopsgatan 8, SF 20500,
Åbo 50, Finland*

1.2.1   Introduction

Chromite ($FeO.Cr_2O_3$) is the only commercially important ore of chromium. The production in 1975 was about 7.9 million tons. The three principal uses of chromite are in metallurgy, refractory and chemical manufacture. The largest chemical uses are in tanning of leather and as pigments [1]. Chromium compounds are used as colouring agents in ceramic glazes and in glassmaking, as fungicides, wood preservatives and as rust inhibitors in cooling-tower recirculating water systems.

The only significant oxidation states of chromium in natural waters are (III) and (VI), the most probable species being $Cr(OH)_2(H_2O)_4^+$ and CrO_4^{2-}. In oxidizing waters, calculation predicts that the stable oxidation state is (VI) [2]. However, significant amounts of chromium(III) have been found. It is probable that chromium(III) is stabilized and solubilized in sea water by complex formation with compounds, such as amino-acids and polybasic organic acids e.g. citric acid [3]. The extent of complex formation of chromium(III) with organic compounds is affected by chloride in acidic conditions and by magnesium and calcium ions in alkaline conditions. From published data it was concluded that sea water (Japan Sea, Pacific Ocean) contains about $0.5 \ \mu g \ dm^{-3}$ dissolved chromium [4], consisting of $\sim 15\%$ inorganic chromium(III), $\sim 25\%$ inorganic chromium(VI) and $\sim 60\%$ organically-bound chromium. A model for the circulation of chromium in sea water has been developed [5]. Inorganic chromium(III) in marine sediments is not oxidized to chromium(VI) by dissolved oxygen, but by the catalytic action of MnOOH. Meanwhile chromium(VI) is reduced biologically after it has been taken up by certain organisms.

Data for the distribution of chromium in the environment are given in Table 1.2 and some industrially important inorganic chromium compounds are listed in Table 1.3.

It is known that trace amounts of chromium play a role in some metabolic processes; in high concentrations the element is toxic. The human nutritional need for chromium is established, but as yet not quantified. The major dietary sources of chromium are beer, meats and yeast. In humans, chromium compounds are essential for glucose metabolism. Chromium is used in the action of formation of insulin via an organic complex with nicotinic acid. Any chromium(VI) that is ingested is irreversibly reduced in the body to the (III) state.

The daily intake of chromium is estimated to be 0.03–0.14 mg. A brief

Table 1.2. Distribution of chromium in the environment [6]

	Range	Average
Crust	80–200 μg g^{-1}	125 μg g^{-1}
Clays	30–590 μg g^{-1}	120 μg g^{-1}
Soils	10–150 μg g^{-1}	40 μg g^{-1}
Coals	10–1000 μg g^{-1}	20–60 μg g^{-1}
Petroleum		0.3 μg g^{-1}
Sea water	1 μg dm^{-3}	
Rivers	1–10 μg dm^{-3}	
Municipal water	0–35 μg dm^{-3}	
Air in urban stations	0.01–0.03 μg m^{-3} (annual mean)	
Air in rural stations	0.01 μg m^{-3} (annual mean)	

Table 1.3. Some industrially important chromium compounds [1, 6]

Formula	Industrial use
$Cr(CH_3COO)_3$	tanning leather, catalyst
$CrBr_3$	in catalysts for polymerization of olefins
Cr_2O_3	pigment, latex paints, catalyst, dyeing polymers
CrO_2	high energy magnetic tapes
CrO_3	corrosion inhibitor, oxidant, catalyst, plating metals
$BaCrO_4$	anticorrosive pigment
$CaCrO_4$	corrosion inhibitor, depolarizer in batteries
$CuCrO_4$	in fungicides, seed protectants, wood preservatives
$K_2Cr_2O_7$	tanning leather, dyeing, painting, corrosion inhibitor, oxidizer
$PbCrO_4$	pigment
$ZnCrO_4$	in anticorrosive pigments

outline of chromium in the environment, in food, plants, animal and man, the role of chromium in metabolism and in biological interactions has been given by Vokal *et al.* [6]. The toxicology of chromium has also been discussed by Carter and Fernando [7].

The fundamental reaction chemistry of chromium has been discussed by Lingane [8]. The analytical chemistry of chromium has been reviewed by Chalmers [9] and Olsen and Foreback [10]. The spectrophotometric determination of chromium has been reviewed in detail by Snell [11].

1.2.2 Spectrophotometric determination of chromium

1.2.2.1 General aspects

Many spectrophotometric methods have been described for the determination of chromium(VI) and (III). Only a few of these methods are suitable for the determination of traces of chromium in natural waters (Table 1.4). Methods 1–6 involve reaction with chromium(VI), whereas 6–13 concern chromium(III). Methods 1, 2, 7 and 8 are the best from the point of view of sensitivity and selectivity and will be discussed in detail.

Table 1.4. Photometric determination of chromium

No. Reagent	Molar absorptivity $(dm^3\,mol^{-1}\,cm^{-1}$ $\times 10^{-4})$	Concentration range $(mg\,dm^{-3})$	Ref.
Chromium(VI)			
1 Diphenylcarbazide	4.0	0.01–0.9	See text
2 Iron(II)–FerroZine	8.3	0.1–0.6	12
3 Catalytic method		0.15–1.2	13
4 Holasol Violet R	2.1	~1	14
5 2,2'-Diquinoxalyl	3.3	0.05–1.0	15
6 Hexacyanoferrate(II)	11.2	0.05–0.7	16
Chromium(III)			
7 TAR	5.0	0.06–1.1	See text
8 PAR	4.7	0.1–0.9	See text
9 Chromazurol S	4.4	0.04–0.4	17
10 Xylenol orange	1.9	0.2–2	18
11 Methyl thymol blue	1.2	0.2–2	19
12 Eriochrome cyanine R	2.6	0.04–2	20
13 Kinetic method		0.1–1	21
14 PAN	1.3	0.3–2	22

In some determinations it is necessary to separate chromium prior to determination by either precipitation, solvent extraction, ion exchange, paper or thin layer chromatography, electrophoresis or coprecipitation [11]. Extraction with methyl isobutyl ketone from acid medium [27] is simple and selective [23]. Ion exchange has been used to eliminate the interfering ions in the determination of chromium in waste water by diphenylcarbazide [26].

Normally total chromium is determined photometrically after oxidation to the (VI) state using permanganate, persulfate, bismuthate or periodate in acid solution [25]. After reaction, the excess of the oxidant is destroyed, for example, by sodium azide [26]. Blundy [27] noted that permanganate oxidation did not give satisfactory results in combination with diphenylcarbazide whereas cerium(IV) was satisfactory. Sodium persulfate has also been used as an oxidant for chromium [28].

1.2.2.2   Procedures

1.2.2.2.1   Spectrophotometric determination of chromium with diphenylcarbazide

(i) *Introductory remarks*

The method using diphenylcarbazide has been used since 1900 [29,30]. The method is very sensitive and has been used in the determination of chromium in many matrixes such as chromium preservative-treated wood, minerals, soils and alloys [31–34].

Since the stability of the reagent solution is poor, it is best to prepare fresh solutions daily. Ethanol, acetone, and water (1:1), methanol, 2-propanol and methyl ethyl ketone have all been recommended as solvents and their effects on stability were studied [35]. The sensitivity of the method is found to decrease with the age of the reagent solution [36]. The most satisfactory solvent appears to be acetone or ethyl acetate. The purity of the commercially available diphenylcarbazide varies and influences the molar absorptivity. Values between 2.6×10^4 and 4.2×10^4 dm^3 mol^{-1} cm^{-1} have been reported. Examination of five samples indicates that there are two groups of reagents with molar absorptivities of 3.5×10^4 and 4.3×10^4 dm^3 mol^{-1} cm^{-1}, respectively [37].

The complex forming reaction with chromium(VI) is pH dependent and the optimum value is between 1 and 2.4 [38]. If the pH value is lower than 1 the maximum absorbance value is not reached and if the pH value is higher than 2.4 the maximum absorbance value is reached very slowly. It has been recommended [38] that the diphenylcarbazide solution should be acidified with sulfuric acid before the reaction with the chromium sample solution and the absorbance should be measured 10–15 minutes after mixing. Any excess of the reagent is not critical.

The reaction between chromium(VI) and diphenylcarbazide has been studied by Bose [39], Marchart [40] and Willems et al. [41]. Bose reported that chromium(VI) formed coloured complexes both with diphenylcarbazide and diphenylcarbazone, but that chromium(III) did not react. However, Pflaum and Howick [42] have shown that chromium(III) reacts with both diphenylcarbazide and diphenylcarbazone in dimethylformamide. Marchart [40] found that chromium(III) and diphenylcarbazone reacted in water, but only very slowly.

From these results the following reaction schemes were deduced for the reaction between chromium(VI) and diphenylcarbazide [41, 42]:

2Cr(VI) + 3 diphenylcarbazide = 2 Cr(III) + 3 diphenylcarbazone

Cr(III) + diphenylcarbazone = Cr(III)-diphenylcarbazone

The Cr(III)-diphenylcarbazone complex is responsible for the violet colour. Willems et al. [41] have confirmed that the product is a 1 : 1 complex.

If the amount of chromium(VI) is higher than that of the reagent, the violet colour develops, but after a short time the solution turns colourless [43]. Probably the excess of chromium(VI) oxidizes diphenylcarbazone to a product which does not form a coloured species.

(ii) *Limit of detection*

Fors [43] found the detection limit was about 0.01 mg dm^{-3} Cr in sea water when 1 cm cells were used. The Lambert–Beer's law was obeyed over the range 0.01–0.9 mg dm^{-3} Cr.

(iii) *Interferences*

Chromium(III) does not interfere, but iron(III), copper(II), mercury(II), molybdenum(VI) and vanadium(V) react with diphenylcarbazide and hence interfere in the determination. Iron can be masked with phosphoric acid or with EDTA. Vanadium, copper and molybdenum can be extracted into chloroform as oxine complexes and mercury masked with chloride [25, 26].

(iv) *Separation, preconcentration*

If the sample solution contains interfering substances, or if the chromium(VI) concentration is very low, it can be separated by extraction, ion-exchange or ion-flotation. Chromate may be extracted using methyl isobutyl ketone [27] or mesityl oxide [44] (4-methyl-3-pentene-2-one). The chromium was then stripped back into water and determined with diphenylcarbazide. However, it is not necessary to strip the chromium, for example Mann and White [45] used diphenylcarbazide in the organic phase after extraction with trioctyl phosphine oxide into benzene. Jirovec and Adam used trioctyl methyl ammonium chloride in chloroform solution [46], followed by addition of diphenylcarbazide; their method is applicable down to 0.02 mg dm^{-3} Cr. Dean and Beverly [47] described the extraction of chromium with methyl isobutyl ketone and development of the diphenylcarbazide complex in the extract.

Cupferron can be used for the separation of microamounts of chromium(VI) and chromium(III) [48]. The chromium(VI) –cupferron complex extracts into chloroform, whereas chromium(III) does not. The chromium(III) remaining in the aqueous layer was then oxidized to the (VI) state and determined with diphenylcarbazide in benzyl alcohol. Total chromium was also determined in the same way after reduction of chromium(VI) with ascorbic acid. Iron(II), mercury(II) and molybdenum(VI) did not interfere.

Yamashige *et al.* [24] used an anion-exchange resin column for the separation of chromium(VI). The method is suitable in waste water analysis. Chuecas and Riley [49] described the development and evaluation of a method for the determination of chromium down to the level of even 0.4 μg dm^{-3} Cr in natural waters, concentrating the chromium by coprecipitation with iron(III) hydroxide after reduction of chromium(VI) with sulfur dioxide. Since iron(III) interferes when chromium is determined with diphenylcarbazide, the two ions were separated by ion-exchange. Shigetomi *et al.* also used an ion exchange column (Dowex 1 × 4) [50]. Ions other than chromium(VI) were removed by washing the column with water. Chromium(VI) was then reduced with hydroxylammonium chloride solution and eluted by 0.1 mol dm^{-3} sulfuric acid/0.001 mol dm^{-3} hydroxylamine. The eluted chromium(III) was oxidized with hydrogen peroxide and determined with diphenylcarbazide. Using this procedure, chromium(III) can be determined in the presence of chromium(VI).

Yoshimura *et al.* [51] described the direct measurement of the resin-phase absorbance for the determination of chromium in sea water with diphenylcarbazide. The calibration curve was reasonably linear in the concentration range 0.003–0.08 mg dm^{-3} Cr. The method has since been developed further [52] to cover the concentration range 0.52–15.6 μg dm^{-3} Cr using 1 dm^3 of sample solution.

Aoyama *et al.* [53] used ion-flotation to concentrate chromium(VI) in the range 3–70 μg dm^{-3} from 1 dm^3 samples. The chromium(III)-diphenylcarbazone complex was floated with sodium lauryl sulfate, and the absorbance of the subsided foam measured after dilution. Both batch and continuous flotation procedures were described.

(v) Reagents

Diphenylcarbazide solution, 0.25% w/v. Dissolve 50 mg of diphenylcarbazide in 25 cm^3 of pure acetone. Prepare this solution freshly each day.

Sulfuric acid solution, 1 mol dm^{-3}.

Standard chromium(VI) solution, 1.000 g dm^{-3}. Dissolve 2.82 g of reagent grade potassium dichromate in 1 dm^3 of water.

(vi) Procedure

Place 2 cm^3 of diphenylcarbazide solution in a calibrated 100 cm^3 flask, add 10 cm^3 of the sulfuric acid solution and, depending on the chromium concentration, a 2–10 cm^3 aliquot of the sample solution. Dilute the resulting solution to the mark, wait for 5 minutes and measure the absorbance at 540 nm. Prepare a blank solution in the same manner and measure its absorbance against water. Prepare a calibration graph using aliquots of the standard chromium(VI) solution to which interfering ions have been added in the same amounts as in the sample solution.

1.2.2.2.2   Determination with iron(II)–FerroZine

(i) Introductory remarks

Bet-Pera and Jaselskis' [12] indirect method for the determination of microamounts of chromate in the presence of chromium(III) is based on reduction of chromium(VI) to chromium(III) by iron(II) and determination of excess iron(II) with FerroZine using the difference in absorbance values for samples containing iron(II) alone and those containing iron(II) in the presence of chromium(VI). It is essential to keep the pH in the range 1–2. Cobalt(II), nickel and permanganate ions interfere if their concentration is greater than 0.2 mg dm^{-3}. Vanadium(V) can be tolerated by up to five times the amount of chromium. The molar absorptivity is 8.3×10^4 cm^3 mol^{-1} cm^{-1} at 562 nm; the Lambert–Beer's law is obeyed in the range 0.1–0.6 mg dm^{-3} Cr.

(ii) Reagents

Ammonium iron(II) sulfate solution. 40 mg of analytical grade salt is dissolved in 100 cm^3 of 0.05 mol dm^{-3} sulfuric acid. Prepare this solution freshly each day.

Sodium monochloroacetate solution, 0.1 mol dm^{-3}. Dissolve 9.5 g of monochloroacetic acid in 600 cm^3 of water, adjust the pH to 5.8 using 3 mol dm^{-3} sodium hydroxide solution and dilute to 1 dm^3.

FerroZine solution, 0.012 mol dm^{-3}. Dissolve 0.6125 g of monosodium 3-(2-pyridyl)-5,6-bis(4-phenylsulfonic acid)-1,2,4-triazine monohydrate in 100 cm^3 of deionized water.

Standard chromium(VI) solution, 5.0 mg dm^{-3} Cr. Dissolve 1.414 g of reagent grade potassium dichromate in 1 dm^3 of water. Dilute 1 cm^3 of this stock solution to 100 cm^3 with deionized water.

(iii) *Procedure*

Add into a calibrated 50 cm^3 polythene flask depending on the chromium concentration, 0.5–6 cm^3 aliquot of the sample solution. Then add exactly 2 cm^3 of the ammonium iron(II) sulfate solution followed by 1 cm^3 of the FerroZine solution. After 2 minutes add 6 cm^3 of the monochloroacetate solution and dilute the resulting solution to the mark. After 3 minutes measure the absorbance at 562 nm. Prepare a calibration graph using aliquots of the standard chromium solution.

The colour of the iron(II)-FerroZine complex is stable.

1.2.2.2.3 Spectrophotometric determination of chromium(III) with TAR

(i) *Introductory remarks*

The total chromium content in a solution can be determined after reduction of all chromium to the (III) state. Sulfite in 1–2 mol dm^{-3} sulfuric acid [21, 58], warm hydrochloric acid [54], ascorbic acid and hydroxyl ammonium chloride [43] have all been used as reducing agents.

The red 1:2 complex, which is formed between chromium(III) and TAR [4-(2′-thiazolyazo)-resorcinol] at pH 5 after heating the solution to 90°C for 45 minutes, can be used for the determination of chromium [55]. The molar absorptivity is 5.0×10^4 dm^3 mol^{-1} cm^{-1} at 525 nm. The reagent is added to the sample solution dissolved in t-butyl alcohol. Lambert–Beer's law is obeyed in the range 0.06–1.1 mg dm^{-3} Cr.

The method is very sensitive. Fors [43] has applied it to the determination of chromium(III) in sea water. Since TAR reacts with chromium(III) [55], solutions of chromium(VI) have to be reduced.

EDTA decreased the absorbance even when it is added last. Chromium(VI) is reduced to chromium(III) when the solution is heated in the presence of TAR at pH 5. The method can thus also be used for the determination of the total amount of chromium in a solution.

For the determination of total chromium in sea water the methanol content had to be increased to ~ 40 per cent of the sample volume. Lambert–Beer's law was obeyed in the range 0.06–1.05 mg dm^{-3} Cr. The molar absorptivity is about 3.9×10^4 dm^3 mol^{-1} cm^{-1} at 531 nm.

(ii) *Reagents*

TAR solution, 0.1 % w/v. Dissolve 100 mg 4-(2′-thiazolylazo)-resorcinol (TAR) in 100 cm^3 of methanol.

Buffer solution, pH 5. Dissolve 41 g of sodium acetate and 14 cm^3 of glacial acetic acid in water and dilute to 1 dm^3.

Methanol standard chromium solution, 1.0 g dm^{-3} Cr. Dissolve 2.829 g of reagent grade potassium dichromate in 1 dm^3 of water. Solutions to be used in the determination are obtained by dilution of this stock solution.

(iii) *Procedure*

Take an aliquot of the chromium solution containing about 70 μg of chromium, add 20 cm^3 of the buffer solution and 4 cm^3 of the TAR solution. Heat the mixture in a boiling water bath for 45–60 minutes and allow to cool to room temperature. Add 16 cm^3 of methanol and transfer the mixture to a 100 cm^3 volumetric flask and dilute to the mark. Measure the absorbance at 531 nm against a reagent blank prepared under similar conditions.

1.2.2.2.4 Spectrophotometric determination of chromium(III) with PAR

(i) *Introductory remarks*

PAR [4-(2-pyridylazo)-resorcinol] is a sensitive reagent for chromium(III). PAR also reacts with many other metals, the selectivity for chromium may be improved by extraction of the chromium complex Cr(HL)$_3$ into chloroform at pH 5. Yotsuyanagi *et al.* [56] determined the distribution constant, D, the cumulative stability constant, β_3 and the dissociation constant, K_a of the acidic hydrogen atom in the complex:

$D = [\text{Cr(HL)}_3]_0/[\text{Cr(HL)}_3] = 1.17$
$\beta_3 = [\text{Cr(HL)}_3]/([\text{Cr}^{3+}][*\text{HL}^-]^3) = 10^{38.4}$
$K_a = [\text{Cr(L) (HL)}_2^-] [\text{H}^+]/[\text{Cr(HL)}_3] = 10^{-5.49}$

where *HL$^-$ is a hypothetical ligand formed from H$_2$L through dissociation of the *o*-hydroxyl proton of the resorcinol group, and o refers to the organic phase.

PAR has also been used in aqueous solution. Akhmedov *et al.* [57] found a 2:1 complex was formed with chromium at 80°C, but not at room temperature. The optimum pH was 5 and λ_{max} 540 nm. The measurement is made on ~ 0.6 mol dm^{-3} sulfuric acid solutions to destroy other metal-PAR complexes.

When the anionic Cr-PAR complex is extracted as an ion association complex with tetradecyl dimethylbenzyl ammonium ions into chloroform, a highly sensitive and selective spectrophotometric method for chromium results [56]. The stoichiometry of the extraction reactions can be described as follows:

$K_{ex}^R = [\text{HL}^-.\text{T}^+]_0/([\text{HL}^-] [\text{T}^+]) = 10^{5.41}$
$K_{ex}^C = [\text{Cr(L) (HL)}_2^-.\text{T}^+]_0/([\text{Cr(L) (HL)}_2^-] [\text{T}^+]) = 10^{6.93}$

where T$^+$ represents the tetradecyl dimethylbenzyl ammonium ion and o refers to the organic phase.

The λ_{max} was 540 nm in chloroform, the molar absorptivity was $4.7 \times 10^4 \, dm^3 \, mol^{-1} \, cm^{-1}$ and Lambert–Beer's law was obeyed in the range 0–0.9 mg dm^{-3} Cr. A disadvantage of the method is that the formation of the Cr-PAR complex in water requires boiling of the solution under reflux, for 135 minutes.

(ii) *Reagents*

PAR solution, 0.1% w/v. Dissolve 100 mg of 4-(pyridylazo)-resorcinol (PAR) in 10 cm^3 of 0.1 mol dm^{-3} sodium hydroxide solution and dilute to 100 cm^3 with water.

Tetradecyl dimethylbenzyl ammonium chloride solution (T$^+$), 0.005 mol dm^{-3}*. Dissolve 1.84 g of tetradecyl dimethylbenzyl ammonium chloride in 80 cm^3 of water and dilute to 100 cm^3.

Acetate buffer solution, pH 5. Dissolve 110 g of sodium acetate and 37.6 cm^3 of glacial acetic acid in water and dilute to 1 dm^3.

EDTA solution, 0.05 mol dm^{-3}. Dissolve 18.61 g of ethylenediamine tetraacetic acid, disodium salt, dihydrate in water and dilute to 1 dm^3 with water.

Chloroform*.

Sodium sulfate anhydrous*.

Standard chromium solution, 1.0 g dm^{-3} Cr. Dissolve 2.829 g of reagent grade potassium dichromate in 1 dm^3 of water. Solutions to be used in the determination are obtained by dilution of this stock solution.

(iii) *Procedure*

(a) Non-extractive. Place an aliquot containing 0–50 μg chromium into a 50 cm^3 volumetric flask. Add, in the following order and with mixing, 2 cm^3 of 2 mol dm^{-3} acetate buffer pH 5 solution and 1 cm^3 of the PAR solution, and dilute to about 45 cm^3 with water. Transfer the mixture to a 100 cm^3 flask equipped with a ground-glass joint. Attach a reflux condenser to the flask and heat under reflux for 135 minutes. After cooling to room temperature, add 2 cm^3 of 0.05 mol dm^{-3} EDTA solution and allow to stand for 2 minutes. Then, dilute to 50 cm^3 with water and measure the absorbance at 530 nm against a reagent blank.

(b) Extractive. Place 50 cm^3 of solution containing 0–9 μg of chromium(III), prepared as described above, into a 100 cm^3 separating funnel. Add 1 cm^3 of the T$^+$ solution, shake with 10 cm^3 of chloroform for 10 minutes, and allow the phases to separate. Run the chloroform layer into a test tube containing a small amount of anhydrous sodium sulfate crystals to remove droplets of water. Measure the absorbance at 540 nm against a reagent blank.

* Reagents used in the extraction procedure only.

18 SECTION 1

1.2.3 References

1 L. Fishbein, *Chromatography of Environmental Hazards*, Vol. II, Elsevier, Amsterdam p. 17 (1973).
2 R.E. Cranston and J.W. Murray, *Anal. Chim. Acta*, **99**, 275 (1978).
3 E. Nakayama, T. Kuwamoto, S. Tsurubo, H. Tokoro and T. Fujinaga, *Anal. Chim. Acta*, **130**, 284 (1981).
4 E. Nakayama, T. Kuwamoto, H. Tokoro and T. Fujinaga, *Anal. Chim. Acta*, **131**, 247 (1981).
5 E. Nakayama, T. Kuwamoto, S. Tsurubo and T. Fujinaga, *Anal. Chim. Acta*, **130**, 401 (1981).
6 H. Vokal, E. Hellsten, A. Henriksson–Enflo and M. Sundbom, *Chromium*, USIP Report 75–21, Stockholm, (1975).
7 D.E. Carter and Q. Fernando, *J. Chem. Educ.*, **56**, 490 (1979).
8 J.J. Lingane, *Analytical Chemistry of Selected Metallic Elements*, Reinhold, New York, p. 39 (1966).
9 R.A. Chalmers, in *Comprehensive Analytical Chemistry*, Vol IC, C.L. Wilson and D.W. Wilson (Eds), Elsevier, Amsterdam, p. 581 (1962).
10 E.D. Olsen and C.C. Foreback, *Encyclopedia of Industrial Chemical Analysis*, Vol 9, F.D. Snell and L.S. Ettre (Eds), Interscience, New York, p. 632 (1970).
11 F.D. Snell, *Photometric and Fluorometric Methods of Analysis. Metals, Part I*, Wiley, New York, p. 703 (1978).
12 F. Bet–Pera and B. Jaselskis, *Analyst*, **106**, 1234 (1981).
13 T.P. Hadjiioannou, *Talanta*, **15**, 535 (1968).
14 Z. Gregorowicz, P. Górka and S. Kowalski, *Z. anal. Chem.*, **294**, 285 (1979).
15 I. Baranowska, *Microchem. J.*, **26**, 55 (1981).
16 S.A. Rahim, H. Abdulahed and N.E. Milad, *J. Indian Chem. Soc.*, **52**, 853, (1975); *Z. Anal. Chem.*, **280**, 239 (1976).
17 M. Malat and M. Hrachovcová, *Collect. Czech. Chem. Commun.*, **29**, 2484 (1964).
18 K.L. Cheng, *Talanta*, **14**, 875 (1967).
19 R.R. Abdulaev, O.A. Tataev and A.I. Busev, *Zavod. Lab.*, **37**, 389 (1971); *Anal. Abstr.*, **22**, 797 (1972).
20 E.B. Sandell and H. Onishi, *Photometric Determination of Traces of Metals*, Part I, 4th edn., Wiley, New York, p. 321 (1978).
21 D.P. Nikilelis and T.P. Hadjiioannou, *Mikrochim. Acta*, 105 (1978).
22 B. Subrahmanyam and M.C. Eshwar, *Bull. Chem. Soc. Jpn*, **49**, 347 (1976).
23 A.E. Weinhardt and A.E. Hixson, *Ind. Eng. Chem.*, **43**, 1670 (1951).
24 T. Yamashige, H. Izawa and Y. Shigetomi, *Bull. Chem. Soc. Jpn*, **48**, 715 (1975).
25 Z. Marczenko, *Spectrophotometric Determination of Elements*, Ellis Horwood, Chichester, p. 213 (1976).
26 B. Saltzman, *Anal. Chem.*, **24**, 1016 (1952).
27 P.D. Blundy, *Analyst*, **83**, 555 (1958).
28 R.W. Bane, *Analyst*, **95**, 722 (1970).
29 A. Cazeneuve, *Bull. Soc. Chim.* (Paris), **23**, 701 (1900).
30 A. Moulin, *Bull. Soc. Chim.* (Paris), **31**, 295 (1904).
31 I.P. Alimarin and B.I. Frid, *Quantitative Mikrochemische Analyse der Mineralien und Erze*, T. Steinkopf, Dresden, p. 132 (1965).
32 M. Pinta, *Detection and Determination of Trace Elements*, Israel Programs for Scientific Translations, Jerusalem, p. 133 (1966).
33 A. Hofer and R. Heidinger, *Z. anal. Chem.*, **233**, 415 (1967).
34 E.S. Pilkington and P.R. Smith, *Anal. Chim. Acta*, **39**, 321 (1967).
35 J.F. Ege and L. Silverman, *Anal. Chem.*, **19**, 693 (1947).
36 P.F. Urone, *Anal. Chem.*, **27**, 1354 (1955).
37 H. Onishi and H. Koshima, *Bunseki Kagaku*, **27**, 726 (1978).
38 T.L. Allen, *Anal. Chem.*, **30**, 447 (1958).
39 M. Bose, *Anal. Chim. Acta*, **10**, 201, 209 (1954).
40 H. Marchart, *Anal. Chim. Acta*, **30**, 11 (1964).
41 G.J. Willems, N.M. Blaton, O.M. Pecters and C.J. de Ranter, *Anal. Chim. Acta*, **88**, 345 (1977).
42 R.T. Pflaum and L.C. Howick, *J. Am. Chem. Soc.*, **78**, 4682 (1956).
43 K. Fors, *Examensarbete*, Åbo Akademi, Åbo, (1980).
44 V.M. Shinde and S.M. Khopkar, *Z. anal. Chem.*, **249**, 239 (1970).
45 C.K. Mann and J.C. White, *Anal. Chem.*, **30**, 989 (1958).
46 J. Jirovec and J. Adam, *Hutn. Listy*, **29**, 739 (1974); *Anal. Abstr.*, **28**, 5H39 (1975).

47 J.A. Dean and M.L. Beverly, *Anal. Chem.*, **30,** 977 (1958).

48 K. Fuji, T. Kusuyama and K. Konishi, *Bunseki Kagaku*, **24,** 332, (1975); *Chem. Abstr.*, **84,** 144178w (1976).

49 L. Chuecas and J.P. Riley, *Anal. Chim. Acta*, **35,** 240 (1966).

50 Y. Shigetomi, T. Hatamoto, K. Nagoshi and T. Yamashije, *Nippon Kagaku Kaishi*, 619, (1976); *Anal. Abstr.*, **33,** 6H56 (1977).

51 K. Yoshimura, H. Waki and S. Ohashi, *Talanta*, **23,** 449 (1976).

52 K. Yoshimura and S. Ohashi, *Talanta*, **25,** 103 (1978).

53 M. Aoyama, T. Hobo and S. Suzuki, *Anal. Chim. Acta*, **129,** 237 (1981).

54 M. Yanagisawa, M. Suzuki and T. Takeuchi, *Mikrochim. Acta*, 475 (1973).

55 B. Subrahmanyam and M.C. Eshwar, *Mikrochim. Acta*, 579 (1976).

56 T. Yotsuyanagi, Y. Takeda, R. Yamashita and K. Aomura, *Anal. Chim. Acta*, **67,** 297 (1973).

57 A. Akhmedov, D.A. Tataev and R.R. Abdulaev, *Zavod. Lab.*, **37,** 756 (1971); *Anal. Abstr.*, **22,** 2300 (1972).

58 S.G. Nagarhar and M.C. Eshwer, *Indian J. Technol*, **13,** 377 (1975); *Anal. Abstr.*, **30,** 3B125 (1976).

1.3   Zinc, Cadmium, Mercury and Lead

Prepared by
W. I. STEPHEN
Chemistry Department, University of Birmingham, Birmingham B15 2TT, UK

1.3.1   Introduction

The existence of heavy metals in aquatic environments has led to much concern over their influence on plant and animal life in these environments and indeed on man's need for wholesome water. Technological advances have greatly increased the use of many metallic elements and increasing amounts of these elements are finding their way into the hydrosphere, and augmenting the 'natural' concentrations. The accumulation of these elements, many of which are highly toxic to animal life, by aquatic plant life and the lower forms of marine animals, is one of man's less praiseworthy influences on the biosphere.

The four elements considered in this section are zinc, cadmium, mercury and lead. The detection and determination of these in natural waters are of considerable importance, not only as a means of establishing their influence on various ecosystems, but also for monitoring and controlling the pathways by which they reach the hydrosphere.

Accurate chemical analyses coupled with effective sampling are essential to any scientific study of problems caused by heavy metal pollution.

Cadmium, mercury and lead are major hazards to health. All are cumulative poisons. WHO (World Health Organization) standards [1] require low concentrations of these elements in drinking water; cadmium and lead below 50 $\mu g\,dm^{-3}$ and mercury below 5 $\mu g\,dm^{-3}$. Zinc is not particularly harmful to man, but can be toxic to lower organisms, and can prove troublesome in food chains. It can be present in drinking water at concentrations as high as 5000 $\mu g\,dm^{-3}$ without undue effect [2].

Uncontaminated natural waters contain between 0.01–1.0 $\mu g\,dm^{-3}$ of cadmium, 0.1–2.5 $\mu g\,dm^{-3}$ of lead and 0.2–5.0 $\mu g\,dm^{-3}$ of zinc [2]. These very low concentrations require the use of sensitive analytical reactions for their determination. The following sections deal with the commonly used spectrophotometric methods for the determination of these elements in natural waters, and especially in potable waters.

1.3.2   Analytical methods—general

The importance of methods for the determination of low concentrations of the elements, zinc, cadmium, mercury and lead in water lies largely in the applications of such methods for control of water quality and of pollution. These applications require 'standard methods' which are formally accepted as legally binding in many countries. Before the advent of techniques such as atomic absorption spectrometry [3] (AAS) and anodic stripping voltammetry, most trace analyses of heavy metals in environmental samples were carried out using colorimetric or spectrophotometric procedures. Although numerous colour-forming reagents have been described for the spectrophotometric determination of these elements [4], very few reagents have satisfied the criteria required for standard methods, particularly for application to natural water samples.

For the elements cadmium, lead and mercury, the reagent 1,5-diphenyl-thiocarbazone [5], commonly known as dithizone, is universally accepted. Although dithizone can also be used for the determination of zinc, the trend in recent years has been to use 'zincon', (2-carboxy-2'-hydroxy-5'-sulfofor-mazylbenzene), which is more selective and more stable than dithizone.

In many countries, the problems of water analysis are dealt with by Government agencies. For example in the UK, a collection of standardized methods has existed since 1972, for the analysis of raw, potable and waste waters [6]. This was intended as a reference work for laboratories concerned with the routine analysis of water: the methods rely heavily on spectrophotometric and titrimetric procedures. This work has continued and many new procedures have since been published [7] e.g. by AAS.

1.3.3 Spectrophotometric determination of zinc

1.3.3.1 General aspects

Zinc exists in natural waters mainly as hydroxo-complexes, $Zn(OH)^+$, $Zn(OH)_2$, chloro-complexes $ZnCl^+$, $ZnCl_2$, carbonato-complexes $ZnCO_3$ and zinc aquo ions, depending on the pH. The hydroxo-complexes are likely to be in the form of colloids or adsorbed on particulate matter. There is considerable evidence to suggest that much of the zinc in natural waters is contained in organic complexes [8, 9]. Surface water of the oceans contains between 0.4 and 3.0 μg dm^{-3} with higher concentrations of 0.6 to 12.6 μg dm^{-3} of zinc being present in coastal waters. Drinking waters may be heavily contaminated with zinc. The 'natural' concentration of zinc is 0.2–5.0 μg dm^{-3} in uncontaminated water. The determination of zinc and other heavy metals in sea-water is frequently achieved by anodic stripping voltammetry but the following spectrophotometric methods have been recommended for water analysis.

Fairhall and Richardson [10] titrated the water sample with potassium hexacyanoferrate(II) to give a colloidal precipitate of zinc hexacyanoferrate(II) which could be measured turbidimetrically. The range of this method was 0.5–10 mg dm^{-3} of zinc. The effect of iron can be eliminated by previous extraction as its thiocyanate or cupferron complex [11]. O'Connor and Renn [12] found that the US standard method was not suitable for concentrations below 0.1 mg dm^{-3} and they evaluated four methods, including colorimetric methods with zincon and dithizone. Feldman and Nakhshina [13] also recommended a dithizone procedure for microgram amounts of zinc in natural water, and Sadilkova [14] makes use of an extractive photometric determination with zincon for water samples containing 5–25 μg in 400 cm^3.

Kish and Zimomrya [15] recommend the blue complex of zinc, thiocyanate and 6-methoxy-3-methyl-2[4-(N-methyl-anilino)phenylazo]-benzothiazolinium chloride for the determination of zinc at concentrations of 2 μg–3 mg dm^{-3}. The complex was extracted with a solution of tributylphosphate in benzene and measured at 640 nm. Interferences are minimal. Sulfarsazen has been used for the determination of zinc or lead in water [16]. When lead is determined, the zinc is masked by cyanide. The procedure involves a preliminary extraction of the metals with diethyldithiocarbamate. Evans and Sayers [17] use TAR (4-(2'-thiazolylazo)-resorcinol) in a single reagent solution containing TAR, buffer and masking agents. The method is rapid and gives results comparable with those by AAS. Recently, Ackermann and Köthe [18] have investigated the established spectrophotometric methods for zinc with a number of well-known reagents. The methods involving the reagents 4(2-pyridylazo)resorcinol, 4-(2-thiazolylazo)resorcinol and 1-(2-pyridylazo)-2-naphthol were recommended because of their sensitivity and reproducibility. Significantly, dithizone was omitted from this study. The spectrophotometric method usually recommended involves the reagent, zincon. The procedure below is representative of general procedures with this reagent [19].

1.3.3.2 Spectrophotometric determination of zinc with zincon

(i) *Principle*

A solution of zincon (2-carboxy-2'-hydroxy-5'-sulfoformazylbenzene) in water has an intense brick-red colour. At pH 9 it reacts with zinc ions to form a blue-coloured complex [20]. Interferences due to heavy metals which also form complexes with zincon are removed by masking with cyanide [21]. The addition of chloral hydrate causes immediate decomposition of the zinc cyanide complex permitting the zinc ions to react with zincon, and since the cyanides of the other heavy metals are only very slowly decomposed by chloral hydrate the absorbance of the zinc-zincon complex can be measured. Interference by manganese is prevented by reduction to manganese (II) with ascorbic acid. Interference by copper when present in a significant amount may be removed by ion exchange [20].

(ii) *Interferences*

The following ions interfere at concentrations in mg dm^{-3} exceeding those listed:
Cd^{2+} 1, Al^{3+} 5, Mn^{2+} 5, Fe^{3+} 7, Fe^{2+} 9, Cr^{3+} 10, Ni^{2+} 20, Cu^{2+} 30, Co^{2+} 30, CrO_4^{2-} 50, polyphosphate 20.

(iii) *Reagents*

Zincon solution 0.1%. Dissolve 0.1 g zincon (2-carboxy-2'-hydroxy-5'-sulfoformazylbenzene) in 10 cm^3 NaOH and dilute to 100 cm^3 with ethanol. The solution is stable for some months.

Borate buffer solution. Dilute $213 \, cm^3$ of $1 \, mol \, dm^{-3}$ sodium hydroxide to about $600 \, cm^3$ with water and in this solution dissolve 37.8 g potassium chloride and 31.0 g boric acid. Dilute to $1 \, dm^3$.

Stock zinc solution, $1.000 \, g \, dm^{-3}$. Dissolve 0.440 g zinc sulfate, $ZnSO_4.7H_2O$, in water and make up to $100 \, cm^3$.

Standard zinc solution, $10.0 \, mg \, dm^{-3}$. Dilute $5.0 \, cm^3$ stock zinc solution to $500 \, cm^3$.

Potassium cyanide solution, $10 \, g \, dm^{-3}$. Potassium cyanide is highly toxic. Take great care in the transference of the solid and its solutions. Always use a safety pipette.

Chloral hydrate solution $CCl_2.CH(OH)_2$, $100 \, g \, dm^{-3}$. Filter if necessary.

Sodium ascorbate, $C_6H_7O_6Na.H_2O$. Fine granular powder.

Sodium hydroxide solution, $1 \, mol \, dm^{-3}$.

Hydrochloric acid solution, $1 \, mol \, dm^{-3}$, $0.01 \, mol \, dm^{-3}$, concentrated-spec. mass 1.18.

(iv) *Preparation of calibration graph*

Into a series of $50 \, cm^3$ graduated flasks pipette appropriate volumes of standard zinc solution covering the range up to 50 μg Zn. Add water to bring the volume to $\sim 30 \, cm^3$. Include a flask containing no zinc as a blank. Then follow the procedure as given under (v) ' Procedure' from 'To each flask add in sequence. . . '. Read the directly compensated absorbances of the standard solutions and prepare a graph relating absorbance to μg zinc.

(v) *Procedure*

Transfer an aliquot of $25 \, cm^3$, or less, of the solution, containing not more than 50 μg zinc, to a $50 \, cm^3$ beaker and adjust the test solution and a blank to pH 7.0 by the addition of dilute sodium hydroxide solution, using a pH meter.

If removal of interference by copper and iron is necessary, proceed as under 'Removal of interferences' (vi). Otherwise place the aliquots of sample and blank into $50 \, cm^3$ volumetric flasks and add water, if necessary, to bring the volume to $\sim 30 \, cm^3$. To each flask add in sequence the following reagents, mixing after each addition: 0.5 g sodium ascorbate, $1 \, cm^3$ potassium cyanide solution (CAUTION), $5 \, cm^3$ buffer solution and $3 \, cm^3$ zincon solution. Add $3 \, cm^3$ chloral hydrate solution to each flask. Make up to the mark, stopper and mix. After 5 minutes measure the absorbance using either a spectrophotometer at 620 nm or absorptiometer using a suitable red filter, with 20 mm cells and the blank in the reference cell. A period of 5 minutes is specified to ensure that the zinc-zincon complex is measured before other heavy metals, which would also react with zincon, are released from their cyanide complexes. From the calibration graph read the number of μg of zinc.

(vi) *Removal of interferences*

Copper. Should the sample contain more than $30 \, mg \, dm^{-3}$ of copper, it must be passed through the ion-exchange column and an equal volume of the blank treated in a similar manner. The column consists of a glass tube 70 mm long and 10 mm internal diameter, fitted with a glass wool plug. Mix an ion-exchange resin (De Acidite FF or equivalent) 50–100 mesh, with water and pour the slurry into the tube to give a column 50 mm in length.

Add 1 cm^3 of concentrated hydrochloric acid to aliquots of the sample and blank. Wash the ion-exchange column with 20 cm^3 of 1 mol dm^{-3} hydrochloric acid at a rate of 1 cm^3 min^{-1}, discarding the eluate. Pass the aliquot of the blank through the column at a rate of 2 cm^3 min^{-1}, discarding the eluate. Place a beaker under the outlet of the column and elute with 25 cm^3 of 0.01 mol dm^{-3} hydrochloric acid at a rate of 2 cm^3 min^{-1}.

Repeat the procedure in the last paragraph using an aliquot of the sample. If it is required to remove iron, proceed as below. Otherwise, first adjust the aliquots of the sample and blank eluted from the ion exchange column to above pH 7. Transfer the contents of the beaker quantitatively to a 50 cm^3 volumetric flask and add water, if necessary, to bring the volume to ~ 30 cm^3 Then follow the procedure given under (v) 'Procedure' starting at 'To each flask add in sequence. . .'.

Iron. Add 2 cm^3 of potassium cyanide to the neutralized sample aliquot, mix well and allow the precipitate to settle. Filter into a 50 cm^3 volumetric flask and wash the precipitate with two 2 cm^3 portions of water, collecting the washings in the volumetric flask. Add water, if necessary, to bring the volume in the flask to ~ 30 cm^3. Treat an aliquot of the blank in the same manner. Then follow the procedure given under (v) 'Procedure' starting at 'To each flask add in sequence. . .'.

Organic matter. When the samples of raw and potable waters are analysed usually it is not necessary to use special pretreatment for mineralization of organic substances.

1.3.4 **Spectrophotometric determination of cadmium**

1.3.4.1 General aspects

Cadmium exists in sea-water as simple chloro-complexes, $CdCl^+$, $CdCl_2$, $CdCl_3^-$ [22]. Surface water in the oceans of the world contain between 0.01 and 0.18 μg dm^{-3} of cadmium. As with zinc, coastal waters tend to contain higher amounts, in the range 0.02–0.30 μg dm^{-3}. The most widely used methods for determining cadmium in natural waters are those involving atomic absorption spectrometry and anodic stripping voltammetry; spectrophotometric methods are now of less importance. Dithizone is the preferred chromogenic reagent for cadmium. Most procedures are based on the work of Saltzman [23] in which cadmium ions form a pinkish-red colour with dithizone, which is extracted with chloroform and measured at 518 nm. Rodier's procedure [24] is based on the observation of Ganotes, Larson and Navone [25]. The cadmium is first extracted from an alkaline tartrate–cyanide medium with a chloroform solution of dithizone. The chloroform is then treated with the solution of tartaric acid and this aqueous phase is re-extracted with chloroform–dithizone in the presence of hydroxylammonium chloride, hydroxide and cyanide. Samples can contain up to 10 μg of cadmium in 25 cm^3 samples (0.4 mg dm^{-3}). The detection limit is 0.02 mg dm^{-3}. Interferences are minimal, and even appreciable amounts of lead, zinc and copper do not interfere. Further improvements in selectivity are possible with ion-exchange procedures [26].

A procedure [27] representative of these using dithizone is given below.

1.3.4.2 Spectrophotometric determination of cadmium with dithizone

(i) *Principle*

Cadmium is determined in the form of its red complex with dithizone spectrophotometrically. The method is generally applicable and copper and nickel interferences are overcome by masking with cyanide. A method for overcoming interference by thallium has been described [22].

(ii) *Reagents*

Potassium sodium tartrate solution. Dissolve 25 g sodium potassium tartrate tetrahydrate, $C_4H_4O_6KNa.4H_2O$ in 100 cm^3 water.

Sodium hydroxide–potassium cyanide solution A. Dissolve 40 g sodium hydroxide and 1.0 g potassium cyanide in 100 cm^3 water. Store in a polythene bottle.

Sodium hydroxide–potassium cyanide solution B. Dissolve 40 g sodium hydroxide and 0.05 g potassium cyanide in 100 cm^3 water. Store in a polythene bottle.

Hydroxylammonium chloride solution, 200 g dm^{-3}.

Chloroform. This should give a faint green tint stable for a day when a minute amount of dithizone solution is added to a portion in a stoppered test tube.

Dithizone extraction solution. Dissolve 80 mg dithizone in 1 dm^3 of chloroform. Keep in a brown glass bottle, store in a refrigerator until required and use while still cold.

Dithizone standard solution. Dissolve 8 mg dithizone in 1 dm^3 of chloroform. Keep in a brown glass bottle, store in a refrigerator and allow to warm to room temperature before use. The fresh solution should be aged a day or two, then used to prepare the calibration graph as given below.

Tartaric acid solution, 20 g dm^{-3}. Store in a refrigerator and use cold.

Thymol blue indicator solution. Dissolve 0.04 g of the indicator in 1 cm^3 of 0.1 mol dm^{-3} sodium hydroxide and dilute to 100 cm^3 with water.

Cadmium stock solution, 100 mg dm^{-3}. Dissolve 0.100 g cadmium metal in 50 cm^3 of 10% nitric acid. Boil to expel oxides of nitrogen and dilute to 1 dm^3.

Cadmium dilute stock solution, 10 mg dm^{-3}. Dilute 10.0 cm^3 cadmium standard solution to 100 cm^3 with 1% nitric acid.

Cadmium standard solution, 1.0 mg dm^{-3}. Dilute 10.0 cm^3 cadmium dilute standard solution to 100 cm^3 with 1% nitric acid. Prepare as needed and use the same day.

(iii) *Preparation of calibration graph*

Measure appropriate amounts of cadmium dilute stock and cadmium standard solution, covering the range up to $20 \mu\text{g}$ Cd, into a series of separating funnels. Neutralize excess acid by adding sodium hydroxide solution until added thymol blue indicator turns yellow, and dilute to 25 cm^3 with tartaric acid solution. Include a separating funnel containing no cadmium as a blank. After adding the tartaric acid solution shake and continue as described below beginning with the wash of 5 cm^3 chloroform [see (iv) 'Procedure']. Deduct the absorbance of the blank from those of the standard solutions and prepare a graph relating net absorbance to μg cadmium. The calibration should be checked every few weeks.

(iv) *Procedure*

Measure into a separating funnel a suitable aliquot of the sample containing not more than $20 \mu\text{g}$ of cadmium. Neutralize excess acid by adding sodium hydroxide solution using thymol blue indicator and adjust the volume to 25 cm^3. Carry out the blank determination with all reagents used including those for the preliminary treatment of the sample.

Add reagents in the following order, mixing after each addition: 1 cm^3 of sodium potassium tartrate solution, 5 cm^3 of sodium hydroxide–potassium cyanide solution A and 1 cm^3 of hydroxylammonium chloride solution. Add 15 cm^3 dithizone extraction solution, shake the mixture for 1 minute and drain the lower, chloroform, layer into a second separating funnel containing 25 cm^3 tartaric acid solution.

Extract again, this time with 10 cm^3 chloroform, adding the lower layer to the tartaric acid solution contained in the second separating funnel. Do not permit any of the aqueous layer to enter the second funnel. Vent through the stopper and not through the stopcock in these and all succeeding operations. Keep the time of contact of chloroform with the strong alkali to a minimum by performing the operations without delay after addition of dithizone.

If lead and zinc are present in appreciable amounts the pink colour of their dithizonates persists through several primary extractions, although all the cadmium has already been extracted, giving the misleading impression that the extraction of cadmium is not complete. Absence of the orange colour of excess dithizone in the aqueous layer indicates that too large a portion of sample was taken, and the determination should be repeated with a smaller quantity.

Shake the combined extracts with the tartaric acid solution in the second separating funnel for 2 minutes. Allow to separate and discard the lower layer. Add 5 cm^3 chloroform and shake the mixture for 1 minute, and again discard the lower layer making as fine a separation as possible. Evaporate the last floating drop by gentle blowing of air.

All the cadmium is now in the tartaric acid solution and it should be substantially free from interfering metals.

Add 0.25 cm^3 hydroxylammonium chloride solution and 15.0 cm^3 dithizone standard solution followed by 5 cm^3 sodium hydroxide–potassium cyanide solution B and shake the mixture for 1 minute. Allow the layers to separate and filter the chloroform layer through a small plug of cotton wool in a funnel.

Measure the absorbances of the test and blank solutions either in a spectrophotometer at 518 nm or absorptiometer with a suitable green filter, using 10 mm cells with water in the reference cell. From the calibration graph read the number of micrograms of cadmium.

In the analysis of raw and potable waters containing small amounts of organic matter there is usually no need to use special pretreatment for mineralization of organic substances.

1.3.5 Spectrophotometric determination of mercury

1.3.5.1 General aspects

In many natural waters and particularly in sea water, mercury exists mainly as chloro-complexes such as $HgCl_2$, $HgCl_3^-$ and $HgCl_4^{2-}$ and to a lesser extent the hydroxo-complexes $HgOHCl$, $Hg(OH)_2$, depending on pH and chloride ion concentration. Inorganic mercury can be involved in biological systems resulting in the formation of methylmercury(II)chloride and dimethylmercury(II), both compounds are highly toxic to mammals [28].

Preferred methods of analysis are those involving atomic absorption spectrometry and neutron activation analysis; dithizone is a principal spectrophotometric reagent. Speciation requires care, as mercury may be present not only in soluble organic and inorganic compounds, but also retained on solid, suspended material of organic or inorganic composition. The determination of total mercury necessitates an initial mineralization of unfiltered samples with oxidizing acids. Occasionally, only dissolved mercury is determined on the filtrate, after separation of the particulate matter. Because of the ease, high sensitivity and selectivity of flameless atomic absorption spectrometry for the determination of mercury, spectrophotometric methods are now seldom used. For completeness of this survey, however, these are included.

The method of colorimetric titration given by Rodier [29] is based on that of Truhaut and Boudène [30]. After mineralization of the sample with sulfuric and permanganic acids, the mercury is selectively extracted as its di-β-naphthyldithiocarbazone complex into chloroform at pH 1.5. The blank is treated identically to give a blue chloroform layer, then standard mercury solution, 10 mg dm^{-3} is added until the colour of the chloroform layer matches that given by the sample, the colours being compared in daylight against a white background. The β-naphthyl analogue of dithizone was preferred because of its improved stability and sensitivity. The pH selected allows mercury to be determined in the presence of copper. This method was stated to be suitable for amounts of mercury up to 400 $\mu g\,dm^{-3}$ within $\pm 5\%$.

Various dithizone procedures have been recommended for application to sea water [31], lake water [32], natural waters and waste waters [33]. Dual-wavelength photometry is used after an exchange reaction [34] between solubilized copper(II) dithizone complex and mercury(II) ions in an aqueous medium. The difference in absorbance at 567 and 493 nm is proportional to the mercury concentration, in the range 5–25 $\mu g\,dm^{-3}$. An interesting catalytic method [35] depends on the reaction of hexacyanoferrate(II) with water, catalysed by mercury(II) ions to give aquopentacyanoferrate(II). This rapidly reacts with nitrosobenzene to give a violet complex in amount proportional to the mercury originally present, after a fixed interval of time; the range is 8–800 $\mu g\,dm^{-3}$.

Monothiobenzoyl acetone has been recommended for the determination of mercury in waste water [36]. The mercury chelate is extracted into benzene, the organic layer is washed with alkali to remove excess reagent, and the absorbance is measured at 345 nm. The method is stated to be suitable for mercury in amounts 0.6–12.0 $\mu g\,dm^{-3}$, with minimal interference.

A standard dithizone procedure [37] is given below.

1.3.5.2   Spectrophotometric determination of mercury with dithizone

(i) *Principle*

The destruction of organic matter present in the sample needs to be carefully performed in the recommended apparatus to avoid loss of mercury due to the volatility of the metal and its covalent compounds. The recommended wet digestion method [38] using the apparatus described, allows the use of vigorous oxidizing conditions without risk of loss.

After destruction of organic matter by wet oxidation with nitric and sulfuric acids, and reduction with hydroxylammonium chloride to destroy oxides of nitrogen, the mercury is separated by extraction with excess of a solution of dithizone (diphenylthiocarbazone) in carbon tetrachloride. The mercury is removed from this extract and returned to the aqueous phase by oxidation with sodium nitrite, excess nitrite being destroyed with hydroxyl-ammonium chloride, and any remaining oxides of nitrogen being removed by treatment with urea. After the addition of EDTA, which hinders the reaction of copper with dithizone, mercury is extracted by titration with a solution of dithizone in carbon tetrachloride and finally determined spectrophoto-metrically.

(ii) *Interferences*

The method is specific for mercury in all ordinary circumstances and contents down to 0.5 μg mercury can be measured. At least 60 μg of copper do not interfere if carbon tetrachloride is used, or at least 600 μg of copper if chloroform is substituted in the final colorimetric determination.

In the presence of more than 60 μg of copper the final colorimetric determination should be made with a solution of dithizone in chloroform instead of in carbon tetrachloride. In this case, the combined extract is diluted to 4.0 cm^3 with chloroform and measurement is made at 492 nm. In all other respects the procedure is identical with that described below.

(iii) *Reagents*

Certain of the reagent solutions used must be purified in order to reduce blank values and so increase the accuracy of the method at low levels of mercury. The purification of hydroxylammonium chloride solution is described below.

Hydroxylammonium chloride, 200 g dm^{-3}. Purify as follows: Transfer the solution to the separating funnel and add a few cm^3 dithizone stock solution, shake for 2 minutes, allow to separate and reject the lower organic layer. Repeat the extraction with dithizone until the lower organic layer has the colour of pure dithizone solution. Finally, extract the solution with successive small quantities of chloroform until the extracts are colourless, and discard them.

Sulfuric acid, concentrated, spec. mass, 1.84.

Nitric acid, concentrated, spec. mass 1.42.

Dithizone stock solution, 0.05 g dm^{-3} in chloroform. This solution should be kept in a brown bottle and stored in a refrigerator.

Dithizone dilute solution in carbon tetrachloride. Dilute 2 cm^3 stock solution in chloroform to 100 cm^3 with carbon tetrachloride. Prepare freshly as required.

Dithizone dilute solution in chloroform. Dilute 2 cm^3 stock solution in chloroform to 100 cm^3 with chloroform. Prepare freshly as required.

Hydrochloric acid, 0.1 mol dm^{-3}.

Sodium nitrite solution, 50 g dm^{-3}.

Urea solution, 100 g dm^{-3}.

EDTA solution, 25 g dm^{-3}. *Dissolve 25 g of EDTA disodium salt, dihydrate in water and dilute to 1 dm^3.

Acetic acid solution, approximately 4 mol dm^{-3}*.

Carbon tetrachloride.

Chloroform.

Mercury stock solution, 100 mg dm^{-3}. Dissolve 0.135 g of mercury(II) chloride in 1 dm^3 of 0.1 mol dm^{-3} hydrochloric acid.

Mercury standard solution, 1.0 mg dm^{-3}. Dilute 10 cm^3 stock solution to 1 dm^3 with 0.1 mol dm^{-3} hydrochloric acid. This solution should be prepared freshly as required.

(iv) *Apparatus*

All glassware must be thoroughly cleaned with nitric and sulfuric acids and washed with distilled water immediately before use. Digestion apparatus is

* These solutions can be purified as described for hydroxylammonium chloride solution.

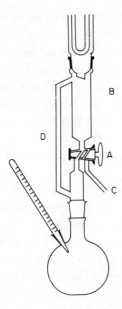

Fig. 1.2. Apparatus for the wet decomposition of organic matter.

shown in Fig. 1.2. The flask has a capacity of 250 cm³ and the reservoir B has a capacity of 150 to 200 cm³. The condenser is a standard double-surface or spiral surface reflux type. The thermometer is graduated up to 200°C and all the connections are made through standard ground-glass-joints.

Separating funnels. 150 cm³, and 1 dm³ pear-shaped separating funnels with well-fitting glass stopcocks and stoppers.

(v) *Preparation of calibration graph*

Pipette into a series of separating funnels, amounts of standard mercury solution to cover the range 0.5 to 10 μg mercury and dilute each to 10 cm³ if necessary, by the addition of 0.1 mol dm⁻³ hydrochloric acid. Transfer to another separating funnel, 10 cm³ 0.1 mol dm⁻³ hydrochloric acid to be used as a blank. Treat each solution as described below.

Add 1 cm³ of sodium nitrite solution and 1 cm³ of hydroxylammonium chloride solution, mix and set aside for 15 minutes. Add 1 cm³ of urea solution and 1 cm³ of EDTA solution and complete the extraction and measurement of each extract as described under 'Determination of mercury' (pp. 32–33).

Deduct the absorbance of the blank from those of the standard solutions and construct a calibration graph relating absorbance to μg mercury.

(vi) *Procedure*

Destruction of organic matter. Use the apparatus shown in Fig. 1.2 for the wet oxidation of the sample. Transfer a suitable aliquot of sample to this oxidation flask, add 5 cm³ of concentrated sulfuric acid, and 10–50 cm³ of concentrated nitric acid, as required. Add a few antibumping granules or glass beads and

31   Zn, Cd, Hg AND Pb

assemble the apparatus as illustrated. Allow any initial reaction to subside and then heat, cautiously at first, collecting the distillate in the reservoir B with tap A closed. When the temperature indicated by the thermometer reaches 116°C, which is close to the boiling point of nitric acid, run the contents of the reservoir through the drain-tube C and collect in a measuring cylinder.

Continue collecting the distillate in the reservoir and when the oxidation mixture darkens, run a little of the distillate from the reservoir to the flask. Continue this procedure, maintaining a slight excess of nitric acid in the oxidation flask, until the solution ceases to darken and fumes of sulfuric acid are evolved.

Allow the mixture to cool, run the contents of the reservoir into the flask and add to the first distillate in the measuring cylinder, the volume of the residue plus distillate usually being $\sim 100\,cm^3$.

Titrate $1\,cm^3$ of this sodium with standard sodium hydroxide solution to determine the concentration of acid present. Dilute with water to produce a solution with a total acidity of $\sim$ pH 0, the volume after dilution being about $400\,cm^3$. Heat to boiling, remove from the source of heat, and add rapidly, with mixing, a volume of hydroxylammonium chloride solution equal to one tenth of the total bulk; then set aside for 15 minutes and cool to room temperature.

Separation of mercury. Transfer the solution to a separating funnel of suitable capacity and extract with carbon tetrachloride, if necessary, to remove any fat. Add $10\,cm^3$ dilute dithizone solution in carbon tetrachloride, shake for 1 minute, allow the layers to separate, and run the lower layer into a $150\,cm^3$ separating funnel. Continue the extraction with successive $1\,cm^3$ portions of dithizone solution until two successive extracts remain green, and combine the extracts in the second separating funnel. If copper is present, a chloroform solution of dithizone should be used, and the final measurement should be at 492 nm.

To these extracts add $10\,cm^3$ of $0.1\,mol\,dm^{-3}$ hydrochloric acid and $1\,cm^3$ of sodium nitrite solution, shake vigorously for 1 minute, allow the layers to separate and discard the lower layer. Add $1\,cm^3$ hydroxylammonium chloride solution and set aside for 15 minutes, shaking occasionally. Add $1\,cm^3$ of urea solution and $1\,cm^3$ of EDTA solution.

Determination of mercury. Add $0.1\,cm^3$ dilute dithizone solution in carbon tetrachloride or chloroform from a $10\,cm^3$ burette. Shake vigorously for 10 s and allow to separate. Run the lower layer into another separating funnel containing $5\,cm^3$ of $4\,mol\,dm^{-3}$ acetic acid** and repeat the operation until the separated layer is greenish-orange. Extend the shaking time to 30 s and reduce the increments of dithizone solution to $0.2\,cm^3$. Continue the titration and separation, combining the extracts, until the organic layer has a greyish

** Solutions of mercury(II) dithizonate in organic solvents are sensitive to light. Exposure to daylight causes the solution to fade, but the original colour is slowly restored if the faded solutions are kept in the dark, and is more rapidly restored on shaking with dilute acids. The light sensitivity is eliminated in the presence of acetic acid.

colour, showing that the mercury has been extracted completely and that the extract contains a slight excess of dithizone. Note the volume of dithizone solution required.

From another $10 \, cm^3$ burette add sufficient carbon tetrachloride or chloroform to adjust the volume of the extract to $4.0 \, cm^3$. Mix, dry the stem of the funnel, and run the lower layer through a small glass-wool plug, supported in a small glass funnel, into a $10 \, mm$ glass spectrophotometer or absorptiometer cell.

Carry out the blank determination on all reagents used in the preparation of the sample.

Measure the absorbance either in a spectrophotometer at 485 nm, or 492 nm if a chloroform solution is used, or in an absorptiometer using a suitable blue filter with either carbon tetrachloride or chloroform in the reference cell. From the calibration graph read the number of μg of mercury in the test and blank solutions and so obtain the net measure of mercury in the sample.

1.3.6 Spectrophotometric determination of lead

1.3.6.1 General aspects

Lead is found as a carbonato complex, $PbCO_3$, chloro-complexes $PbCl^+$, $PbCl_2$, $PbCl_3^-$, hydroxo-complexes, $PbOH^+$, $Pb(OH)_2$, $Pb(OH)_3^-$, $Pb_3(OH)_4^{2+}$, $Pb_4(OH)_4^{4+}$, etc. in natural waters [39]. There can be appreciable absorption on particulate matter and a considerable portion of the lead which is present in natural waters can be found in complexes with organic molecules. Sea water can contain from about $0.001-1.0 \, \mu g \, dm^{-3}$ of lead. Preferred methods of determination are atomic absorption spectrometry and anodic stripping voltammetry.

The usual colorimetric determination of lead in water was based on the formation of the sulfide, but now is based on the dithizone complex. The sulfide method was generally used visually and is subject to considerable interference, apart from lacking sufficient precision for low concentrations of lead. The dithizone method is now accepted as of general application and most countries use an extractive spectrophotometric determination of the lead–dithizone complex for measurement of lead in drinking water. Following the work of Clifford and Wichmann [40], Biefield and Patrick [41] and others, Snyder [42] developed a method involving extraction of the lead–dithizone complex at pH 11.5. At this pH, free dithizone remains in the aqueous phase. Rodier [43] described a method, based on the work of Berger and Pirotte [44], in which lead, after mineralization with sulfuric and nitric acids is determined by extraction with dithizone into chloroform from an ammonia–sulphite–cyanide solution and measurement at 510 nm. The detection limit is $2 \, \mu g$ of lead with an error of 15–20% at the $0.1 \, mg \, dm^{-3}$ level. Bismuth is a major potential interference in dithizone methods for lead, and is eliminated by a prior extraction of bismuth at pH 3.0. Abbott and Harris [45] follow Snyder's practice and extract the lead dithizonate at pH 11.5 in the presence of sodium hexametaphosphate to prevent the precipitation of alkaline earth phosphates. Their proposals form the basis of the recommended method [46] below, which is stated to allow the detection of as little as 1 μg of lead with confidence. Over the range 0–25 μg the method has a precision better than ± 1 μg. Korkisch and Sorio [47] use an ion exchange process for preconcentration of lead in natural waters, followed by a dithizone procedure for its determination. Markova [48] used sulfarsazen, after a preliminary separation with dithizone, for the determination of lead in fresh and mineral waters. The limit of determination of this procedure was stated to be 1–1.5 μg dm^{-3} of lead.

1.3.6.2 Spectrophotometric determination of lead with dithizone [45]

(i) *Principle*

Although dithizone (diphenylthiocarbazone) is not a specific reagent for lead, in a basic cyanide medium containing a reducing agent, the only significant interfering elements are bismuth and tin which, like lead, form dithizonates extractable with chloroform at a pH of 11.5. If these elements are present, particularly bismuth, often associated with lead, they are first removed by the preliminary extraction of their dithizonates into chloroform at pH 2 to 2.5 before proceeding with the lead determination. In the presence of an alkaline cyanide solution the excess of green coloured free dithizone is not extracted by chloroform. If bismuth and tin are absent, preliminary extraction at pH 2 to 2.5 may be omitted.

The absorbance of the red solution of lead dithizonate is determined spectrophotometrically. Since the method is a sensitive one, precautions must be taken to prevent contamination of reagents and apparatus. As bright sunlight tends to decompose dithizone and dithizonates, the determinations should be carried out in diffused light.

(ii) *Reagents*

Alkaline cyanide solution. Dissolve 3.0 g sodium sulfite, $Na_2SO_3.7H_2O$, in a mixture of 340 cm^3 ammonium hydroxide, (specific mass 0.88) and 680 cm^3 water. Add 30 cm^3 of 10 g dm^{-3} potassium cyanide and mix well.

Sodium hexametaphosphate solution, 100 g dm^{-3}. Dissolve 10 g sodium hexametaphosphate in 100 cm^3 water. Remove traces of lead by extraction with 1 g dm^{-3} dithizone solution at pH 9, adjusting with ammonium hydroxide. Make just acid and extract with chloroform to remove traces of dithizone. Adjust to pH 9.5 for maximum stability during storage.

Hydroxylammonium chloride solution, 10 g dm^{-3}.

Ammonia solution, 0.5 mol dm^{-3}. Dilute 3.5 cm^3 ammonia solution, (specific mass 0.88) to 100 cm^3 with water.

Dilute hydrochloric acid, (1 + 1).

Chloroform.

Dithizone stock solution, $1\,g\,dm^{-3}$ in chloroform. Dissolve $0.10\,g$ diphenylthiocarbazone in $100\,cm^3$ chloroform. Store in a refrigerator.

Dithizone working solution in water. Transfer $6\,cm^3$ dithizone stock solution in chloroform to a $100\,cm^3$ separating funnel. Add $10\,cm^3$ $0.5\,mol\,dm^{-3}$ ammonia and shake. Allow to separate, reject the lower chloroform layer and filter the aqueous layer through a wetted filter paper to remove droplets of chloroform.

Lead stock solution, $100\,g\,dm^{-3}$. Dissolve $1.599\,g$ of lead nitrate $Pb(NO_3)_2$ in water, add $10\,cm^3$ of nitric acid, (specific mass 1.42) and dilute to $1\,dm^3$.

Lead standard solution, $10\,mg\,dm^{-3}$. Dilute $10.0\,cm^3$ lead stock solution to $1\,dm^3$. Prepare freshly as required.

(iii) *Preparation of calibration graph*

Into a series of $100\,cm^3$ short-stemmed separating funnels place $50\,cm^3$ lead-free water and introduce appropriate volumes of lead working solution to cover the range up to $20\,\mu g$ Pb. Include a separating funnel containing no added lead. Proceed as described below. Deduct the absorbance of the blank from those of the standard solutions and construct a calibration graph relating absorbance to μg lead. If during the analysis of the sample it is proposed to use the preliminary extraction procedure at pH 2–2.5 to eliminate possible interference by tin and/or bismuth, then a similar step must be incorporated in the preparation of the calibration graph.

(iv) *Procedure*

Into a $100\,cm^3$ short-stemmed separating funnel measure $50\,cm^3$ of sample, or a smaller volume diluted to $50\,cm^3$, containing not more than $20\,\mu g$ of lead. If bismuth and tin are known to be absent, omit the procedure in the next paragraph.

Using a pH meter and dilute hydrochloric acid, adjust the sample to pH 2–2.5. Add $20\,cm^3$ dithizone stock solution in chloroform, shake for 2 minutes, allow the phases to separate and run the chloroform layer to waste. Repeat the extraction with a further $5\,cm^3$ dithizone stock solution, and again run the chloroform layer to waste, Wash the aqueous solution with $5\,cm^3$ chloroform, allow to separate and discard the washings (chloroform layer).

To the aqueous solution add, mixing between each addition, $1.0\,cm^3$ sodium hexametaphosphate solution, $1.0\,cm^3$ hydroxylammonium chloride solution, $30\,cm^3$ alkaline cyanide solution, $0.5\,cm^3$ dithizone working solution (in water), conveniently added from a burette with $0.02\,cm^3$ divisions, and $10\,cm^3$ chloroform. Shake the funnel vigorously for 1 minute and allow the phases to separate. Dry the stem of the separating funnel with filter paper and insert into a plug of cotton wool. Allow about $2\,cm^3$ of the chloroform solution to run to waste. Carry out a blank determination on all reagents used.

Measure the absorbances of the chloroform solution and blank, either in a spectrophotometer at 510 nm or absorptiometer with a suitable green filter using 10 mm cells with chloroform in the reference cell. From the calibration graph read the number of μg of lead of the test and blank solutions and so obtain the net measure of lead in the sample.

1.3.7 References

1 *International Drinking Water Standards*, WHO, Geneva (1972).
2 *European Drinking Water Standards*, WHO, Geneva (1970).
3 D.C. Burrell, *Atomic Spectrometric Analysis of Heavy Metal Pollutants in Water*, Ann Arbor Science, Ann Arbor (1974).
4 I.U.P.A.C. *Tables of Spectrophotometric Absorption Data of Compounds used for the Colorimetric Determination of the Elements*, Butterworths, London (1963).
5 H.M.N.H. Irving, *Dithizone*, The Chemical Society, London, (1977).
6 Department of the Environment, *Analysis of Raw, Potable and Waste Waters*, H.M. Stationary Office, London, (1972).
7 e.g. in the series *Methods for the Examination of Waters and Associated Materials—Lead in Potable Waters by AAS*, H.M. Stationary Office, London, (1976).
8 J.C. Duinker and C.J.M. Kramer, *Marine Chem.*, **5**, 763 (1977).
9 R. Chester and J.H. Stoner, *Marine Chem.*, **2**, 17 (1974).
10 L.T. Fairhall and R. Richardson, *J. Am. Chem. Soc.*, **52**, 938 (1930).
11 J. Rodier, *Analysis of Water*, Halsted Press, Wiley, New York, p. 320 (1975).
12 J.T. O'Connor and C.E. Renn, *J. Am. Water Works Assn.* **55**, 631 (1963).
13 M.B. Feldman and E.P. Nakhshina, *Gidrobiol. Zhur.*, **3**, 86 (1967).
14 M. Sadilkova, *Mikrochim. Acta*, 934 (1968).
15 P.P. Kish and I.I. Zimomyra, *Zavod. Lab.*, **35**, 541 (1969).
16 M.A. Yagodnitsyn, *Gig. Sanit.*, No **11**, 62 (1970).
17 W.H. Evans and G.S. Sayers, *Analyst*, **97**, 453 (1972).
18 G. Ackermann and J. Köthe, *Talanta*, **26**, 693 (1979).
19 Department of Environment, *Analysis of Raw, Potable and Waste Waters*, H.M. Stationary Office, London, p. 212, (1972).
20 R.M. Rush and J.H. Yoe, *Anal. Chem.*, **26**, 1345 (1954).
21 J.A. Platte and V.M. Marcy, *Anal. Chem.*, **31**, 1226 (1959).
22 J. Asplund, *Heavy Metals in Natural Waters—A Literature Survey*, Statens Naturwardsverk, Stockholm p. 89 (1979).
23 B.E. Saltzman, *Anal. Chem.*, **25**, 493 (1953).
24 J. Rodier, *Analysis of Water*, Halsted Press, Wiley, New York, p. 232 (1975).
25 J. Ganotes, E. Larson and R. Navone, *J. Am. Water Works Assn*, **54**, 852 (1962).
26 J. Korkisch and D. Dimetriadis, *Talanta*, **20**, 1295 (1973).
27 Department of the Environment, *Analysis of Raw, Potable and Waste Waters*, H.M. Stationary Office, London, p. 179 (1972).
28 J. Asplund, *Heavy Metals in Natural Waters—A Literature Survey*, Statens Naturwardsverk, Stockholm, p. 99 (1979).
29 J. Rodier, *Analysis of Water*, Halsted Press, Wiley, New York, p. 263 (1975).
30 R. Truhaut and C. Boudène, *Ann. Fals. Exp. Chim.*, 225 (1963).
31 H. Hamaguchi, R. Kureda and K. Hosohara, *J. Chem. Soc. Jpn, Pure Chem. Sect.*, **82**, 347 (1961).
32 C.T. Elly, *J. Water Pollution Control Fed.*, **45**, 940 (1973).
33 N. Ota, M. Terae and M. Isokawa, *J. Chem. Soc. Jpn, Pure Chem. Sect.*, **91**, 351 (1970).
34 K. Ueno, K. Shiraishi, T. Togo, T. Yano and H. Kobayashi, *Anal. Chim. Acta*, **105**, 289 (1979).
35 H. Schurig and H. Mueller, *Z. Chem. Lpz.*, **15**, 286 (1975).
36 M.V.R. Murti and S.M. Khopkar, *Bull. Chem. Soc. Jpn*, **50**, 738 (1977).
37 Department of Environment, *Analysis of Raw, Potable and Waste Waters*, H.M. Stationary Office, London, p. 198 (1972).
38 Analytical Methods Committee, *Analyst*, **90**, 515 (1965).
39 J. Asplund, *Heavy Metals in Natural Waters—A Literature Survey*, Statens Naturwardsverk, Stockholm, p. 106 (1979).
40 P.A. Clifford and H.J. Wichmann, *J. Assoc. Off. Agri. Chem.*, **19**, 130 (1936).
41 L.P. Biefield and T.M. Patrick, *Ind. Eng. Chem., Anal. Edn*, **14**, 275 (1942).
42 L.J. Snyder, *Ind. Eng. Chem., Anal. Edn*, **19**, 684 (1947).
43 J. Rodier, *Analysis of Water*, Halsted Press, Wiley, New York, p. 281 (1975).

44 A. Berger and J. Pirotte, *C.B.E.D.E.*, III, 13 (1951).
45 D.C. Abbott and J. R. Harris, *Analyst*, **87,** 387 (1962).
46 Department of the Environment, *Analysis of Raw, Potable and Waste Waters*, H.M. Stationary Office, London, p. 194 (1972).
47 J. Korkisch and A. Sorio, *Talanta*, **22,** 273 (1975).
48 A.I. Markova, *Zh. Anal. Khim.*, **17,** 952 (1962).

1.4　Nickel

Prepared by
G. ACKERMANN* AND L. SOMMER**

*Lehrstuhl für Analytische Chemie, Section Chemie, Bergakademie Freiberg, Leipziger Strasse, DDR-9200 Freiberg, German Democratic Republic, **and Katedra Analytiché Chemie, Prirodovedecke Fakulty, Univerzity J.E. Parkyne, Kotlarska 2, CS-611 37 Brno, Czechoslovakia*

1.4.1　Introduction

Nickel is in general not present in spectrophotometrically detectable amounts in potable waters, but it has been observed in certain mineral waters $(20 \, mg \, dm^{-3})$ [1]; the amounts in sea water range from 0.12 to $60 \, \mu g \, dm^{-3}$ [2]. Large amounts of nickel may occur in mine drainage waters. Nickel is always present in the divalent state.

Nickel can be preconcentrated from water samples and separated from some interfering substances by a variety of methods. Some of the more important are summarized below.

(a) Nickel and other metal traces can be preconcentrated on a chelating resin containing iminodiacetic acid groups (e.g. Dowex A 1, Na^+-form or NH_4^+-form) at pH 4.6–6.5 with acetate buffer [3], after which Ni(II) can be separated from the other elements on a strongly basic anion exchanger (e.g. Dowex 1-X10, Cl^--form) by elution with $12 \, mol \, dm^{-3}$ HCl [2].

(b) Nickel(II)-hydroxide can be coprecipitated on calcium or magnesium carbonate using 5% (w/v) sodium carbonate as precipitant [2].

(c) Copper(II) and iron(III) can be removed as cupferronates by extraction from slightly acidic solutions in the presence of tartrate [1] into chloroform prior to determination of nickel.

Water samples should be adjusted to pH 1.5–2.0 with $6 \, mol \, dm^{-3}$ HCl just after sampling and may be stored for 24 h before analysis.

1.4.2　Spectrophotometric determination of nickel

1.4.2.1　General aspects

α-Dioximes, especially dimethylglyoxime [2–6], cyclohexanedione dioxime, furildioxime and cycloheptanedione dioxime [7, 8], various N-heterocyclic 2-hydroxyazo dyes, dithiocarbamates [9] or pyridine-2-aldehyde-2-quinolyl-hydrazone [9–11] are the most frequently mentioned reagents for the spectrophotometric determination of nickel. 4-(2-Pyridylazo) resorcinol (PAR) is recommended herein because of the low limit of detection for nickel and because of its ready availability.

1.4.2.2 Procedures

1.4.2.2.1 Spectrophotometric determination of nickel with PAR (4-(2-Pyridylazo)resorcinol)

(i) *Introductory remarks*

Nickel(II) forms a highly absorbing red NiL_2^{2-} chelate with PAR at pH 9–10 (λ_{max} 493 nm; $\varepsilon = 7.55 \times 10^4$; 7.34×10^4: 500 nm [6] or 6.84×10^4 dm^3 mol^{-1} cm^{-1}: 495 nm [12]). This reaction is sensitive, but not selective. The formation of PAR chelates of a number of interfering ions can be masked by addition of citrate and EDTA at pH 9–10, exceptions being those of Co(III) Fe(II), Fe(III), Pd(II), UO_2^{2+} and V(v).

(ii) *Preconcentration and separation*

Nickel and other trace elements can be preconcentrated from 1 dm^3 water samples on a 10×1 cm column of chelating resin (e.g. DOWEX A 1, 30–70 mesh, Na^+-form) at pH 6.5. Fe(II) is first oxidized by addition of 30% H_2O_2 to the acidic sample, the excess of H_2O_2 is removed by boiling. The pH is adjusted to 6.5 by 2 $mol\,dm^{-3}$ sodium acetate and the sample solution transferred to a prepared chelating resin column (flow rate 3.5 $cm^3\,min^{-1}$). The column is then washed with three 10 cm^3 portions of distilled water and nickel and other trace elements eluted with 50 cm^3 2 $mol\,dm^{-3}$ HCl (flow rate 0.5–1.0 $cm^3\,min^{-1}$). The column effluent is evaporated in a silica dish, the residue is dissolved with 2 cm^3 12 $mol\,dm^{-3}$ HCl and placed on a column of a strongly basic anion exchanger (e.g. Dowex 1-X8, 50–100 mesh, Cl^--form) in a 11×0.7 cm column pretreated by 12 $mol\,dm^{-3}$ HCl. The nickel is eluted with 30 cm^3 of 12 $mol\,dm^{-3}$ HCl (flow rate 0.5–0.7 $cm^3\,min^{-1}$). This effluent is evaporated in a silica dish, the residue is dissolved in 10 cm^3 distilled water, transferred to a volumetric flask and nickel is determined. Copper, iron and cobalt are not eluted by 12 $mol\,dm^{-3}$ HCl.

(iii) *Interferences*

If no previous separation of nickel is made, V(v), Fe(II), Fe(III), Co(II), Pd(II), UO_2^{2+} interfere strongly. Limited amounts of Ti (5:1), UO_2(9:1), Pb(36:1), Cu(50:1), Hg(100:1), Zn(100:1), Cd(200:1) do not interfere with the determination of 0.36 mg dm^{-3} Ni. Thiourea masks limited amounts of copper, hydrogen carbonate limited amounts of UO_2^{2+} and fluoride large amounts of Ti(IV) and Al(III). These masking agents must be present in the solution before PAR is added.

(iv) *Reagents*

PAR solution, 0.002 $mol\,dm^{-3}$. Dissolve 50 mg of pure (or doubly recrystallized from methanol) 4-(2-pyridylazo)-resorcinol monosodium salt monohydrate in water containing a few drops of $2 \, mol \, dm^{-3}$ NaOH and dilute to 100 cm³.

EDTA solution, $0.25 \, mol \, dm^{-3}$. Dissolve 46.6 g of ethylenediamine tetraacetic acid, disodium salt, dihydrate in 400 cm³ water, adjusted to pH 9 with dilute sodium hydroxide solution and make up with water to 500 cm³.

Potassium citrate solution, $2.5 \, mol \, dm^{-3}$. Dissolve 96.06 g of citric acid in 200 cm³ water, add 84.0 g of potassium hydroxide whilst stirring and cooling, adjust the pH to 9.5 and dilute with water to 250 cm³.

Borate buffer solution, pH 9.6. Dissolve 61.0 g of boric acid in water adjust to pH 9.6 with 15% potassium hydroxide solution and dilute with water to 1 dm³.

Standard nickel solution, $50 \, mg \, dm^{-3}$ Ni. Dissolve 0.2500 g of pure nickel nitrate hexahydrate in 1 dm³ water. Determine the nickel concentration by EDTA titration using murexide as indicator.

Thiourea solution, $0.5 \, mol \, dm^{-3}$. Dissolve 19.03 g of thiourea in water and dilute to 500 cm³. For preparation of all solutions and during the procedure doubly distilled water should be used.

(v) *Procedure*

Transfer the slightly acidic solution, or the evaporated residue from the column, dissolved in not more than 10 cm³ water to a volumetric flask. Add 1 cm³ thiourea solution (if copper is present) and 5 cm³ PAR solution. Wait 10 minutes and then add successively 5 cm³ potassium citrate solution, 5 cm³ borate buffer solution (the final pH must be between 9 and 10) and 20 cm³ EDTA solution (instead of borate buffer, the phosphate buffer, pH 9, may be used). Dilute to 50 cm³ with water and read absorbance at 500 nm against water, subtract the reagent blank. Determine the nickel content from the calibration graph. The calibration graph is prepared using 0.15–10 cm³ of tenfold diluted standard nickel solution.

(vi) *Characteristics of the procedure*

Range of application: 50–$1000 \, \mu g \, dm^{-3}$ Ni [12].
Standard deviation: $(n = 21)$ $s = 7 \, \mu g \, dm^{-3}$ Ni (for a standard solution containing $50 \, \mu g \, dm^{-3}$ Ni) [12].
Detection limit: $8 \, \mu g \, dm^{-3}$ Ni [6, 12].

1.4.3 **References**

1   A. Nevoral and A. Okáč, *Česk. Farm.*, **17**, 478 (1968).
2   W. Forster and H. Zeitlin, *Anal. Chim. Acta*, **35**, 42 (1966).
3   F.D. Snell, *Photometric and Fluorometric Methods of Analysis of Metals*. Part I, Wiley, New York p. 879 (1978).
4   M. Kenigsberg and L. Stone, *Anal. Chem.*, **27**, 1339 (1955).
5   E. Kentner, D.B. Armitage and H. Zeitlin, *Anal. Chim. Acta*, **45**, 343 (1969).
6   Z. Šimek, M. Langová and L. Sommer, to be published.
7   V.M. Peshkova and N.G. Ignatieva, *Zh. Anal. Khim.*, **17**, 107 (1962).

8 E.B. Sandell and R.W. Perlich, *Ind. Eng. Chem.*, Anal. Edn, **11,** 309 (1939).

9 O.G. Koch and G.A. Koch-Dedic, *Handbuch der Spurenanalyse*, Springer-Verlag, Berlin (1974).

10 S.P. Singhal and D.E. Ryan, *Anal. Chim. Acta*, **37,** 91 (1967).

11 B.K. Afghan and D.E. Ryan, *Anal. Chim. Acta*, **41,** 167 (1968).

12 G. Ackermann, J. Köthe and H. Thor, unpublished investigations.

1.5 Cobalt

Prepared by
G. ACKERMANN* AND L. SOMMER**

*Lehrstuhl für Analytische Chemie, Section Chemie, Bergakademie Freiberg,
Leipziger Strasse, DDR-9200, Freiberg, German Democratic Republic **and Katedra
Analytické Chemie, Prirocovedecke Fakulty, Univerzity J.E. Parkyne, Kotlarska 2,
CS-611 37 Brno, Czechoslovakia.*

1.5.1 Introduction

No evidence has been found of detectable concentrations of cobalt in potable waters, but cobalt is often present in mineral waters and mine drainage waters. Concentrations observed in mineral waters are $0.02–20\ \mu g\ dm^{-3}$ Co, in such cases the concentration ratio of Co:Ni is usually 1:3–1:5 [1]. Small amounts of cobalt ($\leqslant 2\ \mu g\ dm^{-3}$ Co) can be present in surface waters. Cobalt will in general be present in the divalent state.

Cobalt can be preconcentrated from water samples and separated from interfering substances by a variety of methods. The most important are summarized below.

(a) Cobalt and other trace elements are retained on chelating resins containing iminodiacetic acid groups (e.g. Dowex A 1, NH_4^+-form, 30–70 mesh) at pH 6–7.6 with acetate buffer. Cobalt and other elements are eluted from the column with $2\ mol\ dm^{-3}$ HCl. The effluent is evaporated, the residue is dissolved in $12\ mol\ dm^{-3}$ HCl and transferred to a column of a strongly basic anion exchanger (e.g. Dowex 1-X10 in Cl^--form). Nickel is first eluted with $12\ mol\ dm^{-3}$ HCl and thereafter cobalt and copper are eluted with $4.25\ mol\ dm^{-3}$ HCl. More than 0.1 mg Cu interferes with the separation of cobalt. Copper can, however, be separated from cobalt on a cation exchanger (e.g. Dowex 50-XS in Na^+-form by a thiosulfate solution [2, 3]).

(b) Cobalt can be separated from iron and nickel on a strongly basic anion exchanger with mixtures: acetone–$6\ mol\ dm^{-3}$ HCl $(9+1)$ or acetone–$2\ mol\ dm^{-3}$ HCl $(7+3)$ or with $2\ mol\ dm^{-3}$ HCl [4].

(c) A selective retention of cobalt results from water samples containing ascorbic acid and excess of thiocyanate on a strongly basic anion exchanger in SCN^--form. Any accompanying iron(III) is first eluted with a mixture of tetrahydrofuran, ethyleneglycol monomethylether and hydrochloric acid and then cobalt with $6\ mol\ dm^{-3}$ HCl [5].

(d) Similarly, cobalt and iron(III) are first retained on a strongly basic anion exchanger in SCN^--form from sea water in a thiocyanate–hydrochloric acid medium. The thiocyanate complexes may be eluted with $2\ mol\ dm^{-3}$ $HClO_4$, thiocyanate is destroyed and cobalt separated from iron on a second column of strongly basic anion exchanger in Cl^--form with $6\ mol\ dm^{-3}$ HCl [6].

For storage, water samples have to be acidified to pH 1.5–2.0 by addition of e.g., $6\ mol\ dm^{-3}$ HCl.

1.5.2 Spectrophotometric determination of cobalt

1.5.2.1 General aspects

2-Nitroso-1-naphthol has been selected from the large number of chromogenic reagents for cobalt because it is sensitive and exhibits a considerable selectivity in extraction procedures. Two N-heterocyclic 2-hydroxyazo dyes are also selected because of their high sensitivity, the selectivity can be improved by masking of interfering ions or by previous separation of cobalt using ion exchangers.

1.5.2.2 Procedures

1.5.2.2.1 Spectrophotometric determination of cobalt with 2-Nitroso-1-naphthol [7]–[11]

(i) *Introductory remarks*

An insoluble pinkish-violet or red chelate, $Co(III)L_3$ is quantitatively formed after 20 minutes in solutions containing $Co(II)$ and hydrogen peroxide at pH 4–7. This chelate can be extracted into chloroform or toluene ($\lambda_{max} = 535$ nm, $\varepsilon = 1.44 \times 10^4$ dm^3 mol^{-1}cm^{-1} (in toluene) [8] or $\lambda_{max} = 308$ nm, $\varepsilon = 5.6 \times 10^4$ dm^3 mol^{-1} cm^{-1}; or $\lambda = 365$ nm, $\varepsilon = 3.7 \times 10^4$ dm^3 mol^{-1} cm^{-1} (in chloroform) [12]; or $\lambda = 530$ nm, $\varepsilon = 1.46 \times 10^4$ dm^3 mol^{-1} cm^{-1} (in toluene) [13].

A number of interfering ions can be masked by citrate, otherwise several chelates can be back-extracted from the organic solvent with 2 mol dm^{-3} HCl. The reagent excess can also be back-extracted with 2 mol dm^{-3} NaOH.

(ii) *Reagents*

2-Nitroso-1-naphthol solution, 0.1% in methanol. The solution must be freshly prepared. The reagent can be purified by filtration through a pure alumina column in benzene–dioxan–acetic acid $(90 + 25 + 4)$ mixture. The purity of the reagent can be tested by thin-layer chromatography on alumina sheets pretreated with EDTA with a benzene–methanol–acetic acid $(45 + 8 + 4)$ mixture.

Sodium citrate solution, 40%. Dissolve 325.8 g citric acid in 500 cm^3 water and 186 g sodium hydroxide in 300 cm^3 water. Mix thoroughly and dilute to 1 dm^3 with water.

Mercury(II)-nitrate solution, 1 g dm^{-3}. Dissolve 1.0 g mercury in 5 cm^3 concentrated HNO$_3$ under gentle heating and dilute after cooling with water to 1 dm^3.

Hydrogen peroxide solution, 30%.

Hydrochloric acid, 2 mol dm^{-3}.

Sodium hydroxide solution, 2 mol dm^{-3}.

Toluene.

Standard cobalt solution, 100 mg dm^{-3}. Dissolve 0.1000 g cobalt metal in a suitable amount of concentrated HCl, containing some concentrated HNO$_3$. Expel excess of hydrochloric acid by evaporation, dissolve the residue and dilute with water to 1 dm^3.

For preparation of all aqueous solutions and in the procedure, doubly distilled water should be used.

(iii) *Procedure* [7, 8]

Add 10 cm^3 of sodium citrate solution to the aliquot of preconcentrated water sample containing $\geqslant 5\ \mu$g Co, adjust the pH to 4–7 by addition of diluted hydrochloric acid or sodium hydroxide, then add 2 cm^3 hydrogen peroxide solution, 10 cm^3 reagent solution and 10 cm^3 mercury(II)-nitrate solution (the latter if CN$^-$ is present). Dilute with water to 100 cm^3. Mix thoroughly and leave the solution for 30 minutes. Transfer it into a separating funnel, adjust the volume to ~ 150 cm^3 and shake for 5 minutes with 10.0 cm^3 toluene. After phase separation, discard the aqueous layer, wash the organic layer successively with 20 cm^3 2 mol dm^{-3} hydrochloric acid, 20 cm^3 2 mol dm^{-3} sodium hydroxide solution and finally three times with 100 cm^3 water. Filter the organic layer through a clean filter paper into a dry flask and measure the absorbance at 530 nm against the reagent blank. Determine the cobalt content from the calibration graph. The calibration graph is prepared using 0.5–6.0 cm^3 of the tenfold diluted standard cobalt solution.

(iv) *Interferences*

2 mg Fe(III), Zn, Cd, Mn(II), Pb, Hg(II), Cr, Pt(IV) and large amounts of Ca, Mg, Al do not interfere in the determination of 20 μg Co. In the presence of 500 μg Ni, a 5% increase of the absorbance at 535 nm is observed in the determination of 20 μg Co. CN$^-$ interferes, but can be masked by mercury(II).

(v) *Characteristics of the procedure* [13]

Range of application: 0.5–6.0 mg dm^{-3} Co in the toluene extract.
Standard deviation: ($n=21$) $s=15\ \mu$g dm^{-3} Co in toluene extract (for standard solutions containing 2.0 mg dm^{-3} Co in toluene extract).
Detection limit: 0.08 mg dm^{-3} Co.

1.5.2.2.2 Spectrophotometric determination of cobalt with PAR

(i) *Introductory remarks*

A red chelate, Co(III)L$_2$, $\lambda_{max} = 510$ nm, $\varepsilon = 6 \times 10^4$ dm^3 mol^{-1} cm^{-1} [14, 15], is formed from Co(II) and 4-(2-pyridylazo)-resorcinol (PAR) in aqueous solutions in the optimal pH range 6.8–9.0 in the presence of ammonium acetate or borate buffers. Oxidation of Co(II) to Co(III) occurs by oxygen. This chelate is stable and kinetically inert and a number of interfering metal–PAR chelates can be masked by EDTA after the quantitative formation of the Co(III) chelate, with the exception of those of UO_2^{2+}, Ni(II), V(v), Fe(II), Fe(III) and Hg(II).

(ii) *Reagents*

4-(2-Pyridylazo)-resorcinol (PAR) solution, $0.00125 \, \text{mol dm}^{-3}$. Dissolve 0.270 g of the pure monosodium salt monohydrate in a little dilute sodium hydroxide solution and dilute to $100 \, \text{cm}^3$ with water. The reagent should be recrystallized from methanol and its purity tested chromatographically by TLC on silica gel layers, pretreated by EDTA, using benzene–methanol $(9+1)$ as eluant [16].

Borate buffer solution, pH 8.9. Dissolve 15.46 g of boric acid in such a volume of 30% pure sodium hydroxide solution that a solution of pH 8.9 results and dilute with water to $500 \, \text{cm}^3$.

Sodium citrate solution, $1 \, \text{mol dm}^{-3}$, pH 8.2. Dissolve 96.06 g of citric acid in water, adjust the pH to 8.2 with sodium hydroxide and dilute the solution to $500 \, \text{cm}^3$ with water.

EDTA solution, $0.1 \, \text{mol dm}^{-3}$. Dissolve 37.22 g of EDTA in $1 \, \text{dm}^3$ of water.

Thiourea solution, $0.1 \, \text{mol dm}^{-3}$. Dissolve 7.61 g of thiourea in $1 \, \text{dm}^3$ of water.

Standard cobalt solution, $100 \, \text{mg dm}^{-3}$. Dissolve 0.1000 g of cobalt metal by heating in a mixture of concentrated HCl and HNO_3, expel the excess hydrochloric acid by partial evaporation and dilute the remaining solution with water to $1 \, \text{dm}^3$.

Doubly distilled water should be used for preparation of all solutions and in the procedure.

(iii) *Procedure* [14]

Place the very slightly acidic solution of the preconcentrated water sample or the solution of the residue of the ion exchanger column effluent containing 1.5–45 μg Co in a $25 \, \text{cm}^3$ volumetric flask. Add $2.5 \, \text{cm}^3$ $0.1 \, \text{mol dm}^{-3}$ thiourea, $1.5 \, \text{cm}^3$ sodium citrate solution, $5 \, \text{cm}^3$ borate buffer, $2 \, \text{cm}^3$ reagent solution. Mix, check that the pH is between 8 and 9, leave the solution for 5 minutes to allow the Co(III)–chelate to be formed quantitatively, finally add $2.5 \, \text{cm}^3$ EDTA and dilute to $25 \, \text{cm}^3$ with water. Place the flask with contents in a 80°C water bath for 10 minutes, cool and measure the absorbance at 510 nm against water. Subtract the absorbance of the reagent blank and determine the cobalt content from the calibration graph.

(iv) *Interferences*

Nickel(II) strongly interferes and must be first removed as well as large amounts of UO_2^{2+}, Cu(II), Fe(II), Fe(III), Cr(III), Mn(II) and Hg(II). A reduction in the formation of the Co(III)–chelate is observed in the presence of large concentrations of many transition metal ions.

In the determination of 600 μg dm^{-3} Co(II) a 20-fold amount of Cu(II), a 50-fold amount of Mn(II), a 25-fold amount of Pb(II) and a 15-fold amount of UO_2^{2+} do not interfere with the determination of 15 μg Co in 25 cm^3. UO_2^{2+} in a 150-fold excess can be masked by HCO_3^-.

(v) *Characteristics of the procedure* [15]

Range of application: 50–1800 μg dm^{-3} Co.
Standard deviation: ($n=21$). $s=5$ μg dm^{-3} Co (for a standard solution containing 400 μg dm^{-3} Co).
Detection limit: 10 μg dm^{-3} Co.

1.5.2.2.3   **Spectrophotometric determination of cobalt with 2-(5-Bromopyridylazo)-5-diethylaminophenol (5-Br-PADAP)** [3]

(i) *Introductory remarks*

Cobalt(II) forms a blue Co(III)-chelate with 5-Br-PADAP in the presence of a suitable oxidizing agent in sulfate or nitrate medium in the optimal pH range 4–8 ($\lambda_{max} = 586$ nm, $\varepsilon = 9.21 \times 10^4$ dm^3 mol^{-1} cm^{-1}). This complex is soluble in 10% dimethylformamide in the presence of the non-ionic detergent Triton X-100. The chelate is stable in 3.0 mol dm^{-3} H$_2$SO$_4$ or 1.8 mol dm^{-3} HNO$_3$ under which conditions chelates of many interfering metal ions are decomposed.

(ii) *Preconcentration*

Pass 1 dm^3 or more of the water sample at pH 6.0–6.5 through a column of chelating resin (e.g. Dowex A 1, 30–70 mesh, Na$^+$-form, column 10×1 cm, flow rate 3.5–4.5 cm^3 min^{-1}). Wash the column with 30 cm^3 water and elute cobalt and other trace elements with 50 cm^3 2.5 mol dm^{-3} HCl (flow rate 0.5–1.0 cm^3 min^{-1}). Evaporate the effluent, dissolve the residue in 2 cm^3 concentrated HNO$_3$ to remove hydrochloric acid and to oxidize Fe(II). Re-evaporate and dissolve the residue in 1 drop HNO$_3$(1 + 1) and a few cm^3 water.

(iii) *Reagents*

2-(5-Bromopyridylazo)-5-diethylaminophenol   solution, 7.5×10^{-4} mol dm^{-3}. Dissolve 26.2 mg of pure reagent in 100 cm^3 dimethylformamide.

Ammonium   persulfate   solution,   0.25 mol dm^{-3}. Dissolve   5.71 g (NH$_4$)$_2$S$_2$O$_8$ in 100 cm^3 water.

Triton X-100 solution, 5%. Dissolve 5 g of octylphenylpolyethyleneglycol ether in warm water and dilute to 100 cm^3. The solution should not be kept for longer than 3 weeks.

Ammonium acetate solution, 1 mol dm^{-3}. Dissolve 38.54 g in 500 cm^3 water.

Nitric acid, 8 mol dm^{-3}.

Standard cobalt solution, 100 mg dm^{-3}. Dissolve 0.1000 g of cobalt metal by heating in a mixture of concentrated HCl and HNO_3, expel the excess hydrochloric acid by partial evaporation and dilute the remaining solution with water to 1 dm^3.

Doubly distilled water should be used for preparation of all solutions and in the procedure.

(iv) *Procedure*

Place the slightly acidified preconcentrated water sample or the solution prepared from the ion exchange effluent, containing less than 17 μg Co in a 25 cm^3 volumetric flask. Add 0.5 cm^3 Triton X-100 solution, 0.5 cm^3 ammonium persulfate solution and 2.5 cm^3 ammonium acetate solution. Check that the pH is between 5 and 7 then add 2.5 cm^3 of the reagent solution. Mix thoroughly and leave the solution for 15 minutes. Then add 5.7 cm^3 of 8 mol dm^{-3} nitric acid and dilute to volume with water. Measure the absorbance within 10 minutes, at 586 nm against the sample solution blank. Subtract the absorbance of the reagent blank and determine the cobalt content from the calibration graph.

(v) *Interferences*

The reaction is highly selective for cobalt. Pd(II), Cu(II), V(V), UO_2^{2+}, Hg(II) and Ni(II) can interfere, but a 60-fold amount of Ti(IV), 50-fold amount of Cr(III), 50-fold amount of Mn(II), 10-fold amount of Ni(II), 4-fold amount of UO_2^{2+}, or 2-fold amount of Pd(II) do not interfere with the determination of 300 μg dm^{-3} Co. UO_2^{2+} may be masked with HCO_3^- and Ti(IV) in larger amounts with F^-.

(vi) *Characteristics of the procedure* [3]

Range of application: 10–700 μg dm^{-3}.
Standard deviation: $s = 2$ μg dm^{-3} Co.
Detection limit: 6 μg dm^{-3} Co.

1.5.3 References

1 Z. Okáč and V. Nevoral, *Česk. Farm.*, **19**, 139 (1970).
2 J.P. Riley and D. Taylor, *Anal. Chim. Acta*, **40**, 479 (1968).
3 J. Zbiral and L. Sommer, *Fresenius Z. Anal. Chem.*, **306**, 129 (1981).
4 I. Hazan and J. Korkish, *Anal. Chim. Acta*, **32**, 46 (1965).
5 J. Korkish and D. Dimitriades, *Talanta*, **20**, 1287 (1973).
6 T. Kiriyama and R. Kuroda, *Fresenius Z. Anal. Chem.*, **288**, 354 (1977).
7 A. Bíliková, *Vodni hospodařstvi*, B, Nr 9, 258 (1971).
8 A. Bíliková, *Determination of Cobalt and Chromium in Waters*, Special Publication, Nr. 72, Water Research Institute, Bratislava, (1974).
9 H. Schüller, *Mikrochim. Acta*, 107 (1959).

10 A. Classen and A. Daamen, *Anal. Chim. Acta*, **12**,.547 (1955).

11 W. Nielsch, *Mikrochim. Acta*, 725 (1959).

12 O.G. Koch, and G.A. Koch-Dedic, *Handbuch der Spurenanalyse*, Vol. 1, Springer-Verlag, Berlin, p. 592 (1974).

13 G. Ackermann, J. Köthe and I. Pechfelder, unpublished investigations.

14 J. Zbiral and L. Sommer, *Scripta Fac. Sci. Nat. Univ. Brno*, **12**, 283 (1982).

15 G. Ackermann, J. Köthe and H. Thor, unpublished investigations.

16 M. Langová and L. Sommer, unpublished investigations.

1.6 Copper

Prepared by
G. ACKERMANN* AND L. SOMMER**

* Lehrstuhl für Analytische Chemie, Section Chemie, Bergakademie Freiberg, Leipziger Strasse, DDR-9200 Freiberg, German Democratic Republic **and Katedra Analytiché Chemié, Prirocovedecke Fakulty, Univerzity J.E. Parkyne, Kotlarska 2, CS-611 37 Brno, Czechoslovakia

1.6.1 Introduction

The copper content of potable water is usually in the range $0.001-0.6$ mg dm^{-3}, but normally <0.03 mg dm^{-3}; in sea water the concentration is between 0.05 and 0.01 mg dm^{-3}; mineral waters may contain several mg dm^{-3} Cu, mine drainage waters several hundreds mg dm^{-3} Cu. The recommended concentration of copper in potable water is <0.05 mg dm^{-3}, and the upper limit 1.5 mg dm^{-3} [1]. The taste limit (copper causes a bitter taste in water) is 1 mg dm^{-3}.

Copper is, in general, in the divalent state. Concentrations greater than 0.01 mg dm^{-3} Cu can be directly determined photometrically without preconcentration. Preconcentration of copper can be achieved by evaporation of the acidified water or by ion exchangers. Suitable anion exchangers (e.g. Dowex 1-X8, 100–200 mesh, in Cl$^-$-form) bind copper from hydrochloric acid medium in the presence of ascorbic acid. Copper(I) is then oxidized and eluted with 1 mol dm^{-3} nitric acid [2].

The samples must be conserved by the addition of 2–5 cm^3 concentrated nitric or hydrochloric acid per 1 dm^3 of water (for the dicupral method only hydrochloric acid should be used) and stored up to 24 h before analysis.

1.6.2 Spectrophotometric determination of copper

1.6.2.1 General aspects

A vast number of reagents are available for the spectrophotometric determination of copper [3]. The following reagents were selected on the basis of their selectivity: Neocuproine (2,9-dimethyl-1,10-phenanthroline), Bathocuproine disulfonic acid (2,9-dimethyl-4,7-diphenyl-1,10-phenanthroline disulfonic acid), Cuprizone (bis-cyclohexanone oxalyldihydrazone) and Dicupral (tetraethylthiuramidisulfide).

1.6.2.2 Procedures

1.6.2.2.1 Spectrophotometric determination of copper with tetraethylthiuramdisulfide (Dicupral) [4]–[10]

(i) *Introductory remarks*

Dicupral forms a yellowish-brown chelate CuL$_2$ with copper(II) in hydrochloric acid medium (for ~ 1.2 mol dm^{-3} HCl $\lambda_{max} = 430-435$ nm, $\varepsilon = 2.3$

$\times 10^4 \, dm^3 \, mol^{-1} \, cm^{-1}$). The colour develops within 7 minutes and slowly fades after 15 minutes. The complex is soluble in aqueous ethanol or can be extracted into chloroform. The colour develops faster in less acidic solutions and higher molar absorptivities are obtained ($\varepsilon = 3.75 \times 10^4 \, dm^3 \, mol^{-1} \, cm^{-1}$ for 422 nm) [10]. Dicupral is probably decomposed to diethyldithiocarbamate in acid medium which then reacts with $Cu(II)$. The reaction is quite selective for copper(II); only $Hg(II)$, $Se(IV)$ and $Ag(I)$ interfere seriously.

(ii) Reagents

Dicupral solution, 0.3% (w/v). Dissolve 0.300 g tetraethylthiuramdisulfide (Dicupral) in 95% ethanol and dilute to 100 cm^3 with ethanol.

   EDTA solution, 0.1 mol dm^{-3}. Dissolve 37.22 g EDTA in 1 dm^3 of water.

   Hydrochloric acid, conc., specific mass 1.19 g cm^{-3}.

   Chloroform.

   Petroleum spirit, boiling range 60–80°C.

   Ethanol 95% w/v, may contain 5% (v/v) methanol.

   Sodium fluoride solution, 5% (w/v). Dissolve 5.0 g sodium fluoride in 100 cm^3 water.

   Standard copper solution, 0.200 g dm^{-3} Cu. Dissolve 0.200 g copper metal in dilute nitric acid, add a small amount of sulfuric acid, evaporate the solution to SO_3 fumes and dilute with water to 1 dm^3.

   Doubly distilled water should be used for preparation of all solutions and in the procedure.

(iii) Procedure A (suitable for potable water)

Acidify an appropriate volume of water containing 5–500 μg Cu with 5 cm^3 concentrated hydrochloric acid, evaporate to ~ 20 cm^3, cool and transfer to a 100 cm^3 volumetric flask. Dilute to 50 cm^3 with water, add 1 cm^3 0.1 mol dm^{-3} EDTA solution, 40 cm^3 ethanol, 3 cm^3 of dicupral solution and make up to volume with ethanol. Measure the absorbance of the solution at 435 nm against the blank. Determine the copper content from the appropriate calibration graph.

   A slightly yellowish tint of the water coming from iron(III) salts, can be removed beforehand by addition of sodium fluoride solution.

(iv) Procedure B

Add 1 cm^3 concentrated hydrochloric acid and 3 cm^3 dicupral solution to a 100 cm^3 water sample (containing $\geqslant 10$ μg Cu) in a separatory funnel, mix and leave for 10 minutes. Then add 10.0 cm^3 of chloroform and shake for 1–2 minutes vigorously. Filter the chloroform layer through a clean filter paper and measure its absorbance at 435 nm against the reagent blank. Determine the copper content from the calibration graph.

   This procedure is suitable for clear, coloured or turbid surface waters.

(v) *Interferences*

Procedure A: The following ions interfere: $NO_2^- \geqslant 0.2\,mg\,dm^{-3}$, $CN^- \geqslant 2\,mg\,dm^{-3}$ and $NO_3^- \geqslant 20\,mg\,dm^{-3}$, but they can be removed by evaporation with concentrated HCl. Turbidity, large amounts of organic substances and coloured cationic species also interefere. No interference is given in the determination of 20 μg Cu by 2 mg Ca, Mg, Al, Mn(II), Zn, Pb, Cd, Hg(II), Co(II) Ni or 0.5 mg Fe(III) Cr(III).

Procedure B: NO_2^-, NO_3^- and CN^- also interfere, but up to 3 mg of Fe(III), Mn(II), Cr(VI), Zn, Co, Ni, Cd, Pb, Hg(II) or Ag do not in the determination of 20 μg Cu.

Organic compounds causing turbidity (fats, oil, petrol, detergents, pesticides, etc.) can be removed by extraction with petroleum spirit from the acidic water sample.

(vi) *Characteristics of the procedure* [11]

Range of application: 0.05–5 mg dm^{-3}, procedure A;
 1–10 mg dm^{-3}, procedure B.
Standard deviation: ($n=21$), $s=10$ μg dm^{-3} (for standard solution containing 1.3 mg dm^{-3} Cu) procedure A.
 ($n=21$), $s=45$ μg dm^{-3} (for standard solution containing 2.0 mg dm^{-3} Cu) procedure B.
Detection limit: 30 μg dm^{-3}, procedure A;
 90 μg dm^{-3}, procedure B.

1.6.2.2.2 **Spectrophotometric determination of copper with Neocuproine (2,9-dimethyl-1,10-phenanthroline)[12]–[15]**

(i) *Introductory remarks*

Neocuproine forms a yellow copper(I)-complex CuL_2^+ in neutral or slightly acidic medium (pH 3–9). This complex can be extracted as an ion pair into chloroform, n-pentanol, i-pentanol, chloroform + ethanol mixture and especially n-hexanol in the pH range 4–6 in the presence of a suitable buffer (in chloroform: $\lambda_{max}=457$ nm, $\varepsilon=8.6 \times 10;^3$ dm^3 mol^{-1} cm^{-1}; in n-hexanol: $\lambda_{max}=454$ nm, $\varepsilon=8.0 \times 10^3$ dm^3 mol^{-1} cm^{-1}). Copper(II) must be reduced to copper(I) by hydroxylammonium chloride. The reaction is highly selective for copper in the presence of citrate.

(ii) *Reagents*

Neocuproine solution, 1.0 g dm^{-3}. Dissolve 100 mg of 2,9-dimethyl-1,10-phenanthroline hemihydrate in 100 cm^3 methanol.

Hydroxylammonium chloride solution, 10% (w/v). Dissolve 50 g hydroxylammonium chloride in 450 cm^3 water and dilute to 500 cm^3.

Sodium citrate solution, 37.5% (w/v). Dissolve 150 g of sodium citrate dihydrate in 400 cm^3 water. Remove traces of copper by extraction with chloroform after addition of 5 cm^3 of hydroxylammonium chloride solution and 10 cm^3 of reagent solution.

Ammonia, 25% (w/v).

Chloroform.

Standard copper solution, 0.20 g dm^{-3}. Dissolve 0.2000 g copper metal in dilute nitric acid, boil to expel nitrogen oxides and dilute with water to 1 dm^3.

Doubly distilled water should be used for preparation of all solutions and in the procedure.

(iii) *Procedure*

Place a suitable aliquot of the water sample from a preliminary treatment containing 10–300 μg Cu into a separatory funnel, dilute to 50 cm^3 with water, add 5 cm^3 hydroxylammonium chloride solution, 10 cm^3 of sodium citrate solution and adjust the pH with ammonia to pH 4. Add 10 cm^3 of Neocuproine solution and 10 cm^3 chloroform. Shake vigorously for 30 s or more, run the chloroform layer into a volumetric flask. Repeat the extraction with a further 10 cm^3 chloroform, dilute the combined extracts to 25 cm^3 with methanol and mix thoroughly. Measure the absorbance at 457 nm against the reagent blank. Determine the copper content from the calibration graph.

(iv) *Characteristics of the procedure* [16]

Range of application: 0.4–12.0 mg dm^{-3} Cu in chloroform.

Standard deviation: ($n = 21$), $s = 23$ μg dm^{-3} Cu in chloroform (for standard solution containing 3.6 mg dm^{-3} Cu in chloroform).

Detection limit: 20 μg dm^{-3} Cu in chloroform.

1.6.2.2.3   **Spectrophotometric determination of copper with bathocuproine disulfonic acid (2,9-dimethyl-4,7-diphenyl-1,10-phenanthroline disulfonic acid, disodium salt) [17]–[20]**

(i) *Introductory remarks*

Bathocuproine disulfonic acid forms a water-soluble, red chelate with copper(I), CuL_2^{3-}, in the pH range 3.5–8.0 or optimally at pH 4.3–4.5 ($\lambda_{max} = 484$ nm, $\varepsilon = 1.25 \times 10^4$ dm^3 mol^{-1} cm^{-1}). Copper(II) is reduced first to copper(I) by hydroxylammonium chloride or ascorbic acid in slightly acidic medium. The reaction is highly selective in the presence of citrate. The analogous CuL_2^+ chelate of the non-sulfonated bathocuproine can be extracted in the pH range 1.5–10.0 into chloroform, pentanol, i-pentanol or n-hexanol [21, 22].

(ii) *Reagents*

Bathocuproine disulfonic acid solution, 1.0 g dm^{-3}. Dissolve 1.000 g of bathocuproine disulfonic acid, disodium salt in 1 dm^3 water or 2 mol dm^{-3} sodium acetate solution.

Sodium citrate solution, 30% (w/v). Dissolve 300 g of sodium citrate dihydrate in 1 dm^3 water.

Hydroxylammonium chloride solution, 10% (w/v). Dissolve 100 g hydroxylammonium chloride in 900 cm^3 water and dilute to 1 dm^3.

Hydrochloric acid (1 + 1).

Standard copper solution, 0.200 g dm^{-3}. Dissolve 0.200 g copper metal in dilute nitric acid, boil to expel the nitrogen oxides and dilute with water to 1 dm^3.

Doubly distilled water should be used for preparation of all solutions and in the procedure.

(iii) Procedure [18]

Pipette 50.0 cm^3 water sample or a portion of it (then dilute to 50 cm^3 with water) into a suitable Erlenmeyer flask. Then add 1.0 cm^3 hydrochloric acid (1 + 1), 5.0 cm^3 hydroxylammonium chloride solution, 5.0 cm^3 sodium citrate solution and 5.0 cm^3 bathocuproine disulfonic acid solution, mix thoroughly and measure the absorbance at 485 nm against the blank. Determine the copper content from the calibration graph.

(iv) Interferences

Large amounts of Co(II), Cr(III), Ag, Cd, Hg, Sn, Sb interfere as do CN$^-$, SCN$^-$, EDTA, oxalate, persulfate and oxidants, but they are unlikely to be present in most water samples.

(v) Characteristics of the procedure [23]

Range of application: 0.4–4.0 mg dm^{-3} Cu.
Standard deviation: ($n = 21$), $s = 20$ µg dm^{-3} Cu (for standards solution con-
taining 2.3 mg dm^{-3} Cu)
Detection limit: 20 µg dm^{-3} Cu.

1.6.2.2.4 Spectrophotometric determination of copper with Cuprizone (bis-cyclohexanone oxalyldihydrazone)

(i) Introductory remarks

Cuprizone forms a blue, water-soluble CuL$_2$ chelate with copper(II) ($\lambda_{max} = 595$ nm, $\varepsilon = 1.66 \times 10^4$ dm^3 mol^{-1} cm^{-1})[16] at pH 9 in the presence of citrate.

(ii) Reagents

Cuprizone solution, 10 g dm^{-3}. Dissolve 0.1 g of bis-cyclohexanone oxalyl-dihydrazone (cuprizone) in warm 50% (v/v) ethanol + water mixture and dilute to 100 cm^3 with 50% (v/v) ethanol.

Ammonium citrate solution, 10% (w/v). Dissolve 100 g ammonium citrate in 1 dm^3 of water.

Hydrochloric acid solution, 1 and 6 mol dm^{-3}.

Ammonia solution, 1 and 6 mol dm^{-3}.

Phenolphthalein indicator solution, 0.1%. Dissolve 0.10 g phenolphthalein in 60% ethanol and dilute with ethanol to 100 cm^3.

Standard copper solution, 0.200 g dm^{-3}. Dissolve 0.200 g copper metal in 5 cm^3 concentrated nitric acid, boil to expel the nitrogen oxides and dilute to 1 dm^3 with water.

Doubly distilled water should be used for preparation of all solutions and in the procedure.

(iii) Procedure

Place the slightly acidic sample solution containing 10–135 μg Cu in a 50 cm^3 volumetric flask. Add successively 1 cm^3 ammonium citrate, a few drops of phenolphthalein indicator solution, ammonia to give a pink colour and then 1 mol dm^{-3} hydrochloric acid dropwise till the pink colour disappears. Then add 10 cm^3 curprizone solution and water to volume. Measure the absorbance after 30 minutes at 595 nm against the reagent blank solution. Determine the copper content from the calibration graph.

(iv) Interferences

More than a 100-fold excess of Ni(II), Cr(VI) and Co(II) interfere. In the presence of larger amounts of Fe(III) the concentration of citrate in the solution must be increased to 5 cm^3 ammonium citrate solution per 50 cm^3 sample solution.

(v) Characteristics of the procedure [16, 24]

Range of application: 0.6–5.4 mg dm^{-3} Cu.

Standard deviation: ($n = 21$) $s = 8\,\mu$g dm^{-3} Cu (for standard solution containing 1.8 mg dm^{-3} Cu).

Detection limit: 30 μg dm^{-3} Cu.

.03 ppm

1.6.3 References

1 *International Standards for Drinking Water*, 3rd edn, WHO, p. 40 (1971).
2 J. Korkisch, L. Gödl and H. Gross, *Talanta*, **22**, 289 (1975).
3 F.J. Welcher and E. Boschmann, *Organic Reagents for Copper*, Krieger, Huntington, New York, (1979).
4 J. Michal and J. Zýka, *Chem. Listy*, **48**, 915 (1954).
5 J. Michal and J. Zýka, *Chem. Listy*, **48**, 1043 (1954).
6 A. Bíliková and J. Zýka, *Chem. Listy*, **59**, 91 (1965).
7 A. Bíliková, *Determination of Copper and Zinc in Water*, Special publication Nr. 44, Water Research Institute, Bratislava (1968).
8 A. Bíliková, Proceedings of the 12th Seminar *Hydrochemia*, Bratislava p. 39 (1975).
9 A. Bíliková and V. Bílik, *Vodní hospodařstvi*, **15**, 214 (1965).

10   St. Grys, *Mikrochim. Acta*, **I,** 147 (1976).
11   G. Ackermann, J. Köthe, H. Thor and I. Pechfelder, unpublished investigations.
12   G.F. Smith and W.H. McCurdy, *Anal. Chem.*, **24,** 371 (1952).
13   A.R. Gahler, *Anal. Chem.*, **26,** 577 (1954).
14   *Standard Methods for the Examination of Water and Wastewater*, Am. Publ. Health Assn, 14th edn, Washington, D.C. p. 196 (1975).
15   O.G. Koch and G.A. Koch-Dedic, *Handbuch der Spurenanalyse*, Springer-Verlag, Berlin, p. 672 (1974).
16   G. Ackermann, J. Köthe and U. Bittdorf, unpublished investigations.
17   D. Blair and H. Diehl, *Talanta*, **7,** 163 (1961).
18   *Standard Methods for the Examination of Water and Wastewater*, Am. Publ. Health Assn, 14th edn, Washington D.C., p. 198 (1975).
19   *International Standards for Drinking Water*, WHO, Geneva p. 70 (1978).
20   O.G. Koch and G.A. Koch-Dedic, *Handbuch der Spurenanalyse*, Springer-Verlag, Berlin, p. 676 (1974).
21   G.F. Smith and D.H. Wilkins, *Anal. Chem.*, **25,** 510 (1953).
22   L.G. Borchardt and J.F. Butler, *Anal. Chem.*, **29,** 414 (1957).
23   G. Ackermann, J. Köthe and V. Thrakis, unpublished investigations.
24   O.G. Koch and G.A. Koch-Dedic, *Handbuch der Spurnanalyse*, Springer-Verlag, Berlin, p. 646 (1974).

1.7 Molybdenum

Prepared by
S. B. SAVVIN
V. I. Vernadskii Institute of Geochemistry and Analytical Chemistry, Academy of Sciences of USSR, Kosygin St., 19, SU-117975, SZD-1 Moscow, V-334, USSR

1.7.1 Introduction

Molybdenum is a rather rare element. Its content in the earth's crust is estimated to be about 0.0025%. The chief and most widespread mineral is molybdenite MoS_2. Due to oxidation under natural conditions secondary minerals occur such as ilsemannite, $MoO_2.4MoO_3.nH_2O$, molybdite, $MoO_3.4H_2O$ and molybdic ochre, $Fe_2O_3.7MoO_3.7H_2O$.

Molybdenum, in a similar way to other heavy metals, is only present in natural waters in trace concentrations. In sea water, amounts are in the range $0.5–2.0$ $\mu g\,dm^{-3}$ [1]. The molybdenum content in surface waters varies considerably and amounts of $2–10$ $\mu g\,dm^{-3}$ [2], $0.1–60$ $\mu g\,dm^{-3}$ [3] and up to 4 $\mu g\,dm^{-3}$ [4, 5] have been reported. Higher concentrations may be found in effluents from metallurgical, chemical, electro-engineering, dyes and varnishes, petrochemical, glass-making and ceramic production. For example, molybdenum in copper plant effluent was found to be 47 $\mu g\,dm^{-3}$ and from a non-ferrous metal processing plant, 57 $\mu g\,dm^{-3}$ [6].

Molybdenum is regarded as of low-toxicity; no cases of molybdenum poisoning from drinking water have been reported. A molybdenum concentration of 0.25 in water is harmless to man [7]. However, molybdenum produces strong effects on a variety of farm crops and animals, particularly ruminants. When present in the $0.5–100$ $mg\,kg^{-1}$ range in soils it will cause an abnormal growth of flax [8]. At the 50 $mg\,kg^{-1}$ level, an adverse effect on the growth of oats has been reported [9]. Molybdenum at 10 $\mu g\,dm^{-3}$ concentration in irrigation water to pod-bearing crops accumulates up to 5 $mg\,kg^{-1}$ in crops, a level which produces toxic effects in cattle feeding on these plants.

Molybdenum also affects self-purification processes in water reservoirs. At about 5 $mg\,dm^{-3}$ it inhibits biochemical self-purification processes; at 100 $mg\,dm^{-3}$ it prevents the growth of the appropriate microorganisms [10]. The maximum permissible concentration of molybdenum in drinking water in the USSR is set at 0.5 $mg\,dm^{-3}$ [11]. In the USA [12] for waters being continuously used for irrigation of all types of soils, the maximum permitted concentration is 5 $\mu g\,dm^{-3}$.

A number of studies have been made on the speciation of molybdenum in aqueous solutions [13–17]. It always occurs in the hexavalent form. For molybdenum(VI) a cationic form prevails below pH 2. Above this pH value, several anionic forms occur, many are polymeric. The amounts of these polyanions are pH- and concentration-dependent. Very little work has been done on the speciation of molybdenum in natural waters. In rivers the element mainly occurs in the anionic form, cationic forms have also been observed, but no colloidal species [18].

1.7.2 Spectrophotometric determination of molybdenum

1.7.2.1 General aspects

Numerous reagents have been described for the spectrophotometric determination of molybdenum. In sample solutions molybdenum generally is present as Mo(VI) and in many cases a reduction to the Mo(V) state is necessary before reaction to produce a coloured chelate. Interferences are common and preliminary separation is often necessary, particularly that of tungsten from molybdenum for which many effective procedures are available [19, 20]. Data for a selection of established methods are given in Table 1.5.

Table 1.5. Spectrophotometric methods for molybdenum

Reagent	Molar absorptivity $(dm^3\,mol^{-1}\,cm^{-1} \times 10^{-4})$	Conc. range $(mg\,dm^{-3})$	Ref.
Thiocyanate			
in i-pentanol	2.0	0.4–1	13
in butyl acetate	1.5		21
Dithiol	2.0	0.2–1	22
6,7-Dioxo-2,4-			
diphenylbenzopyrillium chloride	5.0	0.01–3	13
Sulfonitrazo	1.4	1–3	23
Bromopyrogallol red +			
cetylpyridinium chloride	8.0	0.1–4	24
Thioglycolic acid	0.235	1–40	25
Thioxine	0.45	0.2–5	13

Several new reagents have been proposed [26–36]; in many cases the addition of surface-active agents improves the spectrophotometric methods for molybdenum [24, 37–43].

Only very few methods have been published on the spectrophotometric determination of molybdenum in natural waters. The determination with thiocyanate has been applied to drinking water [44] and after preconcentration, to sea water [45]. The thiocyanate complex is extractable by various solvents [13]. 5-Chloro-7-iodo-8-hydroxyquinoline has been used for the determination in sea water [46]. Molybdenum has been determined in natural waters by the ternary compound formed with gallein and papaverine [47]. The most selective and most sensitive spectrophotometric determination of molybdenum in waters, however, is that based on the ternary complex of molybdenum(VI) with bromopyrogallol red and cetylpyridinium chloride.

1.7.2.2 Procedures

1.7.2.2.1 Spectrophotometric determination of molybdenum with bromopyrogallol red and cetylpyridinium chloride [24]

(i) Reagents

Molybdenum stock solution, $1.00 \, g \, dm^{-3}$. Dissolve 1.500 g molybdenum trioxide, MoO_3, in $25 \, cm^3$ of $2 \, mol \, dm^{-3}$ sodium hydroxide solution, acidify the solution slightly with hydrochloric acid, and dilute with water to $1 \, dm^3$.

Molybdenum standard solution, $0.010 \, g \, dm^{-3}$. Dilute $10.0 \, cm^3$ stock solution to $1 \, dm^3$ with water.

Bromopyrogallol red solution, $0.001 \, mol \, dm^{-3}$. Dissolve 0.272 g bromopyrogallol red, analytical grade, in 50% ethanol and dilute to $500 \, cm^3$.

Cetylpyridinium chloride solution, $0.01 \, mol \, dm^{-3}$. Dissolve 1.70 g cetylpyridinium chloride in water with heating, and dilute to $500 \, cm^3$.

Ascorbic acid solution, 5% w/v. Dissolve 50 g ascorbic acid in water and dilute to $1 \, dm^3$.

Sulfuric acid solution, $0.5 \, mol \, dm^{-3}$.

(ii) Procedure

Transfer an aliquot of water sample containing 5–70 μg molybdenum(VI), clarified by electrocoagulation, to a $50 \, cm^3$ volumetric flask, containing 4–$6 \, cm^3$ $0.5 \, mol \, dm^{-3}$ H_2SO_4. Successively add $5 \, cm^3$ 5% ascorbic acid solution, $1 \, cm^3$ bromopyrogallol red solution and $0.5 \, cm^3$ cetylpyridinium chloride solution. Dilute to volume with $0.5 \, mol \, dm^{-3}$ sulfuric acid and mix. After five minutes measure the absorbance at 620 nm against a reagent blank solution. Determine the amount of molybdenum in the aliquot from the calibration graph. Prepare the calibration graph according to the given procedure with 0.50–$7.0 \, cm^3$ aliquots of the standard solution of molybdenum(VI).

(iii) Interferences

Tungsten interferes at all levels with the determination of molybdenum. The interfering effect of iron(III) can be eliminated by the addition of ascorbic acid. Ca, Mg, Zn, Ni, Co and Mn do not interfere.

1.7.2.2.2 Spectrophotometric determination of molybdenum with thiocyanate

(i) Introductory remarks

Molybdenum gives an orange-red complex with thiocyanate. Mo(V) thiocyanate in acid solution in the presence of a reducing agent such as tin(II) chloride can be extracted in organic solvents prior to spectrophotometric determination [21] ($\varepsilon = 1.5 \times 10^4$ dm^3 mol^{-1} cm^{-1}).

(ii) *Reagents*

Molybdenum stock solution, 1.00 g dm^{-3}. Dissolve 1.500 g molybdenum trioxide, MoO$_3$, in 25 cm^3 of 2 mol dm^{-3} sodium hydroxide solution, acidify the solution slightly with hydrochloric acid, and dilute with water to 1 dm^3.

Molybdenum standard solution, 0.010 g dm^{-3}. Dilute 10.0 cm^3 stock solution to 1 dm^3 with water.

Iron(II) sulfate solution, 1% w/v. Dissolve 2 g ammonium iron(II) sulfate in 200 cm^3 water, to which 1 cm^3 concentrated sulfuric acid has been added.

Potassium thiocyanate solution, 10% w/v. Dissolve 50 g potassium thiocyanate in water and dilute to 500 cm^3.

Hydrochloric acid, concentrated.

Sodium sulfate, anhydrous.

Tin(II) chloride solution. Dissolve 350 g SnCl$_2$ in 200 cm^3 hot hydrochloric acid (1 + 1). Leave the solution for 12 h, filter and dilute to 1 dm^3 with water. The solution does not keep longer than 1 week.

Washing solution. Add 100 cm^3 concentrated sulfuric acid to 700 cm^3 water; cool and add 10 cm^3 of both the above SnCl$_2$ and KSCN solutions; make up to 1 dm^3 with water.

Butyl acetate.

(iii) *Procedure*

Add 1–5 cm^3 of the water sample containing 10–50 μg molybdenum to a 125 cm^3 separating funnel. Add successively 2 cm^3 concentrated hydrochloric acid, 1 cm^3 iron(II) sulfate solution, 3 cm^3 potassium thiocyanate solution, 3 cm^3 tin(II) chloride solution, 25 cm^3 water and 10 cm^3 butyl acetate. Shake the mixture for 2 minutes and allow the phases to separate. Run the aqueous layer into a second 125 cm^3 separating funnel and again shake for 2 minutes with 5 cm^3 butyl acetate. To the combined organic phases add 25 cm^3 washing solution and shake for 1 minute. After separation, discard the aqueous phase and transfer the organic phase to a 25 cm^3 volumetric flask containing 0.5 g anhydrous sodium sulfate, make up to volume with butyl acetate. Read the absorbance at 475 nm within 15 minutes following the addition of thiocyanate. Determine the amount of molybdenum from a calibration graph prepared according to the procedure above.

Note: The reagents must be cooled to 15°C before use. At higher temperature the colour intensity is not reproducible.

(iv) *Characteristics of the procedure*

Within the range of application for this method other compounds occurring in natural waters under normal circumstances do not interfere. The range of application is 1–100 mg dm^{-3} Mo in the sample solution; limit of determination: 1 mg dm^{-3} Mo.

1.7.3 References

1 F.M. Perelman and A.Ya. Zvorykin, *Molybdenum and tungsten* (in Russian), Nauka Publishers, Moscow, p. 10 (1968).
2 O.A. Alekin, *Principles of hydrochemistry* (in Russian), Gidrometeoizdat Publishers, Leningrad, p. 442 (1970).
3 T.N. Zhigalovskaya, E.P. Makhonko and A.I. Shilina, *Microelements in natural waters and atmosphere.* In *Pollution of natural waters* Series (in Russian), (Eds. T.N. Zhigalovskaya and S.G. Malakhova) Gidrometeoizdat Publishers, Moscow, p. 86 (1974).
4 L.L. Ciaccio (Ed) *Water and water pollution handbook.* Vol. 1, Marcel Dekker, New York, p. 449 (1971).
5 G.W. Dawson, A.J. Chukrow and W.H. Swift, *Control of spillage of hazardous polluting substances.* Water Pollution Control Series, 15090 F 02, Department of Interior, Federal Water Quality Administration, Washington D.C., p. 55 (1970).
6 I.S. Koryakin et al., in *Central Kazakhstan's Productive Forces*, Vol. 6, *Public Health Service and Agriculture* (in Russian), Publishing House of the Academy of Sciences of the Kazakh SSR, Alma-Ata, p. 51.
7 V.G. Nadeenko, V.G. Lenchenko, A.N. Oshchepkova and N.A. Polkovskaya, *Gig. Sanit.*, **No. 3**, 7 (1977).
8 C.R. Millican, *Proc. R. Soc. of Victoria*, **61**, 25 (1949).
9 A. Petrosek and E. Steven, *Ind. Water Eng.*, **10**(4), 26 (1973).
10 T.A. Asmangulyan, *Gig. Sanit.*, **No 4**, 6 (1965).
11 *Regulations for protection of surface waters against pollution by sewage waters* (in Russian), Publishing House of the USSR Ministry of Amelioration and Water Management, USSR Ministry of Health Service and USSR Ministry of Fisheries, Moscow, (1975).
12 *Water Quality Criteria.* Report of the National Technical Advisory Committee to the Secretary of the Interior, Washington D.C., April 1 p. 234 (1968).
13 A.I. Busev, *Analytical Chemistry of Molybdenum* (in Russian), Publishing House of the USSR Academy of Sciences, Moscow (1962).
14 A.K. Babko and B.I. Nabivantsev, *Zh. Neorg. Khim.*, **2**, 2084 (1957).
15 I.I. Alekseeva, *Zh. Neorg. Khim.*, **12**, 1840 (1969).
16 C. Jander and K. Jater, *Z. Anorg. Allg. Chem.*, **194**, 384 (1930).
17 D.V. Ramano Rao, *Anal. Chim. Acta*, **12**, 211 (1955).
18 V.Ya. Yeremenko, *Gidrokhim. Mater.*, **36**, 125 (1964).
19 V.S. Khristoforov, *Vestnik Leningr. Gos. Univ.*, **No 6** (1947).
20 R.B. Golubtsova and G.V. Myasoedova, *Trudy Komis. Anal. Khim.*, Akad. Nauk SSSR, II(IX, XII), 89 (1958).
21 J. Rodier *Analysis of Water*, Halsted Press, Wiley, New York, (1975).
22 A.I. Busev, V.G. Tiptsova and V.M. Ivanov, *Manual on analytical chemistry of rare elements* (in Russian), Khimiya Publishers, Moscow, p. 236 (1978).
23 M.M. Senyavin, *Methods of analysis of natural and sewage waters* (in Russian), Nauka Publishers, Moscow, p. 117 (1977).
24 S.B. Savvin, R.K. Chernova and G.M. Beloliptseva, *Zh. Anal. Khim.*, **35**, 1130 (1980).
25 Yu.M. Dedkov and M.A. Slotintseva, *Zh. Anal. Khim.*, **28**, 2367 (1973).
26 A.K. Chakrabarti and S.P. Bag, *Talanta*, **23**, 736 (1976).
27 R.S. Kharson, K.S. Patel, K.K. Deb and R.K. Mishra, *Z. anal. Chem.*, **295**, 415 (1979).
28 R.S. Kharson, K.S. Patel and R.K. Mishra, *Mikrochim. Acta*, **I**, 353 (1979).
29 D.A. Williams, I.I. Holcomb and D.F. Boltz, *Anal. Chem.*, **47**, 2025 (1975).
30 Y.K. Agraval, P.C. Maru, T.P. Sharma, S. Patke and P.C. Verma, *Z. anal. Chem.*, **276**, 300 (1975).
31 H. Puzanowska-Tarasiewicz, A. Grudniewska and M. Tarasiewicz, *Anal. Chim. Acta*, **94**, 435 (1977).
32 G. Popa and V. Dumitrescu, *Rev. chim. (Bucharest)*, **26**, 761 (1975).
33 A.I. Busev and A.K. Panova, *C. R. Acad. Bulg. Sci.*, **29**, 225 (1976).
34 K. Lal and S.P. Gupta, *Curr. Sci. (India)*, **45**, 84 (1976).
35 S.K. Jain, D.C. Pandey, K.S. Joseph and Y.K. Satyanara, *J. Indian Chem. Soc.*, **56**, 353 (1979).

36 G. Popa, V. Dumitrescu and I.C. Ciurea, *Rev. chem. (Bucharest)*, **27**, 701 (1976).
37 C.L. Leong, *Analyst*, **95**, 1018 (1970).
38 S. Sakuraba, *Bunsoki Kagaku*, **22**, 270 (1973).
39 I. Honsa and V. Suk, *Coll. Czech. Chem. Commun.*, **35**, 1238 (1970).
40 M. Beguchi, M. Iizuka and M. Yashiki, *Bunseki Kagaku*, **23**, 760 (1974).
41 T. Ozawa, T. Okutani and S. Utsumi, *Bunseki Kagaku*, **22**, 1592 (1973).
42 I. Mori, S. Yamamoto and T. Enoki, *Bunseki Kagaku*, **22**, 1081 (1973).
43 R.K. Chernova and I.V. Lobacheva, *Information Sheet*, TNTI, Saratov, No 3 (1979).
44 N.A. Klimakhin, A.V. Perkov, A.I. Petrushenko, Z.I. Shevchenko, N.F. Zmeikova and A.I. Platonova, Proceedings of All-Union Conference: *The problem of man's interaction with the environment* (in Russian), Kursk, p. 19 (1978).
45 H.M. Nakagawa and F.N. Ward, *Geol. Surv. Bull.*, No 1408, 65 (1975).
46 O. Naoichi, F. Mari and T. Kenji, *Bunseki Kagaku*, **27**, 177 (1979).
47 E.S. Rosinskaya, E.A. Morgen and N.A. Vlasov, In *Formation and physico-chemical regime of Eastern Siberia natural waters and methods of their analysis* (in Russian), Irkutsk, 124 (1974).

Section 2 Atomic-, Mass-, X-Ray-Spectrometric Methods, Electronparamagnetic and Luminescence Methods

Section Editors

Commision V.4

IUPAC

L. R. P. BUTLER

CSIR, PO Box 395
Pretoria 001
South Africa

A. STRASHEIM

Chemistry Department
The University
Pretoria 0002
South Africa

2.0   Introduction

2.0.1   Sampling, transportation and storage of water for spectrometric analysis

Spectrochemical methods are often among the most sensitive available and if trace elements at very low concentrations are to be determined it is necessary that special care must be taken to ensure that no changes in sample composition occur which may affect the results of the analysis.

2.0.1.1   Sampling

Section 3 deals with methods of sampling, etc, in detail, but the matter is of such fundamental importance that it is appropriate to outline the essential principles here at the outset of Section 2.

Samples must be taken away from any source of contamination, e.g. ships or boats. Containers used for taking the sample should be made of clean plastic or glass. Metals should be avoided if trace metal analysis is to be done, even corrosion-free metals such as stainless steel. Samples should be acidified with ultra pure acid to a constant pH and placed in contamination-free containers. In some instances it may be necessary to treat samples to prevent bacteriological action which may cause a change in pH and/or precipitation of salts.

It is best to filter samples immediately after taking them. It has become practice to use filters which hold back all particles larger than 0.42–0.47 micron. Even though certain species may be in a non-ionic form such as colloids, it is generally accepted that they are part of the solution.

2.0.1.2   Transportation

It is advisable to freeze samples as soon as possible after being taken and to transport them in a frozen state. Analyses should be carried out as soon as possible after sampling.

2.0.1.3   Storage

If samples must be stored, this should be done in either a frozen state or at temperatures below 10°C. Once samples have been opened care should be taken that exposure to air is limited to a minimum. When chemical enrichment is necessary and if the final form is more stable than the solution it may be advantageous if this form is stored, e.g. as ion exchange resin.

2.0.2 Sample preparation

The need for sample preparation depends on:

the method being used (i.e., the power of detection, the analytical range), the concentrations of the trace elements to be determined,
the degree of interference by other species present on the analytical determination.

Sample preparation may require enrichment or concentration of the analytical element (analyte) and/or the removal of other elements or compounds.

Ideally the analytical method should be able to analyse the sample with no further treatment. The following steps may be required to treat the sample initially.

2.0.2.1 Particle separation

This includes: Filtration, sedimentation, elutriation, centrifugation, particle electrophoresis and precipitation.

2.0.2.2 Enrichment (to concentrate the analytes)

Includes: evaporation, liquid–liquid extraction, ion exchange, drying.

2.0.2.3 Separation (from other constituents)

Includes: co-precipitation, exclusion, electrode position.

Because most spectrochemical methods are generally very specific, techniques of enrichment are usually sufficient to provide the necessary

Table 2.1. Ranges of application

Method	Useful analytical range (direct %)
X-ray spectrometry	10^{-3}–10^2
Electrometric methods (see Section 3)	10^{-6}–10^2
Mass spectrometry	
(incl. isotope dilution)	10^{-8}–10
Spectrophotometry (see Section 1)	10^{-6}–10
Neutron activation (see Section 4)	10^{-8}–10^{-1}
Atomic emission	
(r.f. source techniques)	10^{-9}–10^{-1}
Atomic emission	
(arc, spark, etc.)	10^{-9}–10^{-2}
Atomic absorption	
(incl. electrothermal, etc.)	10^{-9}–1
Molecular fluorescence	
Electron paramagnetic resonance	Specialized
Luminescence spectroscopy	applications

separation required. Where liquid–liquid techniques are used, the analytes are usually concentrated in a non-aqueous phase. Further treatment will depend on the method of analysis used and will be discussed later.

2.0.3   **Analytical techniques of spectrometric analysis**

Table 2.1 lists the useful analytical ranges of the spectrometric methods of analysis discussed in the following pages.

Those techniques requiring the least sample preparation for a particular sample and able to produce reliable and accurate analyses at lowest cost are the most commonly used and are discussed in this Section, viz. atomic emission using inductively coupled sources, microwave induced plasma sources and arc- and spark-sources; X-ray fluorescence; atomic absorption spectrometry (flame and electrothermal atomizers); spark source mass spectrometry; electron paramagnetic resonance spectrometry and luminescence spectrometry.

2.1 Analysis of Water Using Optical Emission Spectrometry with Arc and Spark Excitation

Prepared by
E. PLŠKO

Geologicky Ustav, Universzity Komenského, Zadunajská 15, CD-85101, Bratislava, Czechoslovakia

2.1.1 Introduction

Optical emission spectrometry (OES) is one of the oldest and most sensitive methods of analysis. It has been applied to the determination of trace elements in water with success for many decades. Earlier techniques for the introduction of the sample to the excitation source (usually an arc or spark) included many ingenious devices, but to a large extent the use of radio frequency plasma sources has superseded these more conventional sources. However, in some types of analysis the arc/spark may still find considerable application. OES may be used because of its remarkably high sensitivity, and wide coverage of virtually all metallic elements. Several textbooks are available describing emission methods [1, 2]. This part will discuss only methods concerned with arc or spark excitation.

2.1.2 Principles of the method

The sample solution must either be introduced directly into the source as a solution, or brought into a solid state and introduced as a powder, or with the aid of an electrode of a spark or arc. When exposed to the relatively high temperature of the excitation source it is evaporated and atomized. Atoms and ions are excited to higher electron energy levels and radiation characteristic of the respective elements is subsequently emitted.

The radiation is dispersed through a spectral apparatus and the characteristic wavelength recorded or the intensities measured directly. When recorded on a photographic plate, qualitative analyses may be carried out by examining the spectral lines, and quantitative analyses by measuring the intensities of the required elemental lines.

2.1.3 Sample introduction

In most cases the sample introduction for arcs or sparks is as one or more of the electrodes (i.e. metal rods, carbon rods, etc.) of an ac or dc arc or a high, medium or low voltage spark. However, electrically non-conductive powders may be introduced into the discharge by various other means, e.g. sifter electrodes [3–5], a tape machine [6] or blown samples [7, 8].

Liquids have in many cases been analysed by allowing them to evaporate on the surface of a solid conducting electrode (metal or graphite) which is subsequently arced or sparked [9–12]. Solutions may also be dried onto carbon or other conductive powders and introduced by these techniques into the sources [13, 14].

Other techniques for introducing liquids include the porous cup [15], the capillary electrode [16], aerosol electrode [17] or the use of rotating disc electrodes [18, 19].

2.1.4 Power of detection

Approximately 70 elements can be determined directly down to levels of $10^{-9}\%$. Naturally, by using preconcentration or enrichment methods, even lower limits of detection are possible.

The limiting factors are generally the effect of other constituents of the sample and the precision obtainable at low concentrations. The detecting power with liquid sample systems is usually much poorer than with solid samples even when concentrated 10–50 times by pre-evaporation.

2.1.5 Precision

Photographic recording and measuring methods give poorer precision of measurement than direct reading methods. The use of internal reference elements [20] improves the precision considerably. The analytical range where the precision is linear, usually covers two orders of magnitude, being limited at the upper concentration by self absorption effects [21] and the lower level by the signal-to-noise of the detector and source as well as impurities in the electrode materials and reagents [22]. Precision of analysis in the analytical range may be between 0.5–3.0% RSD with direct reading detection and 2–15% using photographic detection [23].

2.1.6 Limiting factors

The use of arcs or spark sources for the analysis of water usually requires a fair degree of sample preparation. This is time consuming, but if routine techniques have been developed, the actual analysis time may be as little as 20 s per sample (providing direct reading methods are used).

Certain elements of importance in environmental analysis have poorer powers of detection, e.g. mercury, arsenic, lead, than others such as copper and chromium, so for a fuller analysis, supplementary methods such as atomic absorption spectrometry must be used.

The techniques are better suited to the determination of low concentrations than high concentrations. In addition, the presence of high concentrations of alkali or alkaline earth salts may cause interferences such as spectral line enhancement or depression.

2.1.7 Application to water analysis

Very many papers on OES water analysis have been published. They include river and mineral waters [24–27], sea water [28, 29], tap water [30–32], etc. Several review articles discuss the techniques used [33–35].

Generally, samples are either analysed directly or with various forms of enrichment. Direct analysis [36] involves mixing a carrier solution with the sample. $In_2(SO_4)_3$ or possibly $Ba(NO_3)_2$ may be used with the indium or barium also serving as reference elements. The mixture may be dried onto the anode to be burnt in a DC arc or picked up with the rotating disc/spark method. With the latter method the power of detection is too poor to allow trace analysis in most natural waters.

Concentration may be done by evaporation usually onto a powder mixture of graphite with reference elements [37].

Other concentration or enrichment methods such as precipitation, e.g. with $Al(OH)_3$, has been applied to sea water analysis. Detection limits using this technique are given in Table 2.2.

Table 2.2. Detection limits obtained by different procedures. All values are in $\mu g\, dm^{-3}$

	Direct analysis [36]		After concentration	
Element	Surface	Crater	Dry residue [37] $(g\, dm^{-3})$	Coprecipitation of chelates [38] $(500\ cm^3)$
Ag			0.3	
Al			3.0	
As				0.6
B			30	
Ba			2.0	
Bi				0.2
Cd				2.0
Co	4.0	21	3.0	0.2
Cr	1.0	4.0	2.0	0.2
Cu	0.63	1.5	2.0	0.2
Fe	0.26	0.60		0.02
Mn	0.10	1.9	1.0	0.02
Mo	1.2	1.3		0.2
Ni	9.5	7.0	2.0	0.2
Pb	1.0	1.3	3.0	0.02
Sb				2.0
Sn	40	12		0.05
Sr			2.0	
Ti	10	120	10	
Tl				0.2
V	2.5	5.5	10	0.2
Zn	80	800		0.6

2.1.8 Conclusion

Perhaps the greatest value of an arc/spark method lies in the ability to detect and determine a wide variety of elements over a wide concentration range. The methods are thus pre-eminently suited for doing surveys to establish the presence of various elements.

The power of detection especially with enrichment techniques enables nearly all elements in virtually all types of sample to be determined. OES techniques complement other methods better suited to routine high speed analysis of natural waters.

2.1.9 References

1 E.L. Grove (ed.), *Analytical·Emission Spectroscopy*, Vol. I, M. Dekker Inc. New York, Part 1 (1971); Part 2 (1972).
2 T. Török, I. Mika, and E. Gegus, *Emission spectrochemical analysis*, Akadémiai Kiadó, Budapest (1978).
3 C. Feldman and J.Y. Ellenburg, *Anal. Chem.*, **27**, 1714 (1955).
4 J. Czakow, *Rev. Univ. des Mines*, **15**, 341 (1959).
5 E. Plško, In *Emissionspektroskopie*, Akad. Verlag, Berlin, 225 (1964).
6 A. Daniellson, F. Lundgren and G. Sundkvist, *Spectrochim. Acta*, **15**, 122, 126, 134 (1959).
7 A.K. Russnov, and V.G. Khitrov, *Spectrochim. Acta*, **10**, 104 (1958).
8 A.K. Russnov, and V.S. Vorobév, *Zavod. Lab.*, **30**, 41 (1964).
9 M. Fred, N.H. Nachtrieb and F.S. Tomkins, *J. Opt. Soc. Am.*, **37**, 279 (1947).
10 G. Scheibe, and A. Rivas, *Angew. Chem.*, **49**, 443 (1936).
11 R. Hughes, *Spectrochim. Acta*, **5**, 210 (1952).
12 Kh. I. Zilbershtein, *Zh. Techn. Fiz.*, **25**, 1491 (1955).
13 M. Millet, *Reun. Mensuelle GAMS*, 10 (1957).
14 M. Matherny, XVII *Colloq. Spectr. Int.*, Acta Vol. I. Firenze, p. 33 (1973).
15 C. Feldman, *Anal. Chem.*, **21**, 1041 (1949).
16 T.H. Zink, *Appl. Spectrosc.*, **13**, 94 (1959).
17 L. Erdey, E. Gegus, and E. Kocsis, *Acta Chim. Hung.*, **7**, 343 (1955).
18 W. Guttmann, *Acta Chim. Hung.*, **30**, 385 (1962).
19 M. Pierrucci, and L. Barbanti-Silva, *Nuovo Cimento*, **17**, 275 (1940).
20 E. Plško, and J. Kubová, *Chem. Zvesti*, **32**, 624 (1978).
21 J. Kubová and E. Plško, *Chem. Zvesti*, **32**, 631 (1978).
22 E. Plško, XXI *Colloq. Spectr. Int.*, Cambridge, p. 142. Keynote lectures, Heyden, London, (1979).
23 E. Plško, *Kémiai Közlemények*, **48**, 281 (1977).
24 V.E. Sahini, and A. Crain, *Renue Roum. Chim.*, **19**, 165 (1973).
25 M.M. Moselhy, D.W. Boomer, J.N. Bishop, and P.L. Diosady, *Can. J. Spectrosc.*, **22**, 12 (1977).
26 F.M. Farkan, *Analysis*, **2**, 49 (1973).
27 Ya.V. Eremenko, *Spektrograficheskoie opredelenie mikro-elementov v prirodnych vodkh*, Gidrometeoizdat, Moskva (1969).
28 L.I. Kovalchuk, V.P. Koryukova, L.V. Smirnova and E.V. Shabanov, *Zh. Anal. Chim.*, **34**, 1136 (1979).
29 A. Miyazaki, A. Kimzera and Y. Umezaki, *Anal. Chim. Acta*, **90**, 119 (1977).
30 R. Pruiksma, J. Ziemer and E. Yeung, *Anal. Chem*, **48**, 667 (1976).
31 T.E. Morgulis, A.I. Kuznetsova and Ya.D. Rsikhbaum, *Zavod. Lab.*, **43**, 429 (1977).
32 N.G. Karpel, O.K. Fedorchuk and L.P. Zolotova, *Zavod. Lab.*, **44**, 1081 (1978).
33 G. Baudin, *J. Radioanal. Chem.*, **37**, 119 (1977).
34 M.J. Fishman, and D.E. Eudmann, *Anal. Chem.*, **49**, 139 (1977).
35 M.J. Fishman, and D.E. Eudmann, *Anal. Chem.*, **51**, 317 (1979).
36 A.V. Kariakin, and I.F. Cribovskaya, *Emissionyi spektralnyi analiz obiektov biosfery*, Izd. Chimia, Moskva (1979).
37 Project report II-4-8/01, *Progressive methods of determination of trace elements in natural waters*, Geol. Inst. Komenský Univ. Bratislava (1982).
38 A. Gogala, B. Kidrićs Chemical Institute, Ljubljana, Yugoslavia Private communication.

2.2 Analysis of Water by Emission Spectrometry with an Inductively-coupled Plasma Source

Prepared by
J. ROBIN
Laboratoire de Physicochimie Industrielle, INSA de Lyon, 20 Av Albert Einstein, F 69621, Villeurbanne Cédex, France

Of the many analytical techniques available for the determination of trace elements in solutions, the inductively-coupled plasma source used with optical emission direct reading spectrometry has proved to be one of the most useful. It has been designated as a standard method by several authorities.

2.2.1 General features of the method

Like other emission spectrometry methods, inductively-coupled plasma (ICP) methods of analysis make it possible to determine several elements simultaneously. They differ from other emission spectrometry techniques by the nature of the radiation source operating usually at a fixed frequency in the 27–60 MHz range. The role of any emission source is to transform the sample to be analyzed into atomic vapours and to excite the constituent elements. Excited elements emit characteristic radiation which by optical emission spectrometry makes it possible to carry out qualitative and quantitative analysis. Its high sensitivity usually allows direct analysis of water.

In practice the excitation energy is supplied to an inert but flowing plasma-gas, usually argon, by means of an induction coil or inductor which is part of a tuned circuit supplied by a high-frequency current. Inserted in the inductor is a set of concentric refractory tubes called a torch. The plasma-gas is supplied to the torch at a flow-rate of $\sim 10 \, dm^3 \, min^{-1}$.

The sample is introduced as a fine aerosol transported by a weaker flow of gas, $\sim 1 \, dm^3 \, min^{-1}$. In the case of water samples these may be nebulized directly by means of a pneumatic or an ultrasonic nebulizer.

The argon plasma gas which is partially ionized in the excitation zone of the discharge, exhibits a simple spectrum consisting of atom lines.

The various elements present in the sample aerosol emit both atomic and ionic spectra at the same time. Sequential or simultaneous analyses may be carried out depending on the type of spectrometer used. Measurements of the various spectral lines are relatively independent of time for a given sample where the introduction rate into the plasma is uniform, i.e. intensity of the lines is uniform with time.

The analytical source has four main characteristics described in the following sections.

2.2.1.1 The detection limits are very low

Should a solution contain only one salt of the element to be determined, the detection limits are of the order of $0.1–100 \, \mu g \, dm^{-3}$ when using pneumatic nebulization, and of the order of 0.01–10 μg dm^{-3} with ultrasonic nebulization.

2.2.1.2 There are relatively few inter-element effects

Due to the high temperature of the plasma (gas temperature c. 5000 K), little chemical interference or anion effects are observed. By careful selection of parameters there is also little ionization interference. On the other hand, there may be nebulization effects, desolvation, and volatilization–atomization interferences. The last mentioned are caused by the fact that the spectrometer slit is irradiated by different parts of the plasma from one sample to the next, depending on the absence or presence of interfering species. This type of interference is not very important and becomes less as the power supplied to the plasma is increased. The most troublesome problems are spectral interferences which make it necessary to select the analysis lines of the elements being detected carefully and may, in some cases, make it necessary to make corrections to allow for the presence of interfering species.

2.2.1.3 The analytical range is large

The analytical calibration curves resulting from high frequency-coupled plasma emission sources are in general linear over a wide concentration range, i.e. over 4 or 5 orders of magnitude. Certain authors are of the opinion that this large analytical range is of particular interest in that it makes it possible to determine elements at both high or low concentrations in a single solution. However, the determinations are easier to carry out and are more accurate in dilute solutions where inter-element effects are also decreased. This may make it desirable to vary the conditions to suit the ranges of concentrations of the various elements to be determined [1–3].

2.2.1.4 The method is precise, accurate and fast

A distinction should be made between the precision of this source and the precision achieved from the analysis as a whole. ICPs are in general characterized by good stability when the input-power and the gas flow-rates are constant. However, in the course of spectrochemical analysis, other components of the equipment may be less precise, in particular the system for introducing the sample (nebulization/desolvation) or the electronic measuring equipment. Presently the introduction of the sample still represents the limiting factor in the overall precision which is of the order of a few percent for mid concentration levels (i.e. $\sim$200–1000 times the detection limit).

  The accuracy of analyses using the ICP source depends much on the elements to be determined, their concentration ranges and the presence and concentration of such species which may cause radiation, chemical or other interferences already mentioned. In general it is accepted that the ICP source produces some of the most accurate analytical results available. It is also one of the fastest techniques, lending itself to automatic sample introduction.

According to various authors, 1400 [3] and even 4900 [2] determinations may be accomplished by a single analyst in 8 h, i.e. 180–610 samples h^{-1}.

However, to achieve this it is required that operators who have to use ICP have a sound spectrochemical education in order to be able to do wavelength profiling of spectral lines and background correction and to take interferences into account. Multichannel spectrometers or direct reading polychromators [3–11] must be set up with programmes with spectral lines which have been carefully chosen for the types of samples whose composition varies little. They should be equipped with a rotatory quartz plate or 'spectrum shifter' to enable the operator to make background corrections.

Sequential spectrometers [12–16], also called scanning monochromators, are slower but are particularly convenient for samples of varied composition. The spectral lines are, selected depending on the elements to be determined, the composition of the sample and its impurities. In some cases computer control selects the lines and background positions and it is possible to make many corrections at preselected wavelengths.

2.2.2 Application to the analysis of water samples

2.2.2.1 Selection of analyte wavelengths (see Table 2.3)

The wavelengths given in Table 2.3 are those most frequently used and mentioned in the literature for water analysis. This does not mean that they are totally free from any spectral interference and this point must be checked by the user, as much depends on the samples.

Table 2.3. Recommended spectral wavelengths for water analysis

Element	Line I atomic II ionic	Wavelength (nm)	Possible interferences	Literature references
Ag	I	328.07		4, 6, 7, 8, 11, 16, 17
Al	I	236.7		15
	I	308.22		4, 7, 8, 11, 19
	I	309.3	Pb	18
	I	394.40	Ca	13
	I	396.15	B, Ca	4, 13, 16, 17
As	I	189.04		10
	I	193.76	Fe	4, 7, 11, 16, 20
	I	228.81		20, 21
Au	I	242.8		21
B	I	182.64		10
	I	249.68	Fe	4, 7, 17, 22
	I	249.77	Fe	8
Ba	II	233.53		4, 17
	II	413.07		22
	II	455.40		4, 7, 8, 11, 12, 16, 19, 23
	II	493.41		12

Table 2.3. (*Continued*)

Element	Line I atomic II ionic	Wavelength (nm)	Possible interferences	Literature references
Be	II	234.86		16
	II	313.04		4, 7, 8, 11, 23
Bi	I	223.05		4, 7, 11
C	I	247.86		4
Ca	II	315.89		4, 7, 23
	II	317.93		6, 8, 11, 19
	II	393.37		4, 17
	II	396.85		16, 23
	I	422.67		14
Cd	II	214.44		16, 23
	II	226.50		4, 7, 11, 17, 19
	I	228.80	As	4, 6, 9, 14, 15, 21, 22, 24
Co	II	228.62		8, 7, 11, 14
	II	230.79		16
	II	237.86	Fe, Al	6, 24
	II	238.89	Fe	4, 19, 23
	I	345.35		4
Cr	II	205.55		4, 16
	II	267.72	Mg	4, 6, 7, 11, 17, 19, 24
	II	283.56	Fe	4, 14
	I	357.87		4, 8
	I	359.35		22
	I	425.44		6
Cu	II	213.60	Fe, Al	6, 24
	I	324.75	Fe	4, 6, 7, 8, 9, 11, 14, 15, 17, 19, 22, 23, 24
	I	327.40		16
Fe	II	238.2	Cd, Cu, Co, Mn, Ni, Pb, Zn, V	18
	II	259.94		4, 6, 7, 8, 9, 11, 15, 16, 17, 19, 22, 23
	II	261.19		4
Ga	I	294.36		4
Ge	I	265.12		4
Hg	I	184.95		4
	II	194.23		11, 16
	I	253.65	Fe	4, 7
I	I	178.27		10
	I	183.04		10
K	I	766.49		8, 11, 19
Li	I	670.78		8, 11, 21, 23
Mg	II	279.08		7, 11, 19
	II	279.55		4, 15, 16, 17, 18, 23
	II	280.27		6
	I	285.21		14
	I	383.23		8

Table 2.3. (*Continued*)

Element	Line I atomic II ionic	Wavelength (nm)	Possible interferences	Literature references
Mn	II	257.61	Fe, Al	4, 7, 8, 11, 14, 15, 16, 17, 19, 22, 23, 24
	I	403.08		4
Mo	II	202.03		8, 9, 19
	II	203.85		17, 23
	II	281.61		11, 15
	II	287.15		4, 7
	I	379.82		22
	I	386.41		4
Na	I	330.23	for Na 200 $\mu g\,cm^{-3}$	7, 8, 23
	I	589.00		8, 11, 14, 19, 23
	I	589.59		16
Nb	II	316.34		2
Ni	II	216.56		16
	II	231.60	Ca, Fe, Mg	4, 6, 7, 8, 9, 11, 14, 17, 19, 24
	I	341.48		4, 6
	I	351.51		4
	I	352.45		22
P	I	177.50		10
	I	178.23		10, 11
	I	213.62	Cu, Fe, Zn	4, 8
	I	255.33		4
Pb	II	220.35	Al, Fe	4, 6, 7, 11, 14, 16, 17, 19, 23, 24
	I	283.31	Ca, Fe, Mg	6, 24
	I	405.78	Mg	4
S	I	180.73		10, 11, 25
Sb	I	206.84	Fe	4, 7, 11, 16
	I	217.59		4
Sc	II	361.38		8
Se	I	196.09	Fe	4, 7, 11, 16
Si	I	252.41		8
	I	288.16		4, 23
Sn	II	190.00		10, 11, 17
	I	303.41	Fe	4, 7
	I	317.5		21
Sr	II	338.07		4
	II	407.77		4, 7, 11, 19
	II	421.55		8, 23
Te	I	214.28		7, 11
	I	238.58		4
Th	II	401.91		7, 8
Ti	II	334.67		17
	II	334.94		4, 8
	II	337.28		11
	II	368.52		7
Tl	I	351.92		7, 16
	I	377.57		4

Table 2.3. (*Continued*)

Element	Line I atomic II ionic	Wavelength (nm)	Possible interferences	Literature references
V	II	292.40	Al, Fe, Mg	4, 6, 7, 8, 9, 16, 23, 24
	II	309.31	Mg	4, 17
	II	311.07		4, 11, 19
Y	II	371.03		4, 7, 8
	II	417.8		17
Zn	II	202.55	Al, Cu, Mg	4, 6, 11, 24
	II	206.19		8, 23
	I	213.86	Fe	4, 6, 7, 9, 15, 16, 17, 19, 21, 22, 24
Zr	II	339.20		8
	II	349.62		11

2.2.2.2 Direct analysis of fresh water (drinking and irrigation water)

Table 2,4 lists the detection limits using the method of $2 \times$ standard deviation of background fluctuations for a number of elements in water with little or no interference from other species.

In this group of elements, Fassel and co-workers [4], using pneumatic nebulization, have shown that for 14 out of 19 elements, the detection limits are compatible with the limiting levels laid down by the EPA (Environmental Protection Agency) of the USA.

Many authors have tried to improve limits of detection by using various other nebulizers, e.g. the Babington pneumatic nebulizer [26] or ultrasonic nebulizers with desolvation [4, 6; 7, 19].

Compared results are given in Table 2.5. If ultrasonic nebulization is used such elements as arsenic, cadmium, lead and selenium fall within the detection limits prescribed by the EPA. Mercury, however, still cannot be detected with sufficient sensitivity or precision as it would be necessary to be able to determine at least $2 \mu g \, dm^{-3}$.

Another method that has been used to improve sensitivity is electro-thermal atomization with the ICP [21]. Results are also given in Table 2.4. They were obtained from $5 \mu l$ water samples, simply dried in a graphite furnace and vaporized, without the ashing step, the vapour being carried into the ICP.

2.2.2.3 Sea and high salt content water

Some elements, e.g. arsenic, boron, iodine, phosphorus, sulfur and tin are better determined using spectral lines in the vacuum ultraviolet region [10]. An evacuated or argon-flushed spectrometer must be used, together with an ICP argon-purged channel. Results are included in Table 2.4. Special consideration has been devoted to sulfur [23], whose detection limit is of the order of $80 \mu g \, dm^{-3}$ from sulfate-containing water.

Table 2.4. Detection limits for elements in water ($\mu g\ dm^{-3}$) (pneumatic nebulization)

Elements	3	5	8	10	11	15	16	17	33	(Babington) 26
							References			
Ag		—	2	—	2	—	3	4	4	—
Al	:	12	15	—	50	83	10	7	50	—
As		50	—	13	30	—	30	—	—	—
B		8	4	6	—		—	3	20	—
Ba	2	—	0.3	—	4	—	0.5	1	<2	0.4
Be	0.5	—	0.2	—	0.1	—	0.1	—	<2	0.05
Bi	—			—	30	—	—	—		
Ca	20	1	10		60	—	0.3	<0.5	<20	—
Cd	—	1	—		2	3	1.5	2	—	0.1
Ce	—		25		—	—	—	—	—	—
Co	3	1.5	4		5	—	5	4	—	0.05
Cr	—	—	3		3	—	3	1	—	—
Cu	10	0.5	1		2	1	2	1	5	1
Fe	3	1	2	—	40	4	2	2	5	0.3
Hg	—	—	—		4	—	10	—	—	—
I	—		—	16	—	—	—	—	—	—
K	—		50	—	100	—	—	—	—	—
Li	4	0.2	0.2	—	1	—	—	—	10	5
Mg	4	0.5	25		100	0.2	0.1	<0.05	<2	—
Mn	1	0.2	0.5		10	0.8	1	1	2	0.04
Mo	10	5	5		5	6	—	5	—	4
Na	200	—	10		50	—	7	—	—	—
Nb	—	—	4		—	—	—	—	—	—
Ni	—	5	5		8	—	—	15	—	—
P	—	100	15	8	20	—	7	—	50	—
Pb	10	15	—	—	30	—	25	12	—	1
S	—		—	24	70	—	—	—	—	—
Sb	—		—	—	80	—	40	—	—	—
Sc	—		0.3	—	—	—	—	—	—	—
Se	—		—	—	80	—	30	—	—	—
Si	9	2	50	—	—	—	—	—	20	—
Sn	—	—	—	18	7	—	—	12	—	—
Sr	0.5	2	1	—	2	—	—	—	<2	0.3
Te	—	—	—	—	30	—		—	—	—
Th	—	—	10	—	—	—		—	—	—
Ti	—	1	1	—	60	—		1	—	—
Tl	—	—	—	—	—	—	50	—	—	—
V	6	2	4	—	2	—	5	1	—	0.1
Y	—	—	0.7	—	—	—	—	1	—	0.06
Zn	3	0.5	3	—	7	9	2	1	5	—
Zr	—	—	3	—	3	—	—	—	—	—

The high temperature of an ICP dissociates nearly all molecules. However, certain molecular band-heads may be observed in the plasma such as C_2 (416.5 nm) and CN (388.3 nm) [30, 31] and may be useful for the semi-quantitative determination of total organic compounds in polluted water.

Table 2.5. Comparison of pneumatic (PN) and ultrasonic (US) nebulizations, and electrothermal vaporization (ET)

	References								
	4		6		7		19		21
Elements	(PN)	(US)	(PN)	(US)	(PN)	(US)	(PN)	(US)	(ET)
Ag	2	—	—	—	5	0.8	—	—	—
Al	4–5	0.4	—	—	40	3	17	2.9	—
As	30–50	2	—	—	60	6	—	—	10
B	3–5	—	—	—	9	4	—	—	—
Ba	1	0.1	—	—	0.3	0.09	1.3	0.24	—
Be	0.2–0.4	—	—	—	0.2	0.04	—	—	—
Bi	—	—	—	—	50	9	—	—	—
Ca	0.07	—	—		6	2	2	0.36	—
Cd	2–3	0.07	10.7	1.3	5	0.3	0.16	0.05	1
Co	2	0.1	8	2	6	0.4	2	0.31	
Cr	2–5	0.08	3.3	1.3	5	0.3	1.4	0.27	
Cu	1	0.04	2	0.7	3	0.8	2.8	0.57	
Fe	2–5	0.5	2.7	1.3	2	0.2	1.1	0.19	—
Hg	10–20		—	—	15	4	—	—	—
K	—		—	—	—	—	46	4.8	—
Li	—		—	—	—	—	—	—	3
Mg	0.1–0.7	—	—	—	30	3	4.7	0.97	—
Mn	0.2–0.3	0.01	1.3	0.7	0.45	0.08	0.16	0.03	—
Mo	10	0.3	—	—	12	2	1.4	0.28	
Na	—	—	—	—	1500	240	25	4.6	—
Ni	6–10	0.3	9.3	1.3	30	1	1.5	0.30	—
P	40	—	—	—	—	—	—	—	—
Pb	20–30	1	56	4.7	30	2	2	0.59	—
Sb	—	—	—	—	50	4	—	—	—
Se	30–40	1	—	—	600	3	—	—	—
Si	10	—	—	—	—	—	—	—	—
Sn	—	—	—	—	60	4	—	—	30
Sr	0.05	—	—	—	0.06	0.03	0.16	0.03	—
Te	80	—	—	—	50	1	—	—	—
Th	—	—	—	—	12	1.3	—	—	—
Ti	—	—	—	—	2	0.2	—	—	—
Tl	—	—	—	—	75	20	—	—	—
V	1–2	0.09	3.3	2.7	2	0.3	1	0.22	—
Y	—	—	—	—	0.8	0.08	—	—	—
Zn	2–3	0.1	2	0.7	8	0.8	0.31	0.07	1

Similarly, the ICP may be used to determine traces of ammoniacal nitrogen in aqueous solution. This method has been designed for analysing sediments [32]. The analysis is based on the measurement of the intensity of the 336.0 nm NH band-head which is more intense than the atomic nitrogen line (410.99 nm) and the N_2 band-heads (second positive system 337.13 nm) and N_2^+ band-heads (391.44 nm).

In the case of hard water, the presence of elements such as aluminium, calcium, iron, potassium, magnesium, manganese, sodium and silicon at relatively high concentrations causes 'stray radiation' and/or interferences. Attention to this has been drawn by refs [4, 33, 34]. These elements, which are considered either as matrix or as minor elements, must be determined simultaneously with the analytical trace elements and corrections made accordingly.

In the case of sea water, the presence of sodium and calcium chloride at high concentrations does not affect the detection limits of most metallic analytes [4, 35]. However, many trace elements in sea water are below the normal detection limits and preconcentration techniques must be used [36].

Sea water and, more generally, strongly saline waters (brines) tend to disturb the analysis on account of their tendency to clog the capillaries of the gas nebulization device or the injector of the plasma torch. In order to circumvent this problem, Broekaert and Leis [22] used a method known as the injection method in which they alternatively injected a small volume of sample, 0.5 cm^3, into the plasma and then the same volume of water. The size of the samples used in the method is too small to cause clogging and if a salt deposit does form, it can readily be removed by the water injection (rinsing) step. The analytical result is not provided in the form of a continuous signal but as a series of peaks on a recorder. The injection flow-rate and volume are selected in such a way that the peaks obtained correspond to the maximum deflection of the recorder. This method does not improve the detection limits but makes it possible to work with strong saline solutions, e.g. 20 g nitrates per dm^3. The results obtained agree well with those obtained by the normal continuous process.

2.2.3 Methods using enrichment and ICP emission

Other methods that have been used to increase sensitivity without the need for sophisticated nebulization include those listed previously, viz. concentration, by evaporation, coprecipitation, liquid–liquid extraction, etc. Special consideration must be given to certain techniques, however, when the ICP or other plasma is used for the excitation of concentrated analytes [37].

In ref. [11] a rapid preconcentration by evaporation of 10:1 achieved a lowering of detection limits of the same order, i.e. 10 times. However, impurities are also concentrated and the method is only applicable to relatively pure water. Hard water presents difficulties especially for the determination of such elements as molybdenum and sulfur. For many types of natural water samples, e.g. ocean water, this concentration factor is insufficient.

Hydride forming elements have been determined using the well-known hydride methods of AAS associated with a cold trap [20], or using a direct method whereby the hydride is injected directly into the plasma [27–29].

Coprecipitation has been used successfully using indium hydride [14] and flotation. These techniques have been particularly useful for the analysis of sea water or brine with a high salt content. A concentration factor of approximately 240 is achieved [39]. Other hydrides of elements such as zirconium or gallium [40] have been used. Gallium is attractive because it has a simple spectrum, but has the limitation that elements such as arsenic, molybdenum, antimony and vanadium are not coprecipitated.

Another valuable means of concentration is liquid–liquid extraction. Ammonium tetramethylene-dithiocarbamate or diethylammonium diethyl-dithiocarbamate [9] often give concentration factors of 250 to 500. Being extracted into chloroform at pH 4.1 it is necessary to evaporate the chloroform as this may extinguish the plasma if nebulized directly. Concentrated nitric acid usually dissolves the residue.

Another procedure using silica-immobilized 8-hydroxyquinoline in a chromatographic column works well, giving an enrichment factor of the same order [41–43]. Examples of results are included in Table 2.6.

Table 2.6. Examples of analysis of various sea water samples by ICP emission spectrometry after preconcentration ($\mu g\,dm^{-3}$). (These results illustrate the concentration range which must be reached)

| Element | Reference | | | | | |
	Japan sea water [9]	Pacific [9]	Atlantic [9]	Sea water sample B [38]	Coastal sea water [41]	Sea water 2 [42]
Cd	0.016	0.096	< 0.01	—	< 0.09	0.031
Co	—	—	—	—	—	0.017
Cu	0.308	0.348	0.390–0.392	0.73	1.0	0.11
Fe	1.26	1.05	0.544–0.546	3.2	1.5	7.7
Mn	—	—	—	2.3	0.78	0.98
Mo	7.76	8.36	7.48–7.80	—	—	—
Ni	1.46	0.612	0.164–0.168	0.38	0.3	0.46
Pb	—	—	—	—	< 0.2	—
V	—	1.41	1.33–1.30	—	—	—
Zn	1.42	2.30	0.556–0.544	1.6	2.8	0.4

2.2.4 Conditions for good analytical results at trace and ultratrace levels

Five conditions should be met in order to achieve the best results.

(a) avoid contamination of samples, especially when pretreatment is necessary,

(b) use pure water as a blank,

(c) reference solutions for calibration must be prepared with pure water and high-grade and/or purified reagents,

(d) check the quality of the analytical procedure by recovery tests with spiked samples and by comparison with accepted standard reference materials,

(e) every operation must be done in an air-conditioned 'clean' laboratory specially set aside for the preparation of solutions for trace analysis of elements.

As these conditions are common practice, it is not necessary to go into detail and only specific considerations will be indicated.

2.2.4.1 Pure water

Distillation or deionizing alone does not yield water of consistently high purity, suitable for trace analysis [44] and a combination of distillation and deionizing is necessary [43].

2.2.4.2 Reference solutions for calibration

Their preparation requires the use of pure reagents dissolved in pure water, generally with addition of pure nitric acid. Many elements may be mixed together to form compound reference solutions in order to shorten the calibration procedure. But, care must be taken [44] so that:-
— metal salts which form precipitates when mixed together are not present in the same reference solution and
— two or more elements giving spectral interferences are not present in the same reference solution.

    As an example [44], the following element mixtures are acceptable:
(a) Na, K, Ca, Mg, Ba
(b) Fe, Mn, Cu, Zn, Cr(III)
(c) Ti, V, Mo, Cd, Be
(d) Ni, Co, Pb, Al, Sr
(e) Si, B
(f) Hg, Zr.

2.2.4.3 Methods of determination

In the case of natural soft water or of diluted solutions, analytes may be determined simply using the comparison of the intensity of the spectral lines with the intensity emitted by the relevant elements contained in a reference solution. A standard addition method is often recommended [38].

    The use of an internal reference element is not necessary. However, in the case of the analysis of heavy saline solutions such as brines [45], an internal reference element (cadmium) is used in order to correct for the different densities of brines, because matrix-matched calibration solutions were found to be unsatisfactory. In the case of use of microsamples [46], yttrium as an internal reference element compensates for imprecision of sample size.

2.2.4.4 Memory effects

Usually memory effects are eliminated by a convenient automatic rinsing procedure. However, it has been recorded [2] that certain elements, e.g. boron or zinc, exhibit significantly greater memory effects than most others and caution must be excersised.

2.2.4.5 Water reference samples

Reference samples of different origins are mentioned in the literature of the US Geological Survey, US National Bureau of Standards, EPA (Environment

Protection Agency), Environmental Research Association, Inc. (US), National Research Council of Canada, and IOS (Institute of Oceanographic Sciences) of Great Britain.

2.2.5   References

1   J.W. McLaren, S.S. Berman, V.J. Boyko and D.S. Russel, *Anal. Chem.*, **53**, 1802 (1981).
2   G.F. Larson, R.T. Goodpasture and R.W. Morrow, *Applications of Plasma Emission Spectroscopy*, (Ed. R.M. Barnes) Heyden, Philadelphia, p. 46 (1979).
3   J.R. Garbarino and H.E. Taylor, *Appl. Spectrosc.*, **33**, 220 (1979).
4   R.K. Winge, V.A. Fassel, R.N. Kniseley, E. Dekalb and W.J. Haas, *Spectrochim. Acta. Part B*, **32**, 327 (1977).
5   E.S. Peck, A.L. Langhorst Jr and D.W. O'Brien, *Applications of Plasma Emission Spectroscopy* (Ed. R.M. Barnes) Heyden, Philadelphia, p. 63 (1979).
6   S.S. Berman, J.W. McLaren and S.N. Willie, *Anal. Chem.*, **52**, 488 (1980).
7   C.E. Taylor and T.L. Floyd, *Appl. Spectrosc.*, **35**, 408 (1981).
8   G.F. Larson, R.T. Goodpasture and R.W. Morrow, *Developments Atomic Plasma Spectrochemical Analysis.*, (Ed. R.M. Barnes) Heyden, London, 601 (1981).
9   C.W. McLeod, O. Otsuki, K. Okamoto, H. Haraguchi and K. Fuwa, *Analyst*, **106**, 419 (1981).
10 T. Hayakawa, F. Kijui and S. Ideda, *Spectrochim. Acta, Part B*, **37**, 1069 (1982).
11 M. Thompson, M.H. Ramsey and B. Pahlavanpour, *Analyst*, **107**, 1330 (1982).
12 Y. Mauras and P. Allain, *Anal. Chim. Acta*, **110**, 271 (1979).
13 P. Allain and Y. Mauras, *Anal. Chem.*, **51**, 2089 (1979).
14 M. Hiraide, T. Ito, M. Baba, H. Kawaguchi and A. Mizuike, *Anal. Chem.*, **52**, 804 (1980).
15 J.P. McCarthy, M.E. Jackson, T.H. Ridgway and J.A. Caruso, *Anal. Chem.*, **53**, 1512 (1981).
16 D.D. Nygaard, D.S. Chase and D.A. Leighty, *Appl. Spectrosc.*, **37**, 432 (1983).
17 R.J. Ronan, 27th Pittsburgh Conference 1976, *ICP Inform. Newsletter*, **1**, 164 and 264 (1976).
18 M.H. Moselhy and P.N. Vijan, *Anal. Chim. Acta*, **130**, 157 (1981).
19 P.D. Goulden and H.J. Anthony, *Anal. Chem.*, **54**, 1678 (1982).
20 R.C. Fry, M.B. Denton, D.L. Windsor and S.J. Northway, *Appl. Spectrosc.*, **33**, 399 (1979).
21 K.C. Ng and J.A. Caruso, *Anal. Chem.*, **55**, 1513 (1983).
22 J.A.C. Broekaert and F. Leis, *Anal. Chim. Acta*, **109**, 73 (1979).
23 H.E. Taylor, *Developments Atomic Plasma Spectrochemical Analysis* (Ed. R.M. Barnes) Heyden, London, 575 (1981).
24 S.S. Berman, J.W. McLaren and D.S. Russel, *Developments Atomic Plasma Spectrochemical Analysis*, 586 (1981).
25 D.L. Miles and J.M. Cook, *Anal. Chim. Acta*, **141**, 207 (1982).
26 J.R. Garbarino and H.E. Taylor, *Appl. Spectrosc.*, **34**, 584 (1980).
27 P.D. Goulden, D.H.J. Anthony and K.D. Austen, *Anal. Chem.*, **53**, 2027 (1981).
28 D.D. Nygaard and J.H. Lowry, *Anal. Chem.*, **54**, 803 (1982).
29 E. de Oliveira, J.W. McLaren and S.S. Berman, *Anal. Chem.*, **55**, 2047 (1983).
30 D.L. Windsor and M.B. Denton, *Appl. Spectrosc.*, **32**, 366 (1978).
31 D. Truitt and J.W. Robinson, *Anal. Chim. Acta*, **51**, 61 (1970).
32 J.F. Alder, A.M. Gunn and G.F. Kirkbright, *Anal. Chim. Acta*, **92**, 43 (1977).
33 A.F. Ward, H.R. Sobel and R.L. Crawford, *ICP Inform. Newsletter*, **3**, 90 (1977/78).
34 C. Taylor, *ICP Inform. Newsletter*, **3**, 133 and 417 (1977/78).
35 R.M. Barnes, *Wiss. Z. Karl-Marx Univ. Leipzig, Math. Naturwiss. Reihe*, **28**, 383 (1979).
36 J.W. McLaren, S.S. Berman, R.E. Strugeon and S.N. Willie, *ICP Inform. Newsletter*, **8**, 40 (1982).
37 W.B. Kerfoot and R.L. Crawford, *ICP Inform. Newsletter*, **2**, 289 (1977).
38 R.E. Sturgeon, S.S. Berman, J.A.H. Desaulniers, A.P. Mykytluk, J.W. McLaren and D.S. Russel, *Anal. Chem.*, **52**, 1585 (1980).
39 A. Sugimae, *Anal. Chim. Acta*, **121**, 331 (1980).
40 K. Fuwa, Communication P/I,- *23rd Colloquium Spectroscopicum Internationale*, Amsterdam, Abstracts p. 104 (1983).
41 H. Watanabe, K. Goto, S. Taguchi, J.W. McLaren, S.S. Berman and D.S. Russel, *Anal. Chem.*, **53**, 738 (1981).
42 R.E. Sturgeon, S.S. Berman and S.N. Willie, *Talanta*, **29**, 167 (1982).

43 R.E. Sturgeon, S.S. Berman, S.N. Willie and J.A.H. Desaulniers, *Anal. Chem.*, **53,** 2337 (1981).
44 P.L.Kempster–Report Project N3/0501–Task 2, Hydrogeological Research Institute of Republic of South Africa.
45 J.S. Jones, D.E. Harrington, B.A. Leone and W.R. Bramstedt, *Atomic Spectrosc.*, **4,** 49 (1983).
46 H. Uchida, Y. Nojiri, H. Haraguchi and K. Fuwa, *Anal. Chim. Acta*, **123,** 57 (1981).

2.3 Analysis of Water by Emission Spectrometry with a Microwave Induced Plasma Source

Prepared by
K. LAQUA AND J. A. C. BROEKAERT
Institut für Spektrochemie u, Angewandte Spektroscopie, Postfach 778,
Bunsen-Kirchhoff-Strasse 11, D-4600 Dortmund, Federal Republic of Germany

2.3.1 General features of the method

The method is based on the emission spectrometric determination of elements using a microwave discharge as a radiation source. The microwave energy has a frequency of above 1 GHz (often 2.45 GHz). It is coupled inductively to a gas in a quartz tube (the discharge tube) which is held in a resonance cavity. The sample solution, is converted into an aerosol, introduced into the plasma, atomized and the atomic as well as the ionic species excited to emit radiation.

The gas (usually a noble gas) in the case of early microwave induced plasmas (MIP) was often at reduced pressure, as described in the papers of Dagnall *et al.* [1], Bache *et al.* [2] and Alder *et al.* [3]. For aerosol production, pneumatic nebulization is used with subsequent removal of the solvent vapour (desolvation) [4]. Also possible is the electrothermal evaporation of dry residues [5], the gas chromatographic treatment of the sample and introduction of the eluent [6] in the MIP. It is also possible to introduce evolved volatile elements such as mercury [7] into the MIP, a technique where the sample is introduced into a quartz tube which is subsequently filled with a noble gas at reduced pressure and then sealed. Volatile elements or compounds are vaporized by external heating and excited in the cavity of a microwave cavity. This static method was described by van Montfort *et al.* [8]. For this work the $\lambda/4$ Evenson cavity was often used.

In all cases, the system was complicated and the MIP only became popular when it could be run reliably at atmospheric pressure. This became possible with the TM 010 cavity, introduced by Beenakker *et al.* [9–12]. In the latter, a microwave power of below 100 W, a plasma gas flow of down to $0.2 \, dm^3 \, min^{-1}$ argon or $0.4 \, dm^3 \, min^{-1}$ helium and a discharge tube with an inner diameter of 1 mm is used. All sample introduction methods mentioned above can be used here, with some restriction to pneumatic nebulization without desolvation. With a 'wet' aerosol there is some difficulty in piercing the narrow plasma. In order to solve this problem, attempts to produce a toroidal-structured MIP have been made. Kollotzek *et al.* [13] obtained 3-filament and toroidal MIPs by using a special mounting of a discharge tube with an internal diameter of 2.4–4 mm. Bollo–Kamara *et al.* [14] used a discharge tube with 4 mm i.d. as well as a supplementary inner tube with 1.6 mm i.d. and a tangential gas inlet. In both systems, an aerosol generated by pneumatic nebulization could be introduced without desolvation. Other microwave cavities which have been used are the $\lambda 3/4$ Broida cavity, as described by

Zander *et al.* [15] and the surfatron described by Hubert *et al.* [16] both of which may be operated at atmospheric pressure.

For MIPs operating at atmospheric pressure, excitation temperatures of ~ 4500 K are reported, both in argon as well as in helium. These temperatures were calculated from measurements of rotation spectra. Much lower gas temperatures (2000 K) were found and consequently departures from local thermal equilibrium are reported [17]. The low-pressure MIP is thus not a source which may be said to be in thermal equilibrium [4].

The MIP has the following analytical characteristics (2.3.1.1–2.3.1.4).

2.3.1.1 The power of detection, also for a series of non-metals, is high

As viewed from Table 2.7 it can be seen that this is especially the case when combining appropriate sample introduction devices with MIP excitation.

Table 2.7. Detection limits (c_1) obtained by MIP-OES

Element	Argon MIP using continuous pneumatic nebulization with desolvation [30] (c_1 in $\mu g\,cm^{-3}$)	Argon MIP using continuous pneumatic nebulization without desolvation [10] (c_1 in $\mu g\,cm^{-3}$)	Argon MIP using electrothermal evaporation of 50 μl aliquots [20] (c_1 in $\mu g\,cm^{-3}$)	Argon MIP coupled to gas chromatography [31] (c_1 in $pg\,s^{-1}$)
Ag	0.001	0.006	—	—
Al	0.06	0.4	—	—
As	—	—	—	—
Bi	—	—	—	—
Br	—	—	—	10
Ca	—	—	0.077	—
Cd	0.0006	—	0.035	—
Cl	—	—	—	16
Co	—	0.15	—	—
Cr	—	0.15	0.03	—
Cu	0.001	0.009	0.018	—
F	—	—	—	8.5
Fe	0.01	0.2	0.09	—
Ga	—	0.01	—	—
Hg	0.001	—	—	1.0
I	—	—	—	31
Mg	0.0008	0.005	0.002	—
Mn	—	0.03	0.007	0.25
Ni	—	0.08	0.076	—
P	—	—	—	2.1
Pb	—	0.1	0.05	0.49
S	—	—	—	63
Sb	—	0.4	—	—
Si	—	—	—	29
Se	—	—	—	—
Sn	—	—	—	—
Sr	—	0.005	—	—
Te	—	—	—	—
Ti	—	0.4	—	—
Tl	—	—	0.09	—
Zn	0.0005	—	0.01	—

When the electrothermal atomization of dry samples residues [20] is used, the absolute limits of detection lie in the pg range. When using a sample preconcentration process, e.g. by *in situ* electrolysis on a graphite tube, as described by Volland *et al.* [18], the detection limits are in the sub $ng\,cm^{-3}$ range. For the non-metals phosphorus, sulfur and arsenic and even fluorine and chlorine, a helium MIP is very useful. It is often used in combination with gas chromatographic methods. Methods using direct pneumatic nebulization without desolvation, yield a low power of detection as compared with other plasma sources (e.g. the inductively coupled plasma).

2.3.1.2 The matrix effects, especially those caused by alkalines, are high

This is certainly the case in a low-pressure MIP as described by Kawaguchi *et al.* [19]. Alkalis usually enhance the line intensities and are thus often added to samples in excess, acting as buffers or volatilizers [8].

2.3.1.3 The analytical range is large and simultaneous multi-element determinations are possible

As in all emission spectroscopic methods, the dynamic concentration or analytical range, i.e. the concentration range where a clear functional relationship between the measured intensities and the elemental concentrations in the sample exists, is large. In the case of the MIP it is restricted by the sample introduction. Using electrothermal aerosol generation, a linear dynamic range of three decades of concentration was reported by Aziz *et al.* [20]. In the case of gas chromatographic separation prior to MIP-excitation, a linear dynamic range of 20 000 was reported by McMormack *et al.* [21].

2.3.1.4 Elemental speciation is possible

This may be done based on the different volatility of the elemental compound when performing electrothermal evaporation of the sample [22] as well as by using gas or liquid chromatographic separation prior to MIP-excitation.

2.3.2 Application to the analysis of water samples

2.3.2.1 Direct analysis methods

When using pneumatic nebulization without desolvation, the detection limits, as well as the matrix effects caused by alkalis, are so high that the method is unattractive for direct water analysis. With desolvation, as reported by Beenakker [11], the MIP becomes useful for the direct determination of a series of metals in waters, provided their total salt content, e.g. in the case in drinking water, is not high. The MIP in a sealed quartz tube was used for the direct determination of cadmium and tellurium in sea water, tap water, deionized and distilled water [8] (100 to $< 3\ pg\,cm^{-3}$).

The determination of elements and species by MIP excitation subsequent to gas chromatographic separation

Chiba *et al.* [23] determined fluorine in natural waters by extracting the element with trimethylchlorosilane (TMCS) and converting it to trimethyl-fluorosilane (TMFS) in toluene. The extracted toluene solution was injected into a gas chromatograph and detected using an atmospheric pressure helium microwave induced plasma as a source for emission spectrometry. The emission of fluorine is measured at 685.6 nm. The detection limit and linear dynamic range of the GC-MIP measurement system are 7.5 pg s^{-1} and more than 4 orders of magnitude, respectively. Analytical results obtained by the GC-MIP system are consistent with those where fluoride is determined by the ion selective electrode method.

Talmi and Norvell [24] determined arsenic and antimony in environmental samples by MIP-OES coupled to a gas chromatograph. The analytical procedure is based on the co-crystallization of arsenic (As^{3+}) and antimony (Sb^{3+}) with thionalid, and reaction of the precipitate with phenylmagnesium bromide (PMB). Following the decomposition of excess PMB, the triphenyl arsine and stibine are extracted into ether and separated on a GC column. Atomic emission detection of arsenic and antimony is accomplished with an argon MIP attached to the GC column outlet. The detection limits are 20 and 50 pg, respectively, and the relative sensitivities are 50 and 125 ng dm^{-3} for water samples. The relative standard deviation in the case of fresh and salt water ranges from 0.026 to 0.071.

The method may also be used for speciation determinations. Quimby *et al.* [25] determined trihalogenomethanes in drinking water by using an atmospheric pressure helium MIP coupled to gas chromatography. Using a purge and trap technique with 10 cm^3 samples, the detection limits for the trihalogenomethanes are below 1 ng. A Temax GC trap is used and the different halogenomethanes (chloro-, bromo-, iodo-compounds) are monitored selectively by measuring the Cl 481.0 nm, the Br 470.5 nm and the I 206.2 nm line intensities.

Miller *et al.* [26] determined trichloroacetic acid in water in the ng cm^{-3} range. An extraction of the species with diethylether and derivatization to methyl trichloroacetate with diazomethane was performed. A 200 μdm^3 portion of derivatized extract volume was pre-enriched on a Temax GC column and then introduced on the SE-54 fused silica gas chromatographic column. The effluent was led (without splitting) to a helium MIP. The procedure was used to determine the amount of trichloroacetic acid (detection limit 2 μg dm^{-3}) formed during chlorination of water containing sulfuric acid.

Talmi and Bostick [32] determined alkylarsenic acids in water samples. The procedure involves NaBH$_4$ reduction of the arsenic acids to the corresponding arsines, followed by ether extraction of the arsines and their flushing from the sample solution and collection in cold toluene at $-5°$C. The arsines are separated on a GC column and determined by MIP-OES using the As 228.8 nm line. Since the MIP is very selective to arsenic ($> 10^4$), the absolute power of detection (20 pg of arsenic) is little dependent on the molecular structure of the arsines. The concentration detection limit for water samples is at least 0.25 μg dm^{-3}.

Estes *et al.* [27] determined trialkyl-lead chlorides in tap water. They used a valveless gas switching interface to vent large quantities of eluent solvent which would disrupt the helium discharge as sustained in the TM 010 cylindrical resonance cavity.

Thermally sensitive trialkyl-lead chlorides in spiked tap waters are determinable in the range of 10 to 10000 μg dm^{-3}. The detection system features a low-resolution scanning monochromator with a quartz refractor plate background corrector. The latter is required as the 405.8 nm lead line, may suffer from CN-band interferences.

Tanabe *et al.* [28] determined ultratraces of ammonium, nitrite and nitrate nitrogen by atmospheric pressure helium microwave-induced plasma emission spectrometry coupled to a gas generation technique. The ammonium nitrogen was determined by injecting 0.5 cm^3 of pH-adjusted solution (by adding 0.25 mol dm^{-3} NaOH), purging the air, injecting of 0.1 cm^3 10% NaOBr solution) and subsequent measurement. In the case of nitrite nitrogen 1 mol dm^{-3} HCl was used for pH-adjustment and 0.2 mol dm^{-3} sulfamic acid as reagent. Nitrate is reduced to nitrite on a cadmium-copper column and determined as nitrite. Measurements were done using the 746.8 nm N line (detection limit: 0.009 μg N cm^{-3}), the 337.1 nm N$_2$ band (0.006 μg cm^{-3}), the 391.4 nm N$_2^+$ band (0.004 μg cm^{-3}) and the 336.0 nm NH band (0.02 μg cm^{-3}). The method was successfully used for the determination of nitrite and nitrate in pond water and the results agreed well with those using colorimetric determinations.

2.3.2.3 Treatment of samples

In the case of direct determination in water, the preservation of trace metals in natural water samples necessitates taking a series of precautions for the sampling vessels and the storage. They are well-known in water analysis and, as described by Subramanian *et al.* [29] include the use of teflon containers and acidification to pH < 1.5 with nitric acid as described in Sections 2.0.1 and 3.

2.3.2.4 Analytical accuracy

MIP methods are prone to systematic error because of the nature of excitation and especially from the preliminary chemical procedures.

The analysis of reference samples is mandatory to trace these systematic errors. Unfortunately, at low element concentration levels and especially for the determination of species, there is a lack of well-analysed reference samples. Consequently, the only way is to duplicate analyses using an independent method. For elemental determination, the use of furnace atomic absorption spectrometry, for example, is recommended. In the case of speciation, the use of organic mass spectrometry, as for example, by Talmi and Norvell [24] is possible.

2.3.3 Conclusion

MIP-OES is of potential interest for determining elements in different kinds of water, i.e. natural water, sea water, drinking water.

Direct methods, however, suffer from a low power of detection as compared, for example, with ICP-OES. This problem can be solved by combining MIP-OES with electrothermal evaporation of dry sample residues and with gas chromatography. Then, determination of traces at a favourable cost/performance ratio, as compared with ICP-OES, is possible. Speciation is possible at concentration levels which are toxicologically and environmentally relevant and which could only be carried out alternatively by expensive techniques such as mass spectrometry.

2.3.4 References

1 R.M. Dagnall, T.S. West and P. Whitehead, *Anal. Chim. Acta*, **60**, 25 (1972).
2 C.A. Bache and D.J. Lisk, *Anal. Chem.*, **39**, 786 (1967).
3 J.F. Alder and M.T.C. Da Cunha, *Can. J. Spectrosc.*, **25**, 32 (1980).
4 P. Brassem, F.J.M.J. Maessen and L. de Galan, *Spectrochim. Acta, Part B*, **33**, 753 (1978).
5 H. Kawaguchi and B.L. Vallee, *Anal. Chem.*, **47**, 1029 (1975).
6 J.P.J. van Dalen, P.A. de Lezenne Coulander and L. de Galan, *Anal. Chim. Acta*, **94**, 1 (1977).
7 K. Tanabe, K. Chiba, H. Haraguchi and K. Fuwa, *Anal. Chem.*, **53**, 1450 (1981).
8 P.F.E. van Montfort, J. Agterdenbos, R. Denissen, M. Piet and A. van Sandwijk, *Spectrochim. Acta, Part B*, **33**, 47 (1978).
9 C.I.M. Beenakker, *Spectrochim. Acta, Part B*, **32**, 173 (1977).
10 C.I.M. Beenakker, B. Bosman and P.W.J.M. Boumans, *Spectrochim. Acta, Part B*, **33**, 373 (1978).
11 C.I.M. Beenakker, *Spectrochim. Acta, Part B*, **33**, 545 (1978).
12 C.I.M. Beenakker, P.W.J.M. Boumans and P. J. Rommers, *Philips Tech. Rev.*, **39**, 65 (1980).
13 D. Kollotzek, P. Tschöpel and G. Tölg, *Spectrochim. Acta, Part B*, **37**, 91 (1982).
14 A. Bollo-Kamara and E.G. Codding, *Spectrochim. Acta, Part B*, **36**, 973 (1981).
15 A.T. Zander, R.K. Williams and G.M. Hieftje, *Anal. Chem.*, **49**, 2372 (1977).
16 J. Hubert, M. Moisan and A. Richard, *Spectrochim. Acta, Part B*, **33**, 1 (1979).
17 M.H. Abdallah and J.M. Mermet, *Spectrochim. Acta, Part B*, **37**, 391 (1982).
18 G. Volland, P. Tschöpel and G. Tölg, *Spectrochim. Acta, Part B*, **36**, 901 (1981).
19 H. Kawaguchi, I. Atsuya and B.L. Vallee, *Anal. Chem.*, **49**, 266 (1977).
20 A. Aziz, J.A.C. Broekaert and F. Leis, *Spectrochim. Acta, Part B*, **37**, 381 (1982).
21 A.J. McMormack, S.C. Tong and W.D. Cooke, *Anal. Chem.*, **37**, 1471 (1965).
22 K. Bäckmann, U. Hamm, A. Werner, P. Tschöpel and G. Tölg, in *Developments in Plasma Spectrochemical Analysis*, Heyden, Philadelphia (1981).
23 K. Chiba, K. Yoshida, K. Tanabe, M. Ozaki, H. Haraguchi, J.D. Winefordner and K. Fuwa, *Anal. Chem.*, **54**, 761 (1982).
24 Y. Talmi and V.E. Norvell, *Anal. Chem.*, **47**, 1510 (1975).
25 B.D. Quimby, M. F. Delaney, P.C. Uden and R.M. Barnes, *Anal. Chem.*, **51**, 875 (1978).
26 J.W. Miller, P.C. Uden and R.M. Barnes, *Anal. Chem.*, **54**, 485 (1982).
27 S.A. Estes, P.C. Uden and R.M. Barnes, *Anal. Chem.*, **53**, 1336 (1981).
28 K. Tanabe, K. Matsumoto, H. Haraguchi and K. Fuwa, *Anal. Chem.*, **52**, 2361 (1980).
29 K.S. Subramanian, C.L. Chakrabarti, J.E. Sueiras and I.S. Maines, *Anal. Chem.*, **50**, 444 (1978).
30 R.K. Skogerboe and G.N. Colemann, *Appl. Spectrosc.*, **30**, 504 (1976).
31 B.D. Quimby, P.C. Uden and R.M. Barnes, *Anal. Chem.*, **50**, 2112 (1978).
32 Y. Talmi and D.T. Bostick, *Anal. Chem.*, **47**, 2145 (1975).

2.4 Determination of Trace Elements in Natural Waters by Atomic Absorption Spectrometry

Prepared by
I. RUBESKA
UNDP, Box 7285 ADC, Passay City, Metro Manila, Philippines

2.4.1 Introduction

Among spectrometric methods of elemental analysis, atomic absorption spectrometry (AAS) is particularly suited for the analysis of water. It may handle liquid samples directly with little or no chemical pre-treatment, so that contamination is minimized.

Applications of AAS in water analysis are limited to the determination of metals. Though several indirect methods for non-metal anionic constituents such as Cl^-, CN^-, SO_4^{2-}, etc. have been reported, these are usually applicable only to specific types of sample and in general other analytical methods are preferable for these species.

AAS is suitable for the determination of both major, minor and trace elements using the same basic equipment with the alteration of only instrumental or procedural settings, e.g. analytical line, absorption path-length, dilution, etc. Because of its versatility and simplicity of operation AAS has become the most extensively used method for the determination of metals in water. All agencies concerned with water analysis now include AAS methods in their publications of recommended procedures.

Committees attached to these agencies are developing tentative standard methods for water analysis [1–3] and several inter-laboratory studies have been carried out and the results published [4–7].

AAS may be applied basically with three different ways of atomization, viz. flame atomization, electrothermal atomization and atomization with preceding chemical generation of a volatile analyte species of hydride. This last technique, though applicable to only a rather limited number of elements, attains the lowest relative limits of detection.

AAS is particularly suited for the determination of trace metals, as its power of detection, particularly if electrothermal atomization with the latest equipment is considered, allows the determination of most trace elements of interest at their normal concentration levels in unpolluted natural waters (see Table 2.8). In the few cases where the limits of detection are not sufficient, a simple concentration step may in general be included, e.g. liquid–liquid extraction, ion exchange, electrodeposition, coprecipitation or sorption.

The main limitations of AAS with regard to analytical applications for water analysis therefore stem from the effect of matrix elements on the limits of detection and sensitivity of a particular analyte. For natural waters this comprises mainly chlorides and/or sulfates of the alkali and alkaline earth metals. Carbonate and bicarbonate, if present, are eliminated by the addition

Table 2.8. Some typical trace element concentrations in natural waters and limits of detection by AAS (in $\mu g\,dm^{-3}$)

Element	Ground and surface median [35]	River average [36]	Sea [37]	Values reported for open ocean water [29]	Limits of detection by		
					flame	graphite furnace	hydride, vapour or carbonyl generation
Al	10	400	10		30	0.02	
As	2	2	3		140	0.4	0.02
B	10	10	4600		700	30	
Cd	0.03		0.1	0.03	0.5	0.006	
Co	0.1	0.2	0.1	0.003	6	0.04	
Cr	1	1	0.05		2	0.02	
Cu	3	7	3	0.1	1	0.04	
Hg	0.07	0.007	0.03		170	4	0.02
Mn	15	7	2	0.02	1	0.02	
Mo	1.5	1	10		30	0.04	
Ni	1.5	0.3	2	0.27	4	0.2	0.04
Pb	3	3	0.03	0.09	10	0.1	
Se	0.4	0.2	0.4		70	1	0.02
Zn	20	20	10	0.3	0.8	0.002	

Note: It is quite plausible that the data for sea water are overestimated either because too much weight is given to polluted coastal samples or because data from samples contaminated during sampling, storing and/or analysis were included. Some data for open ocean water samples reported more recently are given in column five.

of acids used for sample conservation. Another limitation is that the instrument usually allows the determination of only one element at a time.

2.4.2 Instrumentation

AAS is based on the absorption of radiation at a particular wavelength by neutral unexcited atoms of the analyte. The narrow band-width at which this occurs ensures a high degree of selectivity of determination. The principles of AAS have been described in many publications and are well known throughout the analytical community. However, the analytical performance, i.e. the limits of detection and freedom from interference of an AAS method, depend naturally also on the instrumental parameters and this may affect the transferability of a particular analytical procedure from one instrument model to another. It is, therefore, necessary to discuss some pertinent instrumental parameters.

In flame AAS (FAAS), the analytical sample is nebulized into a chemical flame which serves as atomizer and the atom reservoir. Instrumental requirements of FAAS are well established and most commercially available instruments fulfill these requirements reasonably well. The analytical performance of flame AA spectrometers of different manufacturers are, therefore, similar to a high degree.

In graphite furnace atomic absorption spectrometry (GFAAS), a definite volume of the water sample is introduced into a graphite tube placed in the optical axis of the monochromator. By passing a low voltage high intensity current through the tube this is gradually or rapidly step-wise heated to increasing temperatures according to a pre-selected programme. The sample is thus successively dried, thermally decomposed and atomized.

Atomic absorption measured during the atomization period is then proportional to the amount of the analyte in the graphite tube. However, several processes taking place before and during atomization as well as the kinetics of free analyte atom formation and decay, all influence the resulting analytical signal. This inherent complexity is accentuated by much stronger matrix effects than in FAAS due to the considerably lower dilution by foreign gases and to the long time available for reactions between the analyte, other constituents of the sample and possibly the environment in which atomization takes place.

The main factors controlling the performance of electrothermal atomizers are:

1 the maximum attainable temperature;
2 the heating rate, dT/dt (where T is the absolute temperature and t the time);
3 the properties of the support surface with respect to interaction with the sample constituents during the temperature programme;
4 the temperature and analyte distribution within the atom reservoir during the atomization step, in particular to what extent thermodynamic equilibrium in the gaseous phase is reached.

The first point is mainly controlled by the material of the support, i.e. graphite, tungsten, etc., and is therefore similar for instruments from various producers. Graphite furnaces (GFAAS) are by far the most broadly used among electrothermal atomizers for AAS and have also widest application with respect to the spectrum of elements and matrices. For water analysis, the semi-enclosed tubular types of furnaces allowing larger sampling volumes are preferable.

Considerable differences between various manufacturers' models may be found for the other factors, which for a particular analytical application may be critical. Adaptation of a procedure may thus require considerable modification and in some cases the limits of detection and freedom of interference may not be reproduced.

The situation has improved considerably with atomization from a platform. This is a small plate of pyrolytic graphite within the tube on which the sample is placed. Because of a limited contact with the tube walls the platform temperature is controlled mainly by radiative heat transfer and lags behind the tube temperature. When eventually atomization starts, the analyte atoms enter an atmosphere which is at a higher temperature and either does not change at all or possibly only changes slightly. The conditions are, therefore, close to constant temperature atomization, i.e. closer to thermodynamic equilibrium. Atomization from a platform does not influence sensitivity markedly but interferences are substantially reduced [8], in particular if intercallation reactions of the sample constituents with graphite are prevented by using carbide coated [8] or pyrolytic graphite coated tubes and platforms [9].

The hydride (or cold vapour) generation technique for the determination of elements such as arsenic, is particularly suitable for water analysis since it involves a highly selective separation of the analyte from the sample matrix, thus avoiding many interferences connected with flame and/or electrothermal atomization. Samples of up to ~ 50 cm^3 volume may be treated so that limits of detection in relative concentrations are of the order of 0.1 μg dm^{-3}.

The sample with a definite analytical acid concentration is placed in a generating vessel and a reducing agent such as NaBH$_4$ in form of a pellet or in solution is added. The evolving gases consisting of covalent hydrides and excess hydrogen are swept with a stream of inert gas (N$_2$, Ar) into the atomizer. The maxima of the resulting transient signals are measured, or the absorbance may be integrated. A hydrogen diffusion flame, a silica glass tube heated by a flame or electrically by resistance wiring or a 'flame in tube' may be used as atomizer. The silica tubes generally improve limits of detection by lowering the noise due to flame flickering and simultaneously by prolonging the residence time of free analyte atoms in the observation zone.

Despite the differing constructions of hydride generating devices, they show remarkably similar performances. This is most probably because the formation of the volatile species by chemical reactions in solution is relatively well controlled and the fractions of the volatilized analytes released are virtually complete and thus independent of the instrument [10].

Analytical procedures for specific elements are therefore easily transferrable between different instrumental models. Different limits of detection may be found depending on whether the generated analyte species is being continuously swept into the atom reservoir or whether it is collected and atomized in separate steps. If atomization proceeds as a separate step, then the hydride generation and collection can be simply viewed as an enrichment procedure. In this form it has been recently extended to carbonyl generation for nickel [11].

2.4.3 Sample collection, preservation and pre-treatment

All types of water sample, surface, ground or rain water including snow, are basically treated in the same manner on the instrument. Differences may be found in the methods of collection, preservation and storage of the samples prior to analysis. The methods of collecting and preserving should always be considered with regard to the object of the study for which the analyses are made, as well as to the analytical procedures. The techniques described in the introduction to Section 2 and in greater detail in Section 3 apply.

Any trace element may be present in the sample in different forms, e.g. dissolved as ions or bound in organic complexes, in colloidal form or sorbed on suspended particles. According to the information sought, the procedure may require full solubilization of the analyte usually by mineral acids, or separation of all suspended material by filtration prior to the addition of acid.

Since water is a highly dynamic system, no rigorous procedures discerning between the different forms of existence are available. However, some procedures based more on rational presumptions than on actual experimental proof have been established [1].

Dissolved metals in water are defined as those constituents which pass through a 0.45 μm membrane filter. For their determination, the sample has to be filtered at the time of collection preferably using a plastic filtering apparatus and discarding the first 50–100 cm^3 of the sample. After filtration the sample should be stabilized by addition of a strong mineral acid, to prevent losses by sorption or co-precipitation. For GFAAS, 3 cm^3 of HNO_3 $(1 + 1)$ dm^{-3} are recommended. The amount added should be increased if the pH is >2.

The total metal content is defined as that constituent which is determined in an unfiltered sample acidified in the same manner at the time of collection. The sample should be hot digested additionally in the laboratory with HNO_3, making the final solution 0.5% in HNO_3.

Some elements, e.g. aluminium, present in surface waters as polymerized hydrated oxides may not be depolymerized by simply adjusting the pH [10]. Though this may not affect the signal in FAAS, it may pose problems with GFAAS because of the small volume sampled (5–50 μdm^3). If suspended material is present, the hot acid digestion is imperative otherwise the content may change with time.

The choice of the container material with respect to possible contamination and/or loss of analytes by release or adsorption on the walls is no less important [12–14].

AAS will generally determine the total metal, whether present in ionic form or organically bound, except when separation based on chemical reactivity, e.g. in hydride or vapour generation, chelation-extraction, etc precedes the actual atomization step. In such cases sample pre-treatment, involving oxidation through digestion with some strongly oxidative reagent ($K_2S_2O_8$, $KMnO_4$) or by UV photolysis [15], should be included.

2.4.4 Enrichment and/or pre-concentration procedures

AAS being a very versatile method, may be easily combined with almost any conceivable enrichment–separation technique, including GC or HPLC for speciation studies. Because of the high selectivity of AAS, a full separation from other minor and trace elements is not necessary if elemental content only is to be determined. On the contrary, the enrichment procedure might preferably include as many trace elements of interest as possible so that these all may be measured from the enriched sample.

Selection of the particular enrichment procedure should proceed with consideration of the following points:

1 the necessary enrichment factor;

2 the required degree of separation from major constituents of the water sample;

3 the danger of saturating and thus blocking the functional capacity of the particular process by the major elements in the sample;

4 the danger of contamination during the enrichment operation which depends *inter alia* on the working environment available.

Solvent extraction with the APDC-MIBK extraction system [16] has found widest use since it enables an enrichment factor of ∼40, it is multi-element and iron, an abundant element in solid samples which consumes the reagent and saturates the organic phase, is generally present in water samples at very low levels [17, 18]. Replacing MIBK with the less soluble DIBK [19] improves the possible enrichment factor even further. Halogenated organic solvents are less often used because of possible interferences during atomization in the graphite furnace [20]. If used, the trace elements are usually back-extracted with acids [21]. Nevertheless the use of APDC-CCl$_4$ extraction for Cd with a GFAAS final determination was found to present no problems.

Similar enrichment factors may be achieved with electro-deposition. The main advantage is that it may be carried out in closed systems and is thus particularly suited if 'clean' laboratory space is not available. Its main disadvantage is that unless a multichannel instrument is used, each sample is good only for a single determination as compared with chelation–extraction enrichment, allowing at least a dozen determinations from 1 cm^3 of the organic phase.

Electro-deposition may proceed on the inner walls of a graphite tube through which, during the electro-deposition, the sample is circulating. The tube is then directly used for GFAAS [23,24]. Alternately, and more conveniently, the elements may be deposited on a tungsten wire which is then used as a wire probe inserted into a flame [25] or a pre-heated graphite furnace [26]. This procedure approaches thermal equilibrium atomization which exhibits considerably lower interference effects. In both cases electro-deposition should take place at a controlled potential.

Higher enrichment factors are possible with ion exchange chromatography using chelating resins. Chelex-100, a resin containing an imidoacetate group, has been widely used [27].

Other active groups that have found use for separation of heavy metals, e.g. the dithiocarbamate group fixed on a resin [30] or immobilized on silica [31], or the polyacrylamidoxime group, with a lower tendency to adsorb alkali earth metals, enable higher enrichment factors.

For transition metals, a hydroxyquinoline group resin is preferable, since under neutral or slightly alkaline conditions the alkali earth metals are not retained. Sferoxin, a resin containing this group, however, did not release cobalt fully even by strong acids [28]. This problem was not as serious with 8-hydroxyquinoline immobilized on silica [29] and enrichment factors of up to 500 for cadmium, lead, zinc, copper, iron, manganese, nickel and cobalt in sea water were achieved [29]. This preconcentration makes determinations with GFAAS at ng dm^{-3} concentration levels possible. The main problems of accuracy and precision are not with the measurement procedure itself, but with the collection, preservation, storing and handling of the sample, as well as the preparation and reliability of reference solutions.

Whereas for samples with high salt content, ion exchange chromatography is best suited, the enrichment procedure may be considerably simplified if the water sample is very low in salts, e.g. water from high pressure steam generators or unpolluted snow and ice. Though in principle such samples could be enriched by simple evaporation, the process is difficult to carry out without danger of contamination at the very low levels determined. Thus other processes, such as adsorption on activated tungsten wire are preferable. The wire may then be inserted into a graphite furnace and the adsorbed elements measured [32,33]. This procedure was used for the determination of cadmium, copper, lead and zinc at $ng\,dm^{-3}$ levels in polar snow [34]. The main advantage is that the enrichment procedure can be carried out in the field with very simple equipment, thus simplifying logistic problems such as sample preservation, transport and storage.

2.4.5 Limitations of AAS in water analysis

2.4.5.1 Flame AAS

The application possibilities of FAAS are controlled mainly by the detecting power of the technique. Limits of detection reported by manufacturers for their instruments are valid for pure aqueous solutions and may be taken as an indication of the analytical possibilities for water with low mineralization. In highly mineralized waters, the main interferences encountered in the determination of trace elements are connected with background absorption. In some samples, e.g. sea water, the background equivalents with FAAS may be quite high particularly if compared with the average content, as seen in Table 2.9.

Table 2.9. Typical background equivalents (BGE) for sea water [35]

Element	Wavelength (nm)	BGE ($\mu g\,dm^{-3}$)
As	193.7	8000
Zn	213.9	100
Pb	217.0	1800
Cd	228.8	80
Ni	232.0	400
Cu	324.7	80

The accuracy of the analysis is thus mainly controlled by the correctness of the background correction applied. Since the spectra causing background absorption in water samples are mainly true continuous spectra, e.g. photodissociation of alkali metal chlorides and sulfur dioxide, or molecular spectra of alkali earth oxides or hydroxides with highly diffuse rotational lines, normal deuterium lamp background correctors are adequate and the background absorbance values are usually within the range that can be handled, i.e. <0.4 absorbance unit.

2.4.5.2 Graphite furnace AAS

By far the broadest applications in water analysis for trace elements belong to furnace AAS. A considerable part of all water samples may be analyzed by sampling the water directly into the furnace. Any additional operations in the laboratory, each of which is a possible source of contamination, are eliminated.

The analytical possibilities of GFAAS in water analysis cannot be simply deduced from the quoted limits of detection because analytical signals are much more strongly dependent on matrix composition than in FAAS. In addition, the matrix effects may vary for furnaces of different construction and it is, therefore, also necessary to take into consideration instrumental parameters.

For natural waters, the main matrix constituents usually are sodium, magnesium and calcium in various combinations with chloride and sulfate anions. Thus the main and possibly limiting interferences are:
1 background absorption, which due to the considerably lower dilution of the sample vapour, is a much more severe problem than in flame AAS, and
2 equilibria in the vapour phase.

Losses by covolatization during thermal decomposition of the samples may indeed be diminished and eliminated by adequate changes in the temperature programmes. These inferences are considered with different instrumental parameters.

2.4.5.3 Background

In high chloride waters the most important contribution to background absorption is the photodissociation continuum of sodium chloride. Light scattering contributes markedly only when using small open ended tubes of the CRA type. It may be lowered by using a hydrogen diffusion flame to shield the atomizer.

Background absorption may easily become the limiting factor when analyzing highly mineralized water, particularly sea water. Analytical procedures usually try to eliminate sodium chloride at least partly during the charring step, i.e. before the atomization proper. According to Nakahara and Chakrabarti, sodium chloride may be volatilized slowly at 950°C [38], and this may be exploited for most moderately volatile and involatile elements. Elimination of most sodium chloride before the atomization step has been used in the determination of manganese [40], and molybdenum [38]. However, care has to be used since volatilization losses may occur as a result of different reactions with the concomitants. Thus, for example, chromium can be lost by volatilization of chromylchloride if oxygen is not thoroughly eliminated from the graphite furnace. Though this has not been reported in the analysis of sea water, it has been observed in analysis of blood [39] which has a similar salt content.

Volatilization of sodium chloride may be facilitated by addition of ammonium nitrate (NH_4NO_3) as a matrix modifier [41]. The NH_4Cl formed sublimes at 335°C and the $NaNO_3$ is decomposed at 380°C. Using this modifier, the background adsorption may be reduced to manageable levels for manganese, chromium, nickel and copper [42]. Unfortunately NH_4NO_3 strongly attacks the pyrolytic graphite coating so that the surface of the tube degenerates rapidly thus affecting the sensitivity of determination. Quite often, therefore, the addition technique has to be used for the measurement [43, 44].

More recently $NH_4H_2PO_4$ or $(NH_4)_2HPO_4$ have been suggested as matrix modifiers for natural freshwater samples [45]. This modifier has the advantage that besides helping to eliminate chlorides, it simultaneously raises the decomposition temperature without losing some more volatile metals, notably lead, cadmium and zinc. Most probably, relatively thermostable pyrophosphates are formed.

When measuring volatile elements with appearance temperatures below 600°C, at which temperature sodium chloride starts to volatilize, it is possible to volatilize the analyte before the salt, thus freeing the analytical signal from the background. This approach has been applied for the determination of cadmium [42, 45] and zinc [46] using organic acids to lower the appearance temperature of the analyte. As modifiers, ascorbic [47, 48], citric or aspartic acids [46] and EDTA [49] are added, usually in 0.1–1% concentrations. Simultaneously they reduce other matrix interferences. The shift of the appearance temperature toward lower temperatures is due to changes in the atomization mechanisms. The higher appearance temperatures of cadmium, zinc and lead correspond to atomization from a metal-oxygen bond, and the lower appearance temperatures to metal atomization. This has been demonstrated for lead by Salmon and Holcombe [50], and may be extended by analogy to other elements which atomize below 950°C and have lower heats of vaporization than dissociation energies of the metal-oxygen bond [51] (see Table 2.10).

Table 2.10. Appearance temperatures from oxidized and metallic forms compared with heats of vaporization (ΔH_M) and metal-oxygen bond energies (D_{MO})

	Tap (MO)	Tap (M)	D_{MO} (J mol^{-1})	ΔH_M (J mol^{-1})
Cd	770	450	280	112
Zn	970	530	284	130
Pb	950	770	378	195

It has been tacitly assumed since the paper of Campbell and Ottaway [52] that there are sufficient carbon atoms available for reduction of the analyte and that it is simply a question of temperature when the free energy of the particular reduction reaction becomes negative for the reduction of the analyte to take place. This however, is not true, since the number of available reactive carbon atoms, particularly in pyrolytic graphite coated tubes, is very limited. Carbon atoms in the hexagonal planes are at least 1000 times less reactive than 'edge atoms' in prismatic graphite planes. Thus in tubes coated with pyrolytic graphite, 'active sites' may be expected only on the boundary between two nucleation zones and these may be blocked by chemisorbed oxygen. If all active sites are blocked, no reduction of the analyte will take place below 950°C at which temperature the chemisorbed oxygen is released [58]. Therefore, under certain conditions, elements atomizing below 950°C show double peaks corresponding to atomization from oxidized and metallic forms respectively. Addition of nitric acid will tend to block the active sites and keep the analyte in the oxide form, whereas addition of reducing agents, such as ascorbic acid, or the use of uncoated graphite tubes with many active sites will facilitate reduction of the analyte. The shift of appearance temperatures is shown in Table 2.10.

When analyzing sulfate waters, the dissociation continuum of sulfur oxides [53] could contribute to the background if overlapping in time with the analyte atomization peak. However, it seems that for this type of water, background is not the limiting factor. The critical parameters are rather dissociation equilibria in the vapour phase, with sulfur generated by reduction of alkali metal and/or calcium sulfate or alternatively by dissociation of sulfur oxides.

2.4.5.4   Vapour phase interferences

Dissociation equilibria in the vapour phase are the most serious interferences in water analysis. To be effective, the interfering species must evidently be present in the vapour phase during atomization of the analyte. Knowing the thermal behaviour of different salts present in the sample it should be possible to eliminate, at least partially, the interferents during the thermal decomposition (charring) step. However, this is complictated by interaction of the salts with the graphite support.

In normal graphite tubes made of unoriented crystals, the sample solution has easy access to prismatic planes of the crystals and any constituent forming lamellar compounds will intercallate during the drying and charring stages. Once lamellar compounds are formed it is very difficult to decompose them. Part of the lamellar compounds is released only well above 1000°C, so that intercallation will broaden the temperature interval during which an interferent is released into the vapour phase. In lamellar compounds the reacting species which are usually negatively charged, enter between the individual hexagonal planes of graphite which act as a positively charged macroion. Alkali metal lamellar compounds, where graphite forms a negative macroion, are much less important from the analytical stand point. The distance between the planes may increase up to 9 Å thus causing the crystal to swell. The bond is partly ionic and partly determined by Van der Waals forces. Most of the intercallating species are partly deprotonated mineral acids or halides of transitional metals in different valency states or simply a halide anion surrounded by neutral metal halides. Intercallation is quite a common phenomenon though the number of lamellar compounds which have been studied and described so far is relatively limited [54]. The studies also did not cover the ratios of intercallating species to graphite nor the high temperature regions involved in GFAAS.

The stability of intercallated species may be quite remarkable. Thus Koirtyohann et al. [55] have shown that the depression caused by $HClO_4$ on a number of trace elements may be fully removed only when heating the graphite tube to 1700°C. This shows that the intercallating species or its precursor is retained in the tube at high temperatures and may affect the analyte dissociation in the gaseous phase.

Intercallation may be prevented if the sample solution has no access to the prismatic planes of graphite crystals, i.e. the tube surface is coated with pyrolytic graphite *in situ*. Intercallation may then proceed only at boundary faults between two nucleation zones. With repeated firings the pyrolytic graphite layer may blister and eventually peel off. For long tube life-time the quality of the pyrolytic coating is thus important.

Another possibility is a metallic carbide coating whereby the layered structure of graphite crystals is replaced by the metallic structure of interstitial carbides. Coating with molybdenum [56], or tantalum [57] carbides has proved highly efficient for lowering interferences of chlorides on lead as well as on other elements [58]. Using tantalum carbide coated tubes and platforms, Koirtyohann *et al.* [8] were able to reduce matrix interferences in natural waters for cadmium, lead, chromium, cobalt, arsenic and tin to below 10% relative signals.

The use of glassy vitreous carbon tubes which do not have a layered structure also prevents intercallation. These will no doubt find wider use in the future if only because of their superior life time [59]. So far, however, no experimental data on interferences in these have been published.

Any of these remedies, i.e. pyrolytic graphite or carbide coating or use of glassy carbon, will prevent intercallation of the most common interferents in waters, i.e. alkali or alkali metal halides, and thus facilitate their removal during a properly selected temperature program.

In some cases the formation of lamellar compounds may be exploited to advantage. Thus arsenic may be stabilized in a normal graphite furnace up to 1400°C by addition of H_2O_2 [60], and selenium by addition of K_2CrO_4 up to 1200°C [61]. Both arsenic and selenic acids are known to form lamellar compounds [62]. Both elements may be equally well stabilized with less deleterious effects on the graphite tube by addition of nickel, copper [63] or silver nitrates [64]. All these metals form relatively stable arsenides and selenides shifting the appearance temperature of arsenic and selenium above 1100°C.

The use of lanthanum as modifier for the analysis of water is quite often referred to as 'carbide coating' [65]. However lanthanum forms an ionic carbide which decomposes in the presence of water, releasing acetylene. Its effect may therefore hardly be compared to that of tantalum or molybdenum carbides. Nevertheless, lanthanum is retained in the graphite tube and on heating will again form lanthanum carbide. But since lanthanum is much more efficient in suppressing sulfate interferences than chloride interferences [66] the mechanism of its action is plausibly different.

It has been shown that in the presence of sodium sulfate, sulfur in the gaseous phase affects the dissociation equilibria of elements such as lead, arsenic and others [67,68]. The effect of sulfates decreases in the order $Na_2SO_4 > CaSO_4 > MgSO_4$ [65,66], suggesting that during the analyte atomization more sulfur enters the gaseous phase through the reduction of Na_2SO_4 or $CaSO_4$ than by dissociation of SO_3 which is released by thermal decomposition of $MgSO_4$ around 1000°C. Because of the low solubility of lanthanum sulfate it will crystallize during the drying step preferentially to calcium or sodium sulfates and will then readily decompose at 1150°C releasing sulfur trioxide. Its effect may thus be regarded as analogous to the releasing effect known from flame AAS. Lanthanum as a modifier has been broadly used for controlling interference effects of sulfates on many chalcophile elements [65,66,69,70], whether by adding lanthanum to the water samples or by treating the graphite tubes.

The amount of sulfate which can be eliminated by the addition of lanthanum to the sample depends naturally on the type of graphite furnace and other experimental conditions. In the presence of 1% nitric acid a 0.1% lanthanum solution will eliminate the interference of an equal amount of sulfate.

Results with the constant temperature furnace indicate that dissociation interferences of both chlorides and sulfates are lower the closer the conditions are to thermodynamic equilibrium [71, 72]. This is supported by experimental data with atomization from a platform or a probe [73]. In the case of chloride interference this is mainly an effect of temperature. The sample vapour enters a gaseous phase at a higher temperature than when vaporizing from the wall and all dissociation equilibria correspond to this temperature. In the case of sulfates, a lower concentration of oxygen in the furnace atmosphere contributes to the removal of interference. As Frech *et al.* have shown, most sulfur in the graphite furnaces should be in the form of CS and CS_2 [67]. Because of the above equilibrium concentration of oxygen in the furnace, free carbon atoms are partly used up forming CO, thus increasing the concentration of free sulfur. The later the analyte atomizes, the lower is the concentration of oxygen in the carbon atmosphere. These observations simultaneously show the importance of low temperature gradients in the furnace for efficient elimination of interferences.

2.4.6 **Conclusions**

An understanding of these processes has been emerging in the last few years and its full impact on analytical applications is still to be ascertained. We may therefore expect that the limits of detection of GFAAS procedures on real samples will improve. The main instrumental requirements are:

1 high heating rates without overshooting the atomization temperature;
2 pyrolytic graphite coated tubes;
3 high quality pyrolytic graphite platforms;
4 use of an efficient background corrector, possibly based on the Zeeman effect, enabling correction of high background absorbance values when necessary.

Besides these, the selection of an appropriate temperature programme and a correct modifier are necessary. With all the conditions optimized, the possible interference should be minimal and the limits of detection for real samples may be expected to approach those for pure aqueous solutions.

Since AAS can determine one element at a time, the conditions may be optimized for each particular element. Multi-element analyses have to be done sequentially and this may be relatively time consuming, particularly with the furnace technique.

The above mentioned attributes make AAS the method of choice for water analysis in all laboratories except where multi-element analyses with a high throughput of samples is required. In such laboratories an ICP-optical emission spectrometer may be more economical.

2.4.7 **References**

1 *Methods for Chemical Analysis of Water and Wastes* (Revised), Report No. EPA 600/4-79-020, U.S. Environmental Protection Agency (1978).
2 *Standard Practice for Measuring Trace Elements in Water by GFAAS*, D3919 under the Jurisdiction of ASTM Committee D-19, 1916, Race St., Philadelphia, Penn. 19103, U.S.A.
3 Department of the Environment and National Water Council (U.K.): *Methods Exam. Waters Assoc. Mater.*, (1982).
4 R.E. Sturgeon, S.S. Berman, J.A.M. Desaulniers, A.P. Mykytiuk, J.W. McLaren and D.S. Russell, *Anal. Chem.*, **52**, 1585 (1980).
5 J. Lamathe, C. Magurno and J.C. Equel, *Anal. Chim. Acta*, **142**, 183 (1982).
6 J.C. Meranger, C. Chalifoux and K.S. Subramanian, *Eau du Quebec*, **12**, 272 (1979).
7 M. Ihnat, *Int. J. Environ. Anal. Chem.*, **10**, 217 (1981).
8 M.L. Kaiser, S.R. Koirtyohann, E.J. Hinderberger and H. E. Taylor, *Spectrochim. Acta. Part B*, **36**, 773 (1981).
9 W. Slavin, D.C. Manning and G.R. Carnrick, *Anal. Chem.*, **53**, 1504 (1981).
10 A.M. Gunn, *WRC Tech. Rep.*, TR-169, 75 (1981).
11 D.S. Lee, *Anal. Chem.*, **54**, 1182 (1982).
12 J.D. Hem, Aluminium Species in Water. In *Trace Inorganics in Water*, Am. Chem. Soc., 98, (1968).
13 D.P.M. Laxen and R.M. Harrison, *Anal. Chem.*, **53**, 345 (1981).
14 J.M. Bewers and H.L. Windom, *Marine Chem.*, **11**(1), 71 (1982).
15 R. Massee and F.J.M.J. Maessen, *Anal. Chim. Acta*, **127**, 181 (1981).
16 G.E. Batley and T.J. Farrar, *Anal. Chim. Acta*, **99**, 283 (1978).
17 J.C. Van Loon, *Anal. Chem.*, **51**, 1139A (1979).
18 A. Tessier, P.G.C. Campbell, and M. Bisson, *Int. J. Enviorn. Anal. Chem.*, **7**, 41 (1979).
19 K.M. Bone and W.D. Hibbert, *Anal. Chim. Acta.*, **107**, 219 (1979).
20 G. Volland, G. Kolblin, P. Tschöpel and G. Tölg, *Z. anal. Chem.*, **287**, 1 (1972).
21 R.G. Smith and H.L. Windom, *Anal. Chim. Acta*, **113**, 39 (1980).
22 K.R. Sperling, *Z. anal. Chem.*, **310** (3.4), 254–256 (1982).
23 G.E. Batley and T.P. Matousek, *Anal. Chem.*, **49**, 2031 (1977).
24 G.E. Batley and J.P. Matousek, *Anal. Chem.*, **52**, 1570 (1980).
25 W. Lund, Y. Thomassen and P. Doule, *Anal. Chim. Acta*, **93**, 53 (1977).
26 E.J. Czobik and J.P. Matousek, *Spectrochim. Acta, Part B*, **35**, 741 (1980).
27 M.I. Abdullah, O.A. El.Rayis and J.P. Riley, *Anal. Chim. Acta*, **84**, 367 (1976).
28 Z. Slovak, S. Slovakova and M. Smrž, *Anal. Chim. Acta*, **75**, 127 (1975).
29 R.E. Sturgeon, S.S. Berman, S.N. Willie and J.A.M. Desaulniers, *Anal. Chem.*, **53**, 2337 (1981).
30 J.F. Dingman, K.M. Gloss, E.A. Miland and S. Sigeia, *Anal. Chem.*, **46**, 774 (1979).
31 T. Yao, M. Akino and S. Musha, *Bunseki Kagaku*, **30**, 740 (1981).
32 M.B. Coleila, S. Siggia and R.M. Barnes, *Anal. Chem.*, **52**, 2347 (1980).
33 M.P. Newton, J.V. Chauvin and D.G. Davis, *Anal. Lett.*, **6**, 89 (1973).
34 Y. Hoshino, T. Utsonomiya and K. Fukui, *Chem. Lett.*, **9**, 947 (1976).
35 J. Willis, *Proceedings of the Int. Conf. on Heavy Metals in the Environment*, Toronto, 69 (1975).
36 A.A. Levinson, *Introduction to Exploration Geochemistry*, 2nd Edn., Applied Publishing, Wilmette, Illinois 43 (*circa.* 1980).
37 B. Mason, *Principles of Geochemistry*, J. Wiley and Sons, New York, 195 (1966).
38 T. Nakahara and C.L. Chakrabarti, *Anal. Chim. Acta*, **104**, 99 (1979).
39 M. Beaty, W. Barnett and Z. Grobenski, *At. Spectr.*, **1**, 72 (1980).
40 G.R. Carnrick, W. Slavin and D.C. Manning, *Anal. Chem.*, **53**, 1866 (1981).

41 R.D. Ediger, *At. Absorp. Newsletter*, **13,** 61 (1974).

42 M. Hoenig and R. Wollast, *Spectrochim. Acta, Part B*, **37,** 399–415 (1982).

43 J.R. Montgomery and G.N. Peterson, *Anal. Chim. Acta*, **117,** 392 (1980).

44 R.E. Sturgeon, S.S. Berman, A. Desaulniers and D.S. Russell, *Anal. Chem.*, **51,** 2364 (1979).

45 R. Guevremont, *Anal. Chem.*, **52,** 1574 (1980).

46 R. Guevremont, *Anal. Chem.*, **53,** 911 (1981).

47 D.J. Hydes, *Anal. Chem.*, **52,** 959 (1980).

48 M. Tominaga and Y. Umezaki, *Anal. Chim. Acta*, **139,** 289 (1982).

49 R. Guevremont, R.E. Sturgeon and S.S. Berman, *Anal. Chim. Acta*, **115,** 163 (1980).

50 S.G. Salmon and J.A. Holcombe, *Anal. Chem.*, **54,** 620 (1982).

51 I. Rubeska, *Recent advances in analytical spectroscopy*, Proceedings 9 ICAS + 22 CSI, Tokyo, 93 (1981).

52 W.C. Campbell and J.M. Ottaway, *Talanta*, **21,** 837 (1974).

53 H. Massmann and S. Gucer, *Spectrochim. Acta, Part B*, **29,** 283 (1979).

54 M.L. Dzurus and G.R. Hennig, *Proceedings Vth Conference on Carbon*, Vol. 1, Pennsylvania, Pergamon Press, (1962).

55 S.R. Koirtyohann, E.D. Glass and F.E. Fichte, *Appl. Spectrosc.*, **35,** 22 (1981).

56 D.C. Manning and W. Slavin, *At. Absorp. Newsletter*, **17,** 43 (1978).

57 P. Hocquellet and N. Labeyrie, *At. Absorp. Newsletter*, **16,** 124 (1977).

58 J.E. Poldoski, *Anal. Chem.*, **52,** 1147 (1980).

59 L. De Galan M.T.C. De Loos and R.A.M. Oosterling, *Analyst*, **108,** 138 (1983).

60 J. Koreckova, W. Frech, E. Lundberg, J.A. Persson and A. Cedergren, *Anal. Chim. Acta*, **130,** 267 (1981).

61 G.F. Kirkbright, S.H. Hsiao-Chuan and R.D. Snook, *At. Spectrosc.*, **1,** 85 (1980).

62 G.R. Hennig, *Interstitial compounds in graphite in progress in inorganic chemistry*, Vol. 1, (Ed., F.A. Cotton), Interscience Publishers, N.Y. (1959).

63 R.D. Ediger, *At. Absorp. Newsletter*, **14,** 127 (1975).

64 R.F. Sanzalone and T.T. Chao, *Anal. Chim. Acta*, **128,** 225 (1981).

65 P. Lagas, *Anal. Chim. Acta*, **98,** 261 (1978).

66 A. Andersson, *At. Absorp. Newsletter*, **15,** 71 (1976).

67 W. Frech and A. Cedergren, *Anal. Chim. Acta*, **88,** 57 (1977).

68 W. Frech, J.A. Persson and A. Cedergren, *Prog. Anal. Atom. Spectroc.*, **3,** 279 (1980).

69 M.P. Bertenshaw, D. Gelsthorpe and K.C. Wheatstone, *Analyst*, **107,** 163 (1982).

70 J.C. Meranger, and K.S. Subramanian, *Can. J. Spectrosc.*, **24,** 132 (1979).

71 L.R. Hageman, J.A. Nichols, P. Viswanadham and R. Woodriff, *Anal. Chem.*, **51,** 1406 (1979).

72 L.R. Hageman, A. Mubarak and R. Woodriff, *Appl. Spectrosc.*, **33,** 226 (1979).

73. D.C. Manning and W. Slavin, *Anal. Chim. Acta*, **118,** 301 (1980).

2.5 Spark Source Mass Spectrometry for the Analysis of Natural Waters

Prepared by
A. M. URE

*Department of Spectrochemistry, Macaulay Institute, Craigiebuckler, Aberdeen,
AB9 2QJ, UK*

2.5.1 Introduction

Spark source mass spectrometry (SSMS) is a multi-element analytical technique in which ions, representative of the sample, are produced by subjecting a solid sample to a pulsed, high-voltage, radiofrequency spark discharge. The method of SSMS is, therefore, only indirectly applicable to water samples since these must first be converted to solid samples by processes of evaporation, freezing or preconcentration. The technique is slow, and because of the high cost of the instrumentation, it is likely to be used for water analysis only in those few laboratories already equipped for SSMS and to a limited extent even there.

2.5.2 Instrumentation and general features of methods

The theory and practice of spark source mass spectrometry is discussed in detail by Ahearn [1] and methods for the analysis of solutions are reviewed briefly by Leyder [2]. Solid electrodes of conducting materials, such as graphite or a metal, on which the sample material is supported, or electrodes briquetted from a mixture of sample and a conducting powder are sparked in a pulsed, high-voltage, radiofrequency discharge to form ions. These are accelerated into a mass spectrometer of double-focussing Mattauch Herzog geometry which incorporates an electrostatic analyser as well as a magnetic analyser to delimit the range of ion energies produced by the action of the spark. An ion mass/charge spectrum is produced at a single focal plane where it is recorded by a photographic plate. For singly-charged ions this is effectively a record of the mass spectrum on which measured isotope line densities can be related to sample elemental concentrations by suitable plate evaluation and by calibration procedures which depend on the use of an internal standard element incorporated in the sample electrode. The fundamental relationship used to calculate the unknown concentration, C_x, of a sample constituent is:

$$\frac{C_x}{C_s} = \frac{I_x}{I_s} \cdot \frac{A_s}{A_x} \cdot \frac{1}{S},$$

where C_s is the concentration of the internal standard element, I_x and I_s the relative intensities of the unknown and internal standard isotopes, A_x and A_s the respective isotopic abundances and S is the Relative Sensitivity Coefficient (RSC).

The RSC is the ratio of the element concentration determined by SSMS to the known or authenticated content of that element and must be determined for each element by the analysis of certified reference materials or a sample of known composition. While semi-quantitative analysis, within a factor of about 5 for most elements, is possible without such standardization by RSC, for quantitative analysis, SSMS is a comparative method whose accuracy depends on the accuracy of standardization.

Electrical detection methods have been used but are limited by their sequential, scanning or stepping mode of operation and by the erratic nature of ion production by the spark source.

Spark source mass spectrometry has the advantage of (1) virtually universal element coverage, (2) uniform sensitivity within a factor of 10 for all element isotopes and (3) high sensitivity, with detection limits of 0.01–0.1 mg kg^{-1} for all elements. A precision of $\pm 10-\pm 15\%$ can be obtained and the accuracy can approach this when suitable standardization has been carried out. Higher precision of $\pm 5\%$ or better can be obtained with the use of isotope dilution techniques but this is laborious and slow and cannot be applied to all elements. Despite its limited usefulness for water samples, their analysis by SSMS carries the usual advantages of solutions over solid and powder samples, viz.

(i) freedom from sample inhomogeneity effects,

(ii) ease of standardization, and

(iii) suitability for isotope dilution methods.

2.5.3 Operating conditions

There are four general methods of preparing water samples for analysis by SSMS. These are:

1 freezing to form solid samples for direct sparking;

2 evaporating the sample to small volume and micropipetting a small sample onto an electrode tip where it is dried before sparking;

3 mixing the sample with a collecting, conducting powder and freeze-drying or evaporating to dryness; the electrode is formed from this dry residue;

4 forming an electrode from a mixture of a solid preconcentrate of the water sample and a conducting powder.

The use of frozen water samples, and samples of natural ice, has been described by Chupakhin et al. [3,4] and others. Because the frozen water method involves no preconcentration, the sensitivity is probably adequate only for major constituents.

An example of the second technique is described by Carter et al. [5] in which 100 cm^3 of water is spiked with 2–4 μg of an enriched stable isotope (for up to 24 elements) and evaporated to a volume of a few cm^3. A 2–10 μl aliquot is evaporated onto a high-purity graphite electrode tip for sparking and analysis. Some twelve elements can be determined at ng cm^{-3} concentrations in distilled and process waters and for a few elements a detection limit of 10^{-12} g is claimed for 100 cm^3 original sample volumes. Because of the isotope dilution procedure the precision is high. RSD of $\pm 5\%$ at 10^{-9} g, and an accuracy of about $\pm 10\%$ is claimed. The volume finally pipetted onto the electrode must be limited to about $10\,\mu l$ since appreciable soaking of solution into the electrode could give rise to chromatographic segregation effects. Similar techniques such as that of Mykytiuk et al. [6] make use of standard solutions instead of isotope dilution to calibrate the analysis. In this example 15 elements were determined in high purity water using a 250 cm^3 original water sample volume and a silver instead of a graphite electrode.

The most commonly used technique of sample preparation is perhaps that of drying a slurry of the conducting material and the water sample and briquetting the dry residue to form an electrode for sparking. The method of Hamilton et al. [7] exemplifies the procedure. In this 1 dm^3 of water, tap, purified tap, river or reservoir water, is filtered through a $0.4\,\mu$ filter into a polyethylene tray, to which carbon powder is added as a collector and freeze dried. The dry residue is pressed to form an electrode. Some 20–30 elements at μg dm^{-3} concentrations could be determined. Other conducting materials used to collect the water residue for electrode manufacture include graphite [8], silver [9] and aluminium [10].

2.5.4 Preconcentration methods

The transition metals and some others have been preconcentrated from waters by chelating with 8-hydroxyquinoline (oxine) at pH 8, adding activated charcoal and filtering. The charcoal is oxidized at low temperature, the residue dissolved, the internal standard element indium added and graphite powder mixed in to form a slurry. The dried slurry is used to make electrodes for analysis. Some 30–40 elements can be nearly quantitatively or substantially recovered by this method [11] which has been applied to tap, ground and river water as well as to the IAEA synthetic water sample W3.

A multielement preconcentration procedure using PAN (1-(2-pyridylazo)-2-naphthol) as co-crystallization agent has been described [12]. The precipitate is collected in a filter paper which is ashed at low temperature, the residue redissolved and slurried with graphite. The dried slurry, including indium as internal standard, is used for making the sample electrodes. Some 30 elements are quantitatively or nearly quantitatively recovered by this technique which has been applied to geothermal waters.

For copper, lead and the noble metals, quantitative preconcentration can be obtained by using spontaneous electrochemical deposition (cementation) of trace elements onto a micro-column of the aluminium powder used to make electrodes. The technique is simple and probably applicable to about 15 elements. It is sensitive enough to make the direct determination of gold in river or sea water at ng dm^{-3} concentrations [13, 14].

A coprecipitation [15] method has been used for U in seawater and an ion-exchange resin [16] (Chelex-100) used to preconcentrate 8 elements in seawater using isotope dilution SSMS.

The potential problems of interference from organic species concentrated in the water residue which are generally and unrealistically ignored are avoided by the incorporation of a stage of low temperature ashing with excited oxygen in the procedure of Crocker and Merritt [17], who determined some 30 elements in lake water. Heating the final electrode to 400°C has also been used to eliminate the organic residue, especially where an organic chelating agent has been used to preconcentrate the sample [18]. The main difficulties of water analysis by SSMS, as with many other methods, lie in the dangers of sample contamination which are magnified by the elaborate evaporation or preconcentration stages necessary to prepare the sample and by the low concentrations of the elements to be determined. Clean laboratory conditions, ultra-pure reagents, filtered air supplies and vigorous cleaning of laboratory ware may all be essential. In addition the purity of the conducting material used to form electrodes must be very high.

2.5.5 References

1 A.J. Ahearn, *Trace Analyses by Mass Spectrometry*, Academic Press, New York, p. 460 (1972).
2 F. Leyder, *Spectra 2000*, **7**, 64 (1979).
3 M.S. Chupakhin, O.I. Kryuchkova and A.D. Semenov, *J. Anal. Chem. USSR*, **27**, 1724 (1972).
4 M.S. Chupakhin, O.I. Kruyuchkova and S.I. Sulimova, *J. Anal. Chem. USSR*, **28**, 1504 (1974).
5 J.A. Carter, D.L. Donohue, J.C. Franklin and R.W. Stelzner, In *Trace Substances in Environmental Health*, (Ed. D.D. Hemphill) University of Missouri, Columbia USA 303 (1975).
6 A. Mykytiuk, D.S. Russell and V. Boyko, *Anal. Chem.*, **48**, 1462 (1976).
7 E.I. Hamilton and M.J. Minski, *Environ. Lett.*, **3**, 53 (1972).
8 M.A. Wahlgren, D.N. Edington and F.F. Rawlings, Argonne Nat. Lab., Report ANL-7860 (Pt. 3), Argonne Nat. Lab. Argonne Ill. USA. 55 (1972).
9 P.J. Paulsen and J.R. Moody, *24th Ann. Conf. Mass Spectrom.*, San Diego USA Abstract X-3, 501 (1976).
10 M. Klose, *Z. anal. Chem.*, **254**, 7 (1971).
11 B.N. Vanderborght and R.E. Van Grieken, *Talanta*, **27**, 417 (1980).
12 W. Blommaert, R. Vandelannoote, L. Van't Dack, R. Gijbels and R. Van Grieken, *J. Radioanal. Chem.*, **57**, 383 (1980).
13 K.H. Welch and A.M. Ure, *Anal. Proc.*, 8 (1980).
14 A.M. Ure, *Proc. VI Czechoslovak Spectrosc. Conf.*, Nitra 61 (1980).
15 E.L. Callis, Report NBL-262 New Brunswick Lab. Atomic Energy Commission N.J. USA 52 (1972).
16 A.P. Mykytuik, D.S. Russell, and R.E. Sturgeon, *Anal. Chem.*, **52**, 1281 (1980).
17 I.H. Crocker and W.F. Merritt, *Water Res.*, **6**, 285 (1972).
18 R.E. Sturgeon, S.S. Berman, J.A.H. Desaulniers, A.P. Mykytiuk, J.W. McLaren and D.S. Russell, *Anal. Chem.*, **52**, 1585 (1980).

2.6 Electron Paramagnetic Resonance (EPR) Spectrometry for the Analysis of Natural Waters

Prepared by
A. M. URE
Department of Spectrochemistry, Macaulay Institute, Craigiebuckler, Aberdeen AB9 2QJ, UK

EPR spectrometry is one of the few spectroscopic methods that is able to provide information on the oxidation states and chemical environments of trace metals in natural systems with sufficient sensitivity and discrimination to possess the potential for general applications. The main drawback is its limitation to paramagnetic materials, with spectra being most straightforward for single electron species and s-state ions. Species with an even number of electrons are usually very difficult to detect. However, even with the most readily-observable species [e.g. $Mn(H_2O)_6^{2+}$ with a detection limit of about 10 ppb, i.e. 1 in 10^8], preconcentration is usually necessary for their identification in natural waters. In addition, any metal of interest will probably occur in more than one chemical form in a particular sample and the limit of detection is determined by the concentration of each species (not their sum as in techniques for elemental analyses). Furthermore, for species in solution the detection limit drops dramatically if the molecular motion is not fast compared to the timescale of the EPR transition (i.e. high molecular weight species have a very much higher detection limit than low molecular weight species).

Some exploratory experiments on natural waters in the Aberdeen area performed at the Macaulay Institute have indicated that concentrations of $Mn(H_2O)_6^{2+}$ can be readily monitored by concentrating $100\ cm^3$ of water to $1\ cm^3$ (which is an adequate volume for EPR measurements), but other paramagnetic species in these samples were best studied as solid residues after evaporating to dryness. The complexity of some of these latter spectra is such that conclusive interpretations were difficult and quantitative information was not readily obtained. It is expected, therefore, that, apart from cases of great economic importance, the routine use of EPR in the study of natural waters will probably be limited to monitoring the concentrations of $Mn(H_2O)_6^{2+}$, and possibly $VO(H_2O)_5^{2+}$ ions. The use of EPR spectrometry for total elemental analyses (i.e. by converting all of the species of a particular metal to a single form which can be detected with a high level of sensitivity) has little to recommend it because of the high cost of instrumentation and the lengthy sample preparation time for such measurements, except perhaps when other analytical methods are unavailable, or when a single technique is required for comparing the concentrations of a particular species with the total concentration of that metal (i.e. $Mn(H_2O)_6^{2+}$/total Mn or $VO(H_2O)_5^{2+}$/total V).

No assessment has been made of contamination problems that might arise during sample handling, but it is expected that these will be much more severe than is encountered with work on straightforward elemental analyses, because of the necessity of excluding organic materials, e.g. plasticizers, that could affect the chemical forms of the metals of interest.

2.7 X-ray Fluorescence Spectrometry in the Analysis of Natural Waters

Prepared by
R. JENKINS

Philips Electronic Instrumentation Inc., 85, McKee Drive, Mahwah, NJ 07430, USA

2.7.1 Introduction

When an atom is bombarded with high energy particles, e.g. photons, protons or electrons, inner orbital vacancies are formed in the target atoms. When these vacancies are filled by outer orbital electrons, X-ray lines may be produced. The wavelengths of the lines are related to the atomic number of the target element. X-ray fluorescence (XRF) analysis is based on the use of these characteristic X-ray emission wavelengths or energies to identify an element or range of elements. The intensities of certain selected line(s) can then be used to quantify the appropriate elemental concentrations [1]. XRF has been used since the early 1970s for the study of trace elements in various water samples, and several hundred papers have been published describing a wide variety of instrumentation and techniques for this purpose.

2.7.2 Scope of analysis

Conventional spectrometers allow measurement of all elements in the periodic table down to and including fluorine ($Z = 9$). The dynamic range of the XRF method is large, covering about six orders of magnitude from the low ppm range to 100%. The inter-element (matrix) effects are predictable and relatively easy to correct, particularly in the case of a reasonably modern computer-controlled system. Matrix effects are minimal in the case of so-called 'thin film' samples, that is, a sample of thickness in the ten micron range. The detection limits obtainable directly on the unprepared sample are generally barely sufficient for the analysis of trace elements in natural waters. It is, therefore, common practice to use a preconcentration step. The sample resulting from preconcentration is usually in the form of a thin sample and as such, meets the thin film criterion. Because of this, the conversion of XRF raw data to element concentration is a rather simple procedure.

X-ray spectrometers are generally not transportable and as a result almost all analyses are performed in the laboratory. This in turn requires the transportation of the samples for analysis from the site of interest to the X-ray spectrometer. The actual volume of water taken for analysis is generally rather large, not just because natural waters are usually plentiful and sample errors can be reduced easily by taking large aliquots, but of equal importance, the very low concentration levels of the elements of interest generally require the preconcentration of a large volume of sample to allow a measurable signal to be obtained. To this end it is common practice to preconcentrate and this can sometimes be conveniently done at the sampling site.

2.7.3 Limits of the method

The sensitivity of the X-ray spectrometer varies by about four orders of magnitude over the measurable element range, when expressed in terms of rate of change in response per rate of change in concentration. The minimum detectable concentration limit is inversely proportional to the sensitivity m (i.e. c/s/%) of the spectrometer and directly proportional to the square root of the background response (R_b in c/s) at the analyte wavelengths. Thus the lower limit of detection LLD is given by the expression

$$LLD = 3/m. \sqrt{R_b/T_b}$$

where T_b is the background counting time. For a fixed analysis time the detection limit is proportional to $m/\sqrt{R_b}$ and this is taken as a 'figure of merit' for trace analysis. The value of m is determined mainly by the power loading of the source, the efficiency of the spectrometer for the appropriate wavelength and the fluorescent yield of the excited wavelength. The value of R_b is determined mainly by the scattering characteristics of the sample matrix and the intensity/wavelength distribution of the excitation source.

In general, the background is low when the sensitivity is low and these factors result in an overall variation in the detection limit of the technique of about two orders of magnitude.

The most sensitive elements are generally the transition elements where direct measurements down to one part per million are possible in analysis times of about 100 s. The least sensitive elements are the lower atomic numbers from fluorine ($Z = 9$) through to silicon ($Z = 14$) where detection limits are around 0.01%. Detection limits at these levels are generally insufficient for the analysis of natural waters and preconcentration techniques have to be employed.

2.7.4 Instrumentation

There is a wide variety of instrumentation available today for the application of X-ray fluorescence techniques and for a more detailed description of the types of X-ray spectrometer and terminology related to application and techniques of data handling refer to the IUPAC document by R. Jenkins: Nomenclature, Symbols, Units and their Usage in Spectrochemical Analysis-IV X-Ray Emission Spectroscopy, *Pure Appl. Chem.*, **52**, 2541 (1980).

For the purpose of this discussion it is useful to classify X-ray spectrometers into three categories:
(a) wavelength dispersive spectrometers;
(b) energy dispersive spectrometers;
(c) particle excited spectrometers.

Wavelength dispersive spectrometers are by far the most commonly employed and these systems employ diffraction by a single crystal to distinguish characteristic wavelengths of X-rays excited in the sample. Energy dispersive spectrometers use the proportional characteristics of a photon detector, typically lithium drifted silicon, to separate the excited characteristic radiation in terms of their energies. Since there is a simple relationship between wavelength and energy, these techniques each provide the same basic type of information. As far as the analysis of trace metals in natural waters is concerned, the characteristics of the two methods differ mainly in their relative sensitivities and the way in which data are collected and presented. Generally speaking the wavelength dispersive system is roughly one to two orders of magnitude more sensitive than the energy dispersive system. Against this, however, the energy dispersive spectrometer measures all elements within its range at the same time, whereas the wavelength dispersive system identifies only those elements for which it is programmed. To this extent, the energy dispersive system is more useful in recognizing unexpected elements. Both of these methods typically employ a primary X-ray photon source operating at 0.5–3 kW. The specimen scatters the bremstrahlung from the source, leading to significant background levels, which tend to become one of the major limitations in the determination of low concentration levels.

Several modifications to these instruments have been employed with success in the area of water analysis and in addition, special instruments have been designed specifically for this application. As an example, a low cost energy dispersive spectrometer-based system has been developed at the US Naval Research Laboratory [2]. This instrument has been used to measure Al, Cr, Mn, Fe, Zn, As, Pb, Se, Cu and Cd in samples taken from the Potomac River. One common variation that is used in the energy dispersive spectrometer is to use a radioisotope source in place of the high power X-ray tube. Such a technique gives a lower cost, transportable, but element-specific system [3]. Another variation of the energy dispersive spectrometer is to utilize the technique of total reflection by means of a sample support in the form of an optical flat. Detection limits down to the pg level have been claimed using such a procedure [4].

The third major category of X-ray spectrometer is the particle source, generally a proton-induced X-ray emission, 'PIXE' system. PIXE differs from conventional energy dispersive spectroscopy in that a proton source is used in place of the photon source. The proton source is typically a Van de Graaff generator or a cyclotron. Protons in the energy range of about 2–3 MeV are typically employed for this type of work. The proton source offers several advantages over the photon one. In addition to being intense relative to the conventional photon source, the proton excitation system generates relatively much lower backgrounds. Also the cross-section for characteristic X-ray production is quite large and a single particle can give rise to several characteristic X-rays as it penetrates the sample. Thus the proton-excited XRF system is well suited for the analysis of trace elements in water and an increasing number of papers applying this system have appeared during recent years [5–7, 9].

2.7.5 Analytical techniques

2.7.5.1 Direct methods

The detection limits directly achievable by the X-ray fluorescence method generally lie in the low ppm range or, in absolute terms, a nominal detection limit of 0.1–0.01 μg on a thin film sample. In favourable cases, such as the transition elements, this limit may go to about 0.1 ppm, and in the most unfavourable case, typified by the lower atomic number elements, to about 50 ppm. One of the special problems encountered in the direct analysis of water samples stems from the need to support the specimen under examination. Most conventional spectrometers irradiate the sample from below and the support film both attenuates the signal from the longer wavelength characteristic lines as well as introducing a significant 'blank'. Absorption by air also becomes an important factor for the measurement of wavelengths longer than ~ 2 Å and the need to work in a helium atmosphere introduces further attenuation of the longer wavelength signals. Thus in almost all cases some preconcentration technique is applied to the water sample before analysis [10, 11].

2.7.5.2 Concentration by evaporation

In principle, one of the simpler methods of bringing an analyte element within the sensitivity range of the spectrometer is to preconcentrate by evaporation. In order to achieve detection limits at the low ppb level this in turn would mean the evaporation of about 100 cm^3 of water. For practical reasons preconcentration by evaporation has not found a great deal of application in the X-ray analysis of water.

On the other hand, the application of evaporation preconcentration techniques does show some promise when combined with special techniques for reducing the relatively high inherent background observed in classical XRF methods. One such method utilizes the total reflection of X-rays from a highly polished surface. Wobrauschek and Aiginger [12] have achieved sensitivities down to the ppb level by evaporation of small (about 5 μl) samples of water onto very flat silica plates. A thin layer of insulin was used to give a good distribution of the evaporated sample across the surface of the optical flat, and an energy dispersive spectrometer was used to measure and analyse the characteristic X-ray emission.

2.7.5.3 Use of ion exchange resins

By far the most studied and apparently the most useful method of preconcentration, is the use of ion exchange resins. The major advantage of most ion exchange methods is that the functional group is immobilized onto a solid substrate providing the potential to batch-extract ions from solution. The sample itself can be either the actual exchanged resin or a separate sample containing the appropriate ions eluted from the resin. The success of the method depends to a large extent on the recovery efficiency of the resin which in turn is determined by the affinity of the ion exchange material for the ions in question and the stability of complexes present in solution. Preconcentration factors of up to 4×10^4 can be achieved from suitable ion exchange material using around 100 mg of resin [13].

One of the most useful ion exchange resins is Chelex-100, which contains iminodiacetic functional groups and is similar to EDTA (ethylenediamine tetra-acetic acid). It shows a good recovery efficiency for a wide range of ions and its chemistry can be predicted from prior experiments with EDTA. Against this, however, it is not too successful in the separation of ions from solution which are high in iron and calcium (e.g. sea water) which elements occupy the available sites with the exclusion of the trace element ions. The exchange resin can also be employed as a membrane through which the solution to be analysed is passed [14]. This is particularly convenient means of separation for the XRF method because the paper can be mounted directly in the spectrometer for analysis following the separation process. As an example of the use of Chelex-100, 200 cm^3 of water were passed through two Chelex-100 membranes for a period of about 20 minutes. The enrichment factors obtained were in excess of 1000 allowing the separation of K, Ca, Mn, Co, Ni, Cu, Zn, Rb, Sr and Pb as chlorides or nitrates at element concentrations in the range 10 ppm to 10 ppb [15]. Similar techniques have been used for the determination of U in ground water [16] of Ba [17] and of Se [18].

2.7.5.4   Co-precipitation methods

Another method which has been employed with some success to preconcentrate elements in natural water samples is co-precipitation. This offers a relatively simple method with the advantage of giving a fairly uniform deposit that can be easily collected. One of the earlier applications of this technique involved the use of iron hydroxide as a coprecipitant for the determination of Fe, Zn and Pb in surface waters [19]. One of the more popular coprecipitants in use today is ammonium pyrrolidine dithiocarbamate (APOX). In the application of this method to the analysis of natural waters, detection limits in the range 0.4–1.2 ppb have been claimed for the elements V, Zn, As, Hg and Pb [20]. Other coprecipitants have been described including the use of iron dibenzyl dithiocarbamate for the determination of U at the ppb level in natural waters [21], zirconium dioxide for the determination of As in river water [22], and polyvinylpyrrolidone-thionalide for the determination of Fe, Cu, Zn, Se, Cd, Te, Hg and Pd in waste and natural water samples [23]. The use of this latter reagent is particularly useful where large concentrations of Mg and Ca may be present.

2.7.5.5   Surface adsorption

A third technique which is somewhat less popular, but still useful, is application of surface adsorption reagents. Activated carbon is generally used as the carrier after conversion of the metal ion(s) of interest to a suitable form. For example, iodide in ground water has been determined by this method after conversion to silver iodide [24]. Neutral 8-hydroxyquinoline complexes have been employed in which again the oxine complex is collected on activated carbon [25]. Use of activated carbon is particularly useful where PIXE is being employed as the analysis technique since the resultant sample is ideal for direct presentation to the instrument [26].

2.7.5.6 Other concentration methods

Other preconcentration methods which have been employed in this area include electro-deposition [27], precipitation chromatography [28], liquid/liquid extraction [10], immobilized reagents [29], plus a variety of other techniques well known to analytical chemists.

2.7.6 Conclusions

The relatively large number of published papers describing use of the XRF technique for the analysis of trace metals in natural waters certainly confirms the potential of this technique. Although the sensitivity obtainable is barely sufficient for direct analysis, a wide range of preconcentration techniques has been developed which bring the concentrations of the required elements well within the range of the system. These preconcentration methods are sufficiently well developed that they do not compromise the inherent speed and accuracy of the XRF method. The absence of geometric constraints in the case of the energy dispersive spectrometer makes it ideal for the development of special dedicated instruments for trace analysis, and a useful outgrowth of this has been the use of particle excited spectrometry PIXE and total reflection XRF spectrometry. This latter technique offers tremendous potential for the trace analysis of limited amounts of material as from a typical preconcentration stage.

2.7.7 References

1   R. Jenkins, R.W. Gould and D.A. Gedcke, *Quantitative X-Ray Spectrometry*, Decker, New York (1981)
2   P.G. Burkhalter, *Trace Metal Water Pollutants Determined by X-Ray Fluorescence*, Rep. NRL Pro 1973 (Oct) 61 (1974)
3   B. Holynska, *Radiochem. Radioanal. Lett.*, **17**, 313 (1974).
4   J. Knoth and H. Schwenke, *Z. anal. Chem.*, **301**, 1, 7 (1980).
5   P. Sioshani, A.S. Lodhi and H. Payrovan, *Nucl. Instrum. Methods*, **142**, 285 (1977).
6   Y.C. Lien, R.R. Zombola and R.C. Bearse, *Nucl. Instrum. Methods*, **146**, 609 (1977).
7   R.A. Richey, P.C. Simms and K.A. Mueller, *IEE Trans. Nucl. Sci.*, **26**, 1347 (1979).
8   Ch.M. Fou, *Nucl. Instrum. Methods*, **186**, 599 (1981).
9   N.S. Saleh, *J. Radiochem. Lett.*, **74**, 257 (1982).
10 D. Leyden, *et al.*, *Pergamon Ser. Environ. Sci.*, **3** (Anal. Tech. Environ. Chem.) 469 (1980).
11 J. Smits, J. Nelissen and R. Van Grieken, *Anal. Chim. Acta*, **111**(1), 215 (1979).
12 P. Wobrauschek and H. Aiginger, *Anal. Chem.*, **47**, 852 (1975).
13 D.E. Leyden, T.A. Patterson and J.J. Alberts, *Anal. Chem.*, **47**, 733 (1975).
14 W. Campbell, E.F. Spano and T.E. Green, *Anal. Chem.*, **38**, 987 (1966).

15 R.E. Van Grieken, C.M. Bresele and B.M. Van der Goe, *Anal. Chem.*, **49**(9), 1326 (1977).

16 P. Minkkinen, *Finn. Chem. L.*, 4–5, 134 (1977).

17 P. Clechet and G. Eschalier, *Analysis*, **9**(4), 125 (1981).

18 H.J. Robberecht and R.B. Van Grieken, *Anal. Chem.*, **52**, 449 (1980).

19 E. Bruninx and E. Van Meijl, *Anal. Chim. Acta*, **80**, 85 (1975).

20 A.H. Pradzynski, R.E. Henry and J.S. Stewart, *Radiochem. Rad.*, **21**(5), 273 (1975).

21 S.S. Caravajal, K.J. Mahan and D.E. Leyden, *Anal. Chim. Acta*, **135**, 205 (1982).

22 T. Katsuno, *Nagano-ken Eisei Kogai Kenkyusho Kenkyu Hokoku*, **3**, 52 (1981).

23 R. Panayappan, D.L. Venezky, J.V. Gilfrich and L.S. Birks, *Anal. Chem.*, **50**, 1125 (1973).

24 P.T. Howe, *Analysis for Iodide in Ground Water by X-Ray Fluorescence Spectrometry After Collection as Silver Iodide on Activated Charcoal*, At. Energy Can Ltd., AECL-6444 11 (1980).

25 B.M. Vanderborght and R.E. Van Greiken, *Anal. Chem.*, **49**, 311 (1977).

26 E.M. Johansson and K.R. Akselsson, *Nucl. Instrum. Methods*, **181**(1–3) 221 (1981).

27 B.H. Vassos, R.F. Hirsch and H. Letterman, *Anal. Chem.*, **45**, 792 (1973).

28 W.P. Zeronsa, G. Dobkowski and S. Siggia, *Anal. Chem.*, **46**, 309 (1974).

29 D.M. Hercules, W.P. Zeronsa, G. Dabkowski and S. Siggia, *Anal. Chem.*, **45**, 1973 (1973).

2.8 Trace Element Analysis by Luminescence Spectrometry

Prepared by
W. H. MELHUISH
Institute of Nuclear Sciences, DSIR, Private Bag, Lower Hutt, New Zealand

2.8.1 Introduction

Luminescence methods of analysis of trace amounts of inorganic and organic substances became possible only after the development of the photomultiplier tube in the late 1930s. A pioneer in this field, C.E. White [1] of the USA, developed methods for determining metal ions by forming chelate-complexes and measuring their fluorescence emission. However, it was not until the appearance of commercial fluorescence spectrometers in the 1950s that this method of analysis became widely used.

Today, luminescence methods of trace element analysis are not much used, despite the fact that the instruments are inexpensive, easy to operate and transportable to the test area. The reason is undoubtedly due to the rather low sensitivity and the problem of interference by other elements in the sample. The method is also restricted by the number of elements that can be determined.

It is the intention of this paper principally to point out that new developments may lead to greatly increased sensitivity and selectivity and that fluorescence analysis could, at least for some elements, become the preferred method of analysis.

2.8.2 Luminescence analysis

The determination of trace elements by measuring the luminescence from, for example, metal-chelates has the virtue of ease of sample preparation and simple instrumentation. In its simplest form the instrument consists of a high pressure mercury arc, a set of excitation filters, a sample cell, a series of emission filters and a photomultiplier tube, with appropriate power supplies and amplifiers. The more expensive instruments use monochromators in place of the set of filters.

The detection limit, without pre-concentration is in the range $0.1–10\ \mu\mathrm{g\,dm^{-3}}$ (depending on the element). Pre-concentration by evaporation, co-precipitation or by the use of ion-exchange columns can improve the detection limit by up to 1000 times. It is often important to remove other elements which interfere with the analysis using extraction or ion-exchange techniques [2].

With the invention of the pulsed laser, there has been a considerable improvement in the limit of detection in fluorescence analysis. The advantage of the laser method is that the detector can be turned on for a definite period at a fixed time after the laser pulse. Thus the apparatus can be tuned to the particular analyte being measured and reject scattered exciting radiation and any impurity luminescence which has a decay time shorter or longer than the analyte decay time [3]. One recent example is that of Eu(III) in which a diketo-complex was formed and analysed by a standard fluorescence spectrometer. The limit of detection was $10 \, ng \, dm^{-3}$. The same complex, when excited with a nitrogen gas laser and detected by a gated photomultiplier used in the photon counting mode gave a limit of $2 \times 10^{-3} \, ng \, dm^{-3}$, some 5000 times lower [4]. It is true that this example is not typical since the emission occurs over a relatively narrow (20 nm) band. Further work needs to be done on other elements to test the method but there is some hope that Al, Se, Cd, Mg, Sn, V and Mn could have limits of detection in the $0.2-10 \, ng \, dm^{-3}$ range.

A technique of very high sensitivity for the lanthanides involves co-precipitation with CaF_2 and ignition to high temperature. The matrix is excited with laser radiation of selected wavelengths. A detection limit $2 - 10^{-5} \, \mu g \, dm^{-3}$ is claimed [5].

The purity of the reagents and solvents is especially important in fluorescence analysis, otherwise blanks are too high. Chemicals from several manufacturers should be tested. The sample of water should only come into contact with polyethylene or PTFE after collection. High purity, de-ionized water from which organics have been removed (e.g. by UV radiation) should be used as a solvent.

2.8.3 Chemiluminescence (CL) analysis

This method, only recently developed, is based on the catalysis or inhibition of the alkaline oxidation of a luminescent reagent by (usually) metal ions. The light radiation from the reaction is detected by a sensitive photomultiplier of a type which can be used as a photon counter (i.e. in which single photons give rise to output current pulses with a narrow distribution). The CL reaction chamber should be closely optically coupled to the photomultiplier cathode, with space left to insert optical filters. The apparatus is simple, easily transportable and samples need a minimum of chemical manipulation. Commercially made chemiluminescence apparatus is available; less satisfactory is a scintillation counter used in the CL mode (i.e. where only one photomultiplier is used to view the cell). One advantage of CL analysis is that continuous, on the spot, analysis of water is possible.

The water sample to be analysed is rapidly mixed with the reagent in a mixer, passed into the sample cell and drained to waste. The reagent solution is usually a buffered solution of luminol in water containing hydrogen peroxide together with additives which may enhance the CL or suppress interfering elements. The use of high purity solvents and reagents is, if anything, more important in CL analysis than in fluorescence analysis. To establish a blank measure, pure water is first passed through the system and injected into the reagent flow. The sample is then injected and the peak CL measured [6] Fig. 2.1(a). When using the inhibition method, the catalysed reaction mixture, is

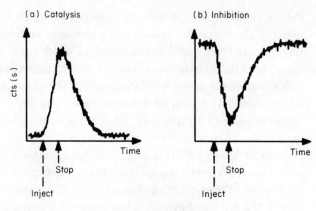

Fig. 2.1. Chemiluminescence analysis.

passed through, then the sample injected; in this case a negative peak is obtained [7] Fig. 2.1(b).

In an attempt to simplify the flow system, a microporous membrane has been used as a mixer, with the reagent being held in one compartment and the sample passed through the other compartment [8].

Table 2.11 shows the limit of sensitivity of some elements measured by CL analysis [9] (elements with sensitivities $> 10 \ \mu g \ dm^{-3}$ are not included). The data are unfortunately not well defined in the published papers so at present they can only be regarded as approximate.

Table 2.11. Trace element analysis by chemiluminescence analysis

Element		Sensitivity ($\mu g \ dm^{-3}$) without pre-concentration
V	(IV)	2.0
Cr	(III)	0.02
Cr	(II)	3.0
Mn	(II)	0.1
Fe	(II)	0.01
Fe	(III)	0.2
Co	(II)	0.01
Ni	(II)	0.1
Cu	(II)	0.01
Zn	(II)	10.0
Ru	(III) (IV)	0.4
Ru	(VI)	1.0
Ag	(I)	10.0
Cd	(II)	10.0
Hf	(IV)	10.0
Os	(VI) (VIII)	0.02
Os	(IV)	2.0
Au	(III)	10.0
Hg	(II)	2.0

2.8.4 References

1 C.E. White, *Anal. Chem.*, **21**, 104 (1949).
2 D.F. Marino and J.D. Ingle Jr., *Anal. Chem.*, **53**, 292 (1981).
3 K. Hiraki, K. Morashige and I. Nishikawa, *Anal. Chim. Acta*, **97**, 121 (1978).
4 S. Yamada, F. Miyoshi and K. Kano, *Anal. Chim. Acta*, **127**, 195 (1981).
5 F.J. Gustafson and J.C. Wright, *Anal. Chem.*, **51**, 1762 (1979).
6 H.K. Schroeder and P.O. Vogelhut, *Anal. Chem.*, **51**, 1583 (1979).
7 J.L. Burguera, M. Burguera and A. Townshend, *Anal. Chim. Acta*, **127**, 199 (1981).
8 D. Pilosof and T.A. Mieman, *Anal. Chem.*, **52**, 662 (1980).
9 D.B. Paul, *Talanta*, **25**, 377 (1978).

References

1. ...

Section 3   Voltammetric Methods

By

Commission V. 5
IUPAC

H. W. NÜRNBERG AND L. MART

Institut für Chemie der Kernforschungsanlage Julich, GmbH,
Institut 4: Angewandte Physikalische Chemie
D-5170, Julich 1
Federal Republic of Germany

3.1 Introduction

Heavy metals of ecotoxic significance exist in natural waters according to the water type and location, over a wide span of trace levels, covering altogether a range from several hundred $\mu g \, kg^{-1}$ to $0.001 \, \mu g \, kg^{-1}$ or sometimes less. The main pathways of input from natural and anthropogenic sources are wet and dry deposition from the atmosphere, run off water from land and discharge of waste water. Wet deposition of ecotoxic heavy metals in rain and snow is the dominant deposition mode in terrestrial ecosystems and contributes, therefore, very significantly to their pollution burden by ecotoxic metals via the atmospheric pathway. For estuarine and corresponding affected coastal waters, fluvial input by polluted rivers will also be a significant contribution. In the water column, heavy metals occur in the dissolved phase and are bound to suspended particulate matter including phytoplankton. The total concentration in the water column and its distribution over the suspended matter and the dissolved phase is governed by a number of parameters such as physical mixing, resuspension and redissolution from bottom sediments, uptake and release by phyto- and zoo-plankton and its detritus, convection, upwelling, recycling of water bodies, currents, sedimentation of suspended particles to bottom sediments, etc.

Accurate analytical procedures which combine high reliability, with high sensitivity, good precision, reasonable determination rate and convenience of operation are, therefore, of crucial significance for monitoring tasks in environmental protection of natural waters and the control of their input into terrestrial ecosystems, e.g. forests and agricultural land, as well as for providing a reliable data base for research on the behaviour, transfer and fate of ecotoxic heavy metals in rivers, ponds, lakes, estuaries and the sea. Special monitoring requirements arise with respect to drinking water because, as a rule, the majority of the population is immediately affected by heavy metal doses taken up with drinking water and food.

The levels of ecotoxic metals in various water types and differences in composition of major constituents and suspended matter content suggests that water types should be divided into the following four classes, particularly with respect to the requirements of sampling, sample pretreatment and voltammetric determination.

Analytical procedures with voltammetric determination are presented for the following trace metals:

(I) Cu, Pb, Cd, Zn—determinable simultaneously by differential pulse anodic stripping voltammetry (DPASV) at a mercury electrode;

(II) Ni, Co—determinable simultaneously by adsorption differential pulse voltammetry (ADPV) at a mercury electrode;

(III) Cu, Hg—determinable simultaneously by differential pulse anodic stripping voltammetry (DPASV) at a gold electrode.

Due to their extraordinary sensitivity, combined with good precision and inherently high accuracy, these voltammetric methods provide one of the most

Table 3.1. Classification of water types according to typical dissolved trace metal levels

Class	Water type	Trace metal concentration range*
1	Water from the open sea, surface and depth; molten snow from unpolluted, e.g. polar, regions	From 0.0001 (Cd in polar snow and surface water of certain oceanic regions) to 0.8 $\mu g \, kg^{-1}$ (Ni in deep sea waters)
2	Water from unpolluted coastal areas, lakes and creeks	From 0.005 to 0.8 $\mu g \, kg^{-1}$
3	Drinking water from fountains and wells; rain and molten snow in Europe and North America	From 0.1 to several hundred $\mu g \, kg^{-1}$, in exceptional cases (Zn) even more
4	Water from coastal areas, estuaries, rivers and lakes with pollution burdens	From 0.1 to several hundred $\mu g \, kg^{-1}$, in exceptional cases even more

* The concentrations are given in trace metal amount per kg water.

useful and convenient trace analytical approaches for determining these heavy metals in all types of natural waters and drinking water [1].

This classification can be only a rough one. The dissolved levels of the various trace metals can be very different in the same class and even in samples from the same location. For example, Pb will be extremely low in deep sea water, but Cd, Cu and particularly Ni are much more elevated than in surface water. Therefore, the dissolved trace metal levels expand altogether in each class over a rather wide range.

3.2 Analytical Procedures

The complete electro-analytical procedure consists of the stages:
sampling;
sample preparation;
determination by voltammetry.

3.2.1 Sampling

Procedures depend on water class and type. They are, however, common to all analytical procedures for trace metal determinations in natural waters irrespective of the analytical method. Therefore, sampling will be treated in detail in Section 3.3. Typical sample volumes collected are 1–2 dm^3, but may be also only 100 cm^3 or less. Substantially larger water volumes tend to introduce contamination.

3.2.2 Sample preparation

Sample preparation procedures frequently have a close relation to the subsequent determination method, in this case voltammetry.

Class 1 water samples

For these water samples usually no preparation except acidification to the optimal pH for voltammetry is required [2]. Open sea water from zones of high biological productivity, i.e. high phytoplankton contents, e.g. upwelling areas, demand a more elaborate preparation procedure such as that applied to water from Class 4.

For the simultaneous determination of Cu, Pb, Cd, Zn at the mercury electrode by DPASV the pH is adjusted to about 2.

Add 1 cm^3 10 mol dm^{-3} HCl (ultrapure grade) to 1 dm^3 water.

For the simultaneous determination of Ni and Co at the mercury electrode by ADPV the pH should be adjusted to ~ 9.2 with an ammonia buffer.

Take 5 cm^3 water sample, add 0.1 cm^3 2.5 mol dm^{-3} NH$_3$/NH$_4$Cl and 5 μl 0.1 mol dm^{-3} ethanolic dimethylglyoxime as chelator for Ni and Co.

For the simultaneous determination of Cu and Hg by DPASV at a gold electrode, an aliquot from the natural water sample adjusted to pH 2 with HCl is taken, but after the plating step the medium is exchanged to 0.1 mol dm^{-3} HClO$_4$ + 2.5 × 10^{-3} mol dm^{-3} HCl solution to avoid damage to the gold electrode and interference problems as a consequence of the oxidation of substantial amounts of chloride.

Class 2 water samples

Usually sample preparation is the same as for Class 1. However, filtration, as for Class 3 and Class 4 water samples, may be necessary, depending on the level of suspended matter.

Class 3 water samples

Frequently simple treatment, as for Class 1, suffices for drinking water. For rain and molten snow, however, suspended matter should be filtered off. Sometimes the filtrate requires decomposition of dissolved organic matter (DOM) by UV-irradiation as for Class 4 samples. Some types of drinking water also require UV-treatment.

Class 4 water samples

For these samples the normal levels of suspended matter and of dissolved organic matter (DOM) require a more elaborate preparation stage.

3.2.2.1 Filtration [3, 4]

Filtration equipment similar to the Sartorius, Göttingen, SM 16511, is recommended. Cellulose acetate membrane filters of 0.45 μm pore size are commonly used. According to common convention in aquatic trace chemistry, materials passing this filter are defined as dissolved. This is a convenient, though in a strict sense, purely empirical definition. For atmospheric precipitates this device is installed in the automated sampler. Subsequently the procedure for collected natural water samples is described.

A fresh filter, cleaned as described in Section 3.4, is taken out of the container and put on the filter holder, which should be attached securely to the filtration system. The filter is handled with plastic pincers which were rinsed well before use. Water from an unpacked 250 cm³ bottle is poured through the central opening into the upper compartment of the filtration system and is forced by nitrogen through the filter to eliminate possible contamination introduced during handling of the filter. After this, the emptied 250 cm³ bottle is attached as a receiver flask to the filtration unit (See Fig. 3.1).

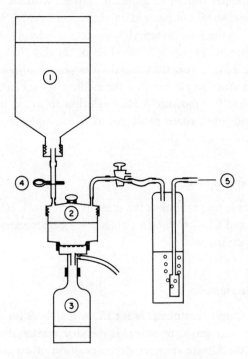

Fig. 3.1. Filtration unit. 1, collected unfiltered water sample; 2, filtration device; 3, receiver polyethylene bottle; 4, clamp; 5, nitrogen inlet.

A 2 dm³ polyethylene bottle containing the unfiltered sample is capped with a stopper and coupled to a silicone tube with a junction to the upper reservoir of the filtration device. This tube is cleaned by opening the clamp and passing about 50 cm³ of water through it. After this, the bottle is inserted into the holder and is connected by a plug-in polycarbonate junction with the upper reservoir of the filtration device. 250 cm³ of water can be drawn from the bottle by squeezing it manually, but it is better to build up a slight overpressure with nitrogen in the bottle by opening the three-way stop-cock and the clamp. Nitrogen bubbles into the bottle, builds up pressure and stirs up suspended matter, thus avoiding its sedimentation. After closing the clamp and releasing overpressure via the three-way stop-cock, water can be drawn into the upper reservoir of the filtration set by opening the clamp. This water is filtered under nitrogen pressure. The first 250 cm^3 are needed to neutralize the filter and to condition the receiver bottle. They are rejected. The next 250 cm^3, depending on the turbidity of the sample, are rejected also or are taken as filtered to be stored for later analysis. Waters with low contents of particulate matter need at least 1 dm^3 of filtered sample in order to get a measurable deposit on the filter. Then, only the fourth aliquot of 250 ml is stored for voltammetric determination. If Hg is to be determined an aliquot of the filtrate has to be transferred to a glass bottle to avoid losses during storage.

The total amount of filtered water must be recorded for later calculation of the heavy metal content of the particulate matter in the water column at the sampled depth level, expressed as ng kg^{-1} water.

The loaded filter is removed with a plastic pincette and stored in a plastic dish, protected by wrapping in a polyethylene bag.

In field missions, e.g. on board a vessel, the ambient atmosphere will be dust-laden. Contamination of the receiver bottles by entry of air during emptying, when rejecting the first portions of the filtrate according to the described procedure, is possible. In order to avoid an uncontrollable blank, the bottle should be shaken vigorously for some seconds with the last remaining 20 cm^3 of water in order to wash the air in the bottle. After this, the bottle is emptied completely and linked immediately to the filtration set.

After filtration, the receiver bottle is unscrewed, (hands should be protected by polyethylene gloves) and air is eliminated by squeezing the bottle, which is then sealed. The bottle is wrapped in the first polyethylene bag which in turn is wrapped in a second bag which is closed by a plastic fastener.

3.2.2.2 Digestion of filtered suspended matter

Two alternative digestion procedures for the filtered suspended matter and the membrane filter are commonly used: –preferentially low temperature ashing and, as second choice, wet digestion with $HClO_4/HNO_3$. This wet digestion procedure provides sufficiently low heavy metal blanks if performed carefully.

Low temperature ashing [4]

Commercial instrumentation such as that available from the International Plasma Corp. Hayward, Calif., or equivalent, may be used.

During missions, filters loaded with particulate matter should be stored deep frozen ($-20°C$) in clean glass or plastic dishes protected from dust by sealed polyethylene bags. The filters and the particulate material are digested by low temperature ashing (LTA) at 150°C in an oxygen plasma induced by microwaves. This method is suitable for trace metals such as Cu, Pb, Cd, Zn, Ni and Co and avoids volatilization losses. Hg, however, requires a wet digestion procedure.

The filters are put into marked silica dishes with plastic pincers. Four of the silica dishes can be loaded in each of the four compartments of the LTA-device from the International Plasma Corp. The pressure is adjusted to $1-5 \times 10^5$ Pa under an oxygen supply flow of $300 \, cm^3 \, min^{-1}$ and a microwave input of 150 W; the digestion is completed within 4 h. The residues are taken up in $1 \, cm^3$ of water. Depending on the ash quantity 50–200 μl HCl, ultrapure grade, are added. After 5 minutes the silica dish is emptied into a voltammetric cell and rinsed with $\sim 40 \, cm^3$ water.

The blank values can be kept low provided appropriately cleaned filters are used, and amount typically to 1 ng Cd, 3 ng Pb and 5 ng Cu per filter.

The LTA-procedure decomposes all biological and organic components of the particulate matter, i.e. algae and detritus, including organic films on silica or clay particles which themselves remain unaffected. Inorganic carbonates are dissolved by subsequent acidification. Thus, a determination includes the trace metal content of the samples which might become remobilized by dissolution and which can enter the aquatic food chain at least partially due to uptake by filter feeders, e.g. mussels.

Wet digestion [5]

Put the loaded filter into a $10 \, cm^3$ precleaned silica beaker with plastic pincers. Add 0.1–$0.5 \, cm^3$ 70% $HClO_4$ and 0.1–$1 \, cm^3$ 65% HNO_3 (both ultrapure grade) and cover with a silica watch glass. Dry at 110°C and maintain at 100–110°C until evolution of nitric oxides almost ceases. Raise the temperature to 200°C and heat until a light yellow or colourless liquid is obtained, then evaporate to dryness. Dissolve the residue in 2–$5 \, cm^3$ deionized water plus an appropriate amount of acid ($10 \, mol \, dm^{-3}$ HCl or 70% $HClO_4$) to adjust the pH to 2 or 1. If necessary for subsequent voltammetric determination, the analyte volume can be increased with deionized water acidified to pH 2 with HCl.

The wet digestion should be performed in a fume hood behind a protective glass wall. Due to the small amounts of acids used no explosion risks exist.

The digestion requires about 2 h.

The wet digestion procedure has proved to be optimal for all types of biological and environmental materials, including food, when heavy metals are to be determined subsequently by voltammetry [5].

3.2.2.3   Acidification and UV-Irradiation [4, 6]

Before UV-irradiation for photolytic decomposition of DOM containing a certain amount of dissolved heavy metals in nonlabile complexes scarcely accessible to voltammetry, the filtered water sample is acidified to pH 2 by addition of $1 \, cm^3$ HCl (ultrapure grade) to $1 \, dm^3$.

Instrumentation. 150 W mercury vapour lamp, e.g. Heraeus, or equivalent, with appropriate radiation from 238–580 nm is used. Voltammetric cells are used as irradiation vessels. A silica beaker should be used for dust protection of the cell during irradiation.

Procedure. UV-irradiation is performed in a silica flask, or if for subsequent determination by DPASV at the mercury film electrode, MFE, in the same teflon flask that is used as the voltammetric cell. The cell is half filled with sample water and its weight is determined on an electrical balance, whose pan is covered by a clean polyethylene sheet. The cell is covered with a clean silica beaker and isolated from the outer atmosphere by the water bath (see Fig. 3.2). All these manipulations are carried out at a clean bench with hands covered by clean polyethylene gloves.

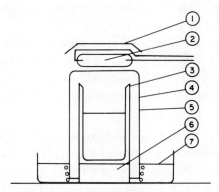

Fig. 3.2. UV-irradiation in a teflon voltammetric cell for subsequent determination by DPASV at the MFE (see text). 1, reflector; 2, mercury vapour lamp; 3, voltammetric cell with filtered and acidified water sample; 4, protective silica beaker; 5, aluminium foil on outer side of silica beaker; 6, cell holder; 7, glass dish filled with water for separation from laboratory atmosphere.

The subsequent UV-irradiation in the quasi-closed device can be performed outside the clean bench. The irradiation from a 150 W mercury vapour lamp, e.g. Hanau, is focussed with the aid of a reflector through the silica cover onto the sample solution. Depending on the origin, the sample is irradiated for 1–4 h. During irradiation the temperature rises to about 100 °C and the solution simmers down to one half or one third of the original volume.

Samples with more elevated DOM-levels, e.g. frequently encountered in in-shore waters, due to the effect of waste water inlets or polluted rivers, are more strongly acidified by addition of 1 cm^3 concentrated ultrapure HCl, to support the photolytic decomposition of organics and leaching of chelated trace metals.

Usually, the additional amount of trace metals released by this treatment is on average 20% of the total dissolved trace metal level, but may go up in certain samples to 60% or more.

To samples from estuaries and polluted rivers or lakes with relatively high levels of dissolved organic matter, an additional 1 cm^3 ultrapure Perhydrol, is added to achieve more rapid and efficient oxidative decomposition. The resulting somewhat higher blank is compatible with the usually higher trace metal levels in this type of sample.

3.2.3 Voltammetric determination

Usually heavy metal concentrations in natural waters (dissolved and suspended matter phase) are at such trace levels that voltammetric determinations require prior preconcentration of the heavy metals. This preconcentration is carried out *in situ* at the working electrode without additional contamination risks.

A very efficient, rapid and convenient photolysis of DOM can be achieved in a new device with high intensity (1200 W) mercury lamps [16].

For the heavy metal concerned the following determination procedures are applied. They are all based for reasons of optimal sensitivity and performance on differential pulse modes [1].

Modern commercial polarographs equipped with the differential pulse mode, can be applied in conjunction with an *XY*-recorder for recording voltammograms. Semi- and fully-automated devices with the ability to store, add, subtract and compare voltammograms improve convenience and precision.

1 (See 3.2.3.1) Cu, Pb, Cd, Zn—simultaneously by differential pulse anodic stripping voltammetry (DPASV) at the HMDE (hanging mercury drop electrode) typically down to levels of 0.1–0.05 $\mu g\,kg^{-1}$.

2 (See 3.2.3.2) Cu, Pb, Cd—simultaneously by DPASV at the mercury film electrode (MFE) typically in the range from $\leqslant 0.001$ to 1 $\mu g\,kg^{-1}$. At higher levels the HMDE should be used while between 1 and 0.001 $\mu g\,kg^{-1}$ the usage of the MFE is optional.

3 (See 3.2.3.3) Ni, Co—simultaneously by adsorption differential pulse voltammetry (ADPV) at the HMDE. For these metals that do not form amalgams and others of analogous behaviour, preconcentration is achieved by adsorption of suitable chelates at the working electrode surface.

4 (See 3.2.3.4) Cu, Hg—simultaneously by DPASV, usually with medium exchange after the plating step, at the gold electrode.

For determinations from group (1), (3) and (4) a precleaned pyrex glass cell is used, equipped with the working electrode. A Pt coil, preferentially contained in a separate compartment and connected to the analyte by a frit is used as auxiliary electrode and a saturated calomel or Ag/AgCl reference electrode. The reference electrode is contained in a separate compartment connected to the analyte by a salt bridge. For ultratrace determinations in group (2), a teflon cell is used.

Deaeration is carried out with 99.999% nitrogen passing through a washing bottle filled with the analyte or analogous medium.

3.2.3.1 Simultaneous determination of Cu, Pb, Cd, Zn by DPASV at HMDE

Put 15–20 cm^3 natural water sample acidified to pH 2, or a filtrate or analyte resulting from suspended matter digestion, as analyte into the voltammetric cell. Deaerate for 10 minutes.

Adjust the plating potential to -1.2 V (Note: more negative potential values are not permissible at pH 2 to avoid interference by hydrogen evolution).

Depending on the concentration levels, plating times (t_d), are 2–5 minutes, while stirring with a teflon-covered magnetic bar at 900 rpm. Subsequently turn the stirring off and allow a quiescent interval of 30 s.

Then perform anodic stripping in the differential pulse mode and record the voltammogram.

Parameters: scan rate 5 mV s^{-1}; pulse height 50 mV; pulse duration 57 ms; clock time of pulses 0.5 s.

For concentration evaluation carry out two standard additions at fresh mercury drops, the use of calibration relationships is, as a rule, not permissible or recommended in natural waters, because varying trace levels of surface active materials might affect electrode processes and, therefore, peak height and will then lower precision.

3.2.3.2 Simultaneous determination of Cu, Pb, Cd by DPASV at the MFE [7]

This more elaborate and delicate determination should usually be applied only in the ultra-trace range below 0.5 μg kg^{-1} down to 0.001 μg kg^{-1} and in exceptional cases less.

The voltammetric cell is a precleaned closed teflon cell with a Pt-coil as auxiliary electrode and an Ag/AgCl reference electrode. Both electrodes are housed in separate teflon tubes to exclude contamination. The tubes are filled with saturated KCl and connected to the analyte by Vycor glass tips (4 mm diameter, Corning Glass Corp., Cleveland) providing low ohmic resistance and very small leakage of KCl solution. All soldering points are shielded by silicone to avoid contamination by solder corrosion products.

The rotatable working electrode is an *in situ* formed mercury film electrode (10–100 nm thickness) on a glassy carbon substrate. The best electrodes require glassy carbon from Tokai, MfG, Tokyo. The production of electrodes is described in ref. [7] in detail. The synchronous motor for rotation of the electrode during plating is capsuled in perspex to avoid any contamination of the analyte. The performance of all manipulations at a clean bench with laminar filtered air-flow and wearing protective polyethylene gloves is mandatory.

Procedure. Put 20–30 cm^3 of sample acidified to pH 2 into the precleaned teflon cell, spike with 50 μl 2.5×10^{-2} mol dm^{-3} ultrapure Hg(NO$_3$)$_2$, to adjust to $\sim 10^{-5}$ mol dm^{-3} Hg(II) analyte for mercury film formation and close the cell. Deaerate with 99.999% N$_2$ for 10–20 minutes.

If several cells are deaerated simultaneously before measurement, each cell should be additionally deaerated for a further 3 minutes before measurement.

Adjust the potential to -1.0 V for plating, while the electrode rotates at 1500 rpm but sometimes also up to 4000 rpm. High rotation speed increases sensitivity due to enhanced mass transfer, but decreases the life span of the electrode (roller bearings wear) and therefore 4000 rpm should be applied only in extreme ultra trace determinations below 0.005 $\mu g \, kg^{-1}$. The negative plating potential is important to obtain a homogeneous mercury film of closely packed microdroplets of mercury of near equal size. Once a thin mercury film has been built up, e.g. after 3–5 minutes plating time, less negative deposition potentials can be applied for further plating.

Simultaneously with mercury film formation, an aliquot of trace metal ions, e.g. Cu, Pb and Cd, is plated as amalgam. This electrochemical preconcentration is achieved *in situ* and, therefore, without any additional contamination risk. Deposition time, t_d, is adjusted between 3 and 12 minutes according to the order of the trace metal bulk concentration. A deposition period t_d of 3 minutes can be regarded as a practical minimum with respect to the recommended mode of standard additions.

At the end of the plating step, the motor is switched off and the solution is allowed to come to a rest within an interval of 30 s. During this step and subsequent stripping, the sample is blanketed with nitrogen.

Subsequently the preconcentrated heavy metal traces are stripped, scanning with 10 mV s^{-1} in the differential pulse mode in the anodic direction, and the voltammogram is recorded.

Adjust the following parameters: pulse height 50 mV; pulse duration 29 ms; clock time of pulses 240 ms.

The resulting peak heights are proportional to the heavy metal concentrations in the analyte solution.

Usually an unknown sample initially requires an exploratory run with a plating time of 5 minutes, yielding an approximate estimate of the different trace metal levels. From this run the appropriate plating time of 3–12 minutes can be deduced. In the subsequent determination the appropriate instrumental sensitivity for the trace metal with the most negative reoxidation potential, usually Cd, is adjusted and the Cd-peak is recorded. If necessary, the scan can be stopped in the valley after this first peak. The instrumental sensitivity is readjusted, according to the results of the exploratory run, in order to record the following Pb peak in suitable height. During this instrumental setting, effected within 10 s, the plating current is negligible, due to lack of convection. The scan is started again and the next peak is recorded. At the end of the potential scan, usually at -0.1 V, the motor is switched on again and during the following 3 minutes the remaining amounts of all the trace metals in the mercury film are reoxidized to clean the mercury film completely. Then, at the same mercury film, two standard additions are carried out for evaluation of the trace metal concentrations. After each standard addition a similar voltammetric procedure is carried out. However, as pointed out in the next section, the plating times are appropriately reduced.

Standard addition and adapted plating times [7]

Making a correct standard addition is a matter of experience. From the first voltammogram a skilled operator can deduce the approximate amount of trace metals necessary roughly to double the bulk concentration by the first standard addition. This same amount is also applied in the second addition.

After the first addition, the plating time is halved and after the second addition it is reduced to one-third of the original. Besides the advantage of reducing the overall analysis time, this practice eliminates slight deviations from the linear relation of peak height to trace metal concentration in DPASV at film electrodes. Each stripping step starts with approximately the same trace metal amount in the mercury film and the recorded peaks are all of the same order of magnitude. It must be emphasized that this method works reliably in analyses of Cd and Pb even if the primary aim of approximately doubling the bulk concentration has not been attained, e.g. by making too high or too low a standard addition.

Determination of copper

The procedure described offers the only convenient way to cope with problems of Cu-analysis at a thin mercury film electrode. Compared to Cd and Pb, the solubility of Cu in Hg (0.002%) is very low. Depending on the bulk concentration, the film thickness and the plating time, only part of the Cu deposited will be soluble as amalgam. Problems arise during anodic stripping as the peak-concentration relation of both plated Cu species is different. This is reflected by constantly decreasing Cu peaks in repetitive analysis, as the Hg-film thickness grows with each plating, thus amalgamating an increasing amount of Cu. The usual standard addition followed by constant plating times will seriously alter the proportion of both deposited Cu species, leading to a completely non-linear bulk concentration–peak height relationship.

To avoid these problems, the determination of Cu is performed after the complete procedure for Cd and Pb, including their two standard additions. As the approximate amount of Cu is known from the exploratory run, an appropriate plating time can be chosen. For the determination of high Cu concentrations, a medium or long plating time and a low instrumental gain is applied. Thus, the non-amalgamated portion of plated Cu is predominant. For low and ultratrace Cu levels, the shortest possible deposition time and high instrumental sensitivity should be chosen. The small amount of deposited Cu is totally amalgamated and gives a defined anodic stripping response. The procedure of standard addition, reducing plating times as outlined above, is applied. If the resulting peaks differ by more than $\sim 30\%$ in height, e.g. the standard addition has not been well adjusted, the plating step should be repeated increasing or decreasing the deposition time according to circumstances.

Determination rate

The daily determination rate by working simultaneously with 2 polarographs and 2–4 electrodes is 8–10 analyses per day. This number will be reduced to 6–8 analyses if a preliminary pretreatment such as UV-irradiation is required. The number of extreme ultratrace analyses should be restricted to 2–4 per day.

Table 3.2. Summarized sequence of main operations during voltammetric analysis for Pb, Cd and Cu with the mercury film electrode

---

1 Sample acidified to pH 2 (ultrapure HCl); spike with $Hg(NO_3)_2$:
   20 mm^3 (5000 mg dm^{-3}) to 50 cm^3 sample.
2 Deaerate with N_2, 10 min.
3 Potential-controlled pre-electrolysis at -1.0 V for a definite plating time, e.g., 6 min;
   glassy carbon electrodes rotate usually at 1500 rpm; $Hg^{2+} + 2e^- \rightarrow Hg^0$ (film formation);
   $Me^{2+} + 2e^- \rightarrow Me^0$; trace metal plated as amalgam.
4 Stop rotation, adjust start potential: -0.8 V; wait 30 s.
5 Start potential scan from -0.8 to -0.1 V; differential pulse mode. Stripping of
   Me-Amalgam $\rightarrow Me^{2+} + Hg^0 + 2e^-$; voltammogram is recorded.
6 Clean Hg-film at -0.1 V; electrode rotates; add known amount of Me^{2+}.
7 Repeat pre-electrolysis: step 3 to step 6. Reduce plating time to one-half, e.g. 3 min.
8 Repeat pre-electrolysis: step 3 to step 6. Reduce plating time to one-third, e.g. 2 min.
9 Evaluation.

Rotating glassy carbon electrode maintenance and storage [7]

Usually the working electrode is in continuous use during a series of analyses. The first step in each analysis has to be devoted to the regeneration of the working electrode. The voltammetric determination of a fresh sample is preferably carried out with a freshly formed MFE. Therefore, the old mercury film used in the previous analysis has to be removed. When working in acidified (pH 2–3) samples, it is not sufficient to wipe it off with wet filter paper. This leaves the glassy carbon surface in an unsuitable state reflected by a progressively decreasing hydrogen overvoltage. The resulting slope of the base line in the voltammogram would soon impede the precise determination of trace metals, particularly those with rather negative reoxidation potentials e.g. Cd. The most severe consequences of this inadequate maintenance of the working electrode occur in the ultra-trace range, when high instrumental gain has to be used.

An adequate maintenance procedure of the working electrode is as follows:

A wet folded filter paper on which some aluminium oxide (0.3 μm grain size) has been spread, is brought for ~ 20 s into contact with the electrode rotating at reduced speed, in order to avoid spilling of the abrasive material by centrifugal forces. The reduced speed is achieved by switching a resistor in series with the motor. The old mercury film is removed and the exposed glassy carbon surface is subjected to brief polishing. Subsequently the re-established glassy carbon electrode is rinsed carefully with deionized water, completely removing all traces of the aluminium oxide which would act as a serious contaminant. A final rinsing step is peformed by rotating the electrode for some minutes in deionized water acidified maximally to pH 1. Higher acidity might have adverse effects on the glassy carbon surface. The working electrode is now ready for fresh formation of a mercury film in a subsequent analysis.

Alternatively, the glassy carbon electrode can be stored. Before storage a small flask filled with mercury is put into the centre of the cell and the remaining volume is filled up with acidified water. By fitting the cell to its cover, the working electrode is plunged into mercury. Storage in water or worse, in acidified water, will quickly cause the glassy carbon surface to deteriorate. The Vycor tips of the counter and reference electrode compartments are protected from drying out by immersion in acidified water. The cell with its electrodes can be stored for longer periods without any deterioration in electrode performance.

According to the maintenance procedure outlined, properly treated glassy carbon working electrodes will keep their high performance qualities as substrates for mercury film electrodes for years.

3.2.3.3 Adsorption differential pulse voltammetry for the simultaneous determination of Ni and Co [8]

This extremely sensitive approach performed at the HMDE can also be used for other metals that do not form amalgams.

Put 5–10 cm^3 prepared sample (pH 9.2, 0.05 $mol\,dm^{-3}$ ammonia buffer, $10^{-4}\,mol\,dm^{-3}$ dimethylglyoxime (DMG) added as chelator) into the voltammetric cell (pyrex glass) and deaerate for 10 minutes with 99.999% N_2.

Adjust the adsorption potential to -0.7 V and stir the solution with a teflon-covered magnetic bar at 800 rpm. The adjusted potential, close to the zero charge potential value of the mercury electrode (HMDE), is optimal for preconcentration by adsorption at the electrode. Depending on the bulk concentration level of the heavy metal, adsorption times of 5–10 minutes should be selected. The ammonia buffer concentration can be increased at higher concentration levels of Ni and Co. In this manner, by appropriate adjustment of the complexing buffer and the adsorption time for DMG-chelates, the basic condition is kept, that full coverage of the electrode surface is avoided and the adsorbed amount of chelates remains in the rising part of the adsorption isotherm and is consequently directly proportional to the Ni and Co bulk concentrations.

Stirring is terminated and a quiescent interval of 30 s is allowed. Subsequently scan $(5\,mV\,s^{-1})$ in the differential pulse mode in a negative direction to -1.15 V, reducing the Ni and Co from the adsorbed DMG-chelates. The (differential pulse voltammetry) DPV-parameters are: pulse height 50 mV, pulse duration 57 ms, clock time of pulses 500 ms. Two standard additions suffice for evaluation of the concentration.

The use of voltammetric devices with components able to store and subtract the response of the blank analyte significantly improves the base line and consequently the detection limit and precision in the extreme ultra-trace range down to 0.001 $\mu g\,kg^{-1}$.

The daily determination rate is about 10 samples per day.

3.2.3.4 Simultaneous determination of Cu and Hg by DPASV at the gold electrode [6]

The working electrode is a gold disc electrode polished first with carborundum and then with diamond paste of progressively decreasing particle size (7, 3, 0.7 mm) until a mirror-like surface is obtained.

The polished electrode needs electrochemical activation before each application. Immerse the electrode in $0.1 \, mol \, dm^{-3}$ $HClO_4 + 2.5 \times 10^{-3} \, mol \, dm^{-3}$ HCl, deaerate 10–15 minutes with 99.999% N_2. Treat at -0.25 V for 20 s and subsequently $+1.7$ V for 20 s several times.

The determination is carried out as follows: Use a pyrex glass cell. As reference electrode only the Ag/AgCl electrode is permissible, housed in a separate compartment connected to the analyte by a Vycor glass tip, e.g. Corning Glass, Cleveland. For ultratrace determinations the Pt-coil auxiliary electrode should be treated in the same manner.

Immerse the activated gold electrode in the sample solution, (pH 2) deaerate for 10–15 minutes, let the electrode rotate at 3000 rpm and adjust the plating potential to -0.2 V for 5 minutes. Terminate the rotation and allow a quiescent interval of 30 s. For subsequent stripping, scan at $10 \, mV \, s^{-1}$ in the differential pulse mode up to $+0.9$ V. Pulse parameters are: pulse height 50 mV, pulse duration 37 ms, clock time of pulses 200 ms.

If the chloride concentration in the sample exceeds $5 \times 10^{-3} \, mol \, dm^{-3}$, interference with the Hg-peak and damage of the gold electrode surface occurs, due to substantial chloride oxidation. Therefore, medium exchange before the stripping step becomes mandatory.

The working electrode is disconnected from the voltage and the sample solution is substituted by the $0.1 \, mol \, dm^{-3}$ $HClO_4 + 2.5 \times 10^{-3} \, mol \, dm^{-3}$ HCl stripping medium used in the prior electrochemical activation of the gold electrode. The small chloride concentration ensures that the reoxidation potential of the plated Hg does not become too positive.

Two to three standard additions, each with preceding electrochemical activation of the gold electrode, suffice for concentration evaluation. The determination of the dissolved mercury content of natural waters by this voltammetric method includes the soluble methylmercury component.

The determination rate is ~ 8 to 10 samples per day.

For the determination of larger Cu-concentrations, the gold electrode is preferred, due to its large dynamic range and because of the difficulties arising from the limited solubility of Cu in mercury.

For extreme ultra-trace determinations of Hg from 0.08 to $0.001 \, \mu g \, kg^{-1}$, a more complicated determination procedure by subtractive differential pulse anodic stripping voltammetry (SDPASV) at a twin gold electrode becomes necessary. In the presence of comparable or high Cu-concentrations, an additional more elaborate plating step with intermediate cleaning by Cu-stripping is required. Cu has then to be determined separately according to the above procedure with a single gold electrode. Details of this extreme ultra trace determination of Hg by SDPASV are outlined in refs [6, 9].

3.3 Sampling

This first step of all analytical procedures in aquatic trace metal chemistry is the most critical one in obtaining meaningful data. Errors made at this stage render the whole analysis useless and create the paradoxical situation that even with great care and effort in the subsequent stages, and applying advanced instrumental determination methods with high detection power and inherently good accuracy, such as differential pulse voltammetry, only meaningless, even if precise, data may be obtained.

Therefore, the essential aspects of accurate sample collection are outlined below for different water types. They apply generally to water sample collection for heavy metal trace analysis irrespective of the subsequent method of determination.

3.3.1 Collection of water samples from the sea, estuaries, rivers and lakes

3.3.1.1 Surface water samples [4, 10]

Principle. Leave the main ship or research vessel and collect samples only from a rubber boat during sampling.

A large ship always constitutes a severe handicap for collection of surface water samples, as it creates a rapidly expanding cloud of polluted water around it. This plume must be avoided during sample collection. Factors that have to be considered when choosing the sampling area are; waters, which have been crossed by the ship, as well as the waters around it when stationary that are also severely polluted. A ship will drift with the wind and thus will leave polluted surface water behind it. Airborne pollution from the ship will also create a large area of polluted water, depending on location and the wind direction and strength. The ship must be left behind and a rubber boat used in order to reach an unaffected area about 0.5 km or more from the ship.

The area for sampling can be reached by a small rubber boat, provided the weather conditions permit it. Of course, this boat will also be inevitably polluted by heavy metals and as soon as it stops a plume of polluted water will spread around it. The only possibility of avoiding this contamination is to move the boat slowly upwind, as far as possible by rowing and to sample water in front of the boat as it moves continuously into uncontaminated water. Airborne contamination resulting from the motor, the boat or crew will also be minimized in this way. When using a small boat, water can be collected directly by hand in a polyethylene bottle by pressing it under the water surface, avoiding the upper surface boundary layer, at which surfactive material, and thus trace metal ions, accumulate. The sampling bottle is filled slowly during progression through waters unaltered by the body of the boat. While using bottles, hands in contact with the sampling bottle must always be protected by polyethylene gloves.

Depending on the size of the boat, the water surface may on occasion only be reached with a plastic telescopic bar extendable to ~ 4 m. The sampling bottle is inserted into a holder at its end. This holder should be made of stainless steel completely coated with plastic (Fig. 3.3). The plastification of the holder with molten Nylon 6.6 under a protective gas gives a resistant covering, clean enough for this purpose considering the short residence times in the water. A polyethylene disk separates the bar into a contaminable zone for handling from the boat and a clean bottle holder zone. In this manner water running along the bar is prevented from coming into contact with the sampling bottle (see Fig. 3.3). The completeness of the plastic coating of the metallic parts can be checked by immersing the coated part of the holder into salt water, using the metal part as an electrode and measuring the electric resistance of the coated part against an auxiliary electrode immersed in the same medium.

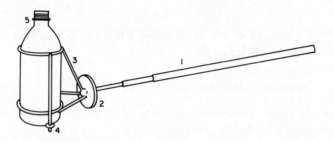

Fig. 3.3. Telescope bar with sampling bottle for surface water collection. 1, telescopic bar; 2, polyethylene disc to protect sampling bottle; 3, nylon coated holder; 4, silicone rubber fastener; 5, polyethylene bottle

Manipulations with the bottle, e.g. inserting it into the holder or closing it, will be a crucial step, as they have to take place on the (usually) polluted boat.

Several bottles necessary for sampling, generally one to three for one station, are packed into double polyethylene bags before the cruise. The bottles should have been cleaned and conditioned as described previously. On reaching the sampling area, the holder also having been protected by a polyethylene bag during transport, is cleaned by immersing it into the water in front of the moving boat. The sampling bottle is then unwrapped from the first polyethylene bag. The bottle is uncapped within the second clean bag and taken out, the cap being left in the bag. This operation is carried out wearing clean polyethylene gloves. This bag and the cap in it must be inserted into the first bag during the sampling procedure in order to minimize its contamination. After fixing the bottle into position using a silicone rubber fastener (see Fig. 3.3), the bottle and the holder are cleaned by submersion into sea water in front of the moving boat. Then the bottle is filled partially at least twice and shaken well before emptying it. Subsequently it is filled by holding it under the surface to a depth of ~ 1 m. While filling the bottle and also during all manipulations, the boat must progress upwind. After sampling, the polyethylene bag is put over the holder and the bottle is capped. The closed bottle is immediately wrapped in both protective bags. It is of utmost importance that the sampling bottle be clean on the outside before placing it into the precleaned bag. If this is not so then contamination from the outer bottle walls or from the inside of the protective bag could later infiltrate the thread of the cap from droplets of water condensing during thawing of a sample that had been stored frozen. Merely opening such a bottle could immediately contaminate the sample in it.

The polyethylene sampling bottles, precleaned according to the procedures outlined in Section 3.4.1 are suitable for samples in which the levels of Cu, Cd, Pb, Zn, Ni, Co, etc. are subsequently to be determined. These polyethylene sampling bottles can also be used for the collection of water samples in which the Hg-level is to be analysed. However, an aliquot of the water sample should soon be transferred into a glass or silica bottle to avoid Hg losses by adsorption on the wall.

In general, care has to be taken that all operations with sampling bottles are performed in the bow of the boat and within the wind stream, trying to avoid turbulence that could induce dust contamination from the boat or clothing. All items for sampling, such as the holder or a set of cleaned polyethylene gloves and the bagged sampling bottles, have to be protected by at least one polyethylene bag. It has to be carefully rinsed or better replaced by a fresh clean bag after each sampling mission. Hitherto the significance of such manipulations has often been underestimated with regard to their significance in reducing contamination during sampling to a negligible extent.

In open sea areas, where larger scale currents occur and even in upwelling regions, heavy metal levels are fairly constant, so that the number of samples taken at one station can remain quite limited. Depending on the measuring capacity and the number of stations spread over a certain area or distance, 1–3 samples should be collected in one run.

In contrast to this, coastal waters with complex pollution patterns resulting from river inputs and dispersal by tidal currents need a rather more intensive net of sampling stations. The same applies to estuaries along the decreasing salinity gradient upstream.

3.3.1.2 Deep water samples

One of the best commercially available samplers for intermediate waters up to 200–300 m is the so called GoFlo from General Oceanics, Miami. Proposed modifications: replace all O-rings by silicone O-rings; replace water outlet by an all teflon outlet, working without O-rings.

For intermediate depths in the sea or in lakes, the GoFlo may be used dynamically, i.e. the GoFlo is placed at the end of the hydroline and is moved down during sampling. A weight should be centred at the middle of the GoFlo, so that its lower opening is at the deepest point (remaining thus uncontaminated by upper parts of the system) during dynamic sampling.

For trace metal analysis, conventional deep sea water sampling by clamping samplers to a hydroline is not recommended. Even a rather well designed system, e.g. the GoFlo is bound to fail, because there might be severe contamination by the hydrowire itself. This can only be avoided by using a plastic coated hydrowire and a number of precautions in handling it such as a special winch, special all plastic pulleys, etc.

Only sophisticated samplers, e.g. the so called CIT sampler designed by Schaule and Patterson [11] and the Jül sampler, also called 'Moonlander', designed by Mart [12], are appropriate for collection of uncontaminated samples, particularly for Pb-analysis from the sea to a depth of several thousand meters.

3.3.1.3 Drinking water

The general precautions in handling of bottles, that have been described before, must also be used for collection of drinking (tap) water, as once again, levels of some trace metals in drinking water can be very low. If the drinking water is tap water, the first water running from the tap must be avoided because there will be a high accumulation of trace metals stemming from the pipes, soldering and welds. The best way is to let water run at least for 2 minutes before collecting the sample. For continuous drinking water control in water works, voltammetric automates can be coupled via a by-pass to the water mains [13].

3.3.2 Collection of rain and snow samples [14]

Samples are collected with an automated sampler which restricts sampling to precipitation periods, thus eliminating interference from dry deposition. The principle of the sampler designed for continuous weekly sampling as a rule is depicted in Fig. 3.4. The sampler is placed in a rig locating the opening about 2 m above ground to eliminate contamination by soil splashing during intensive rainfall.

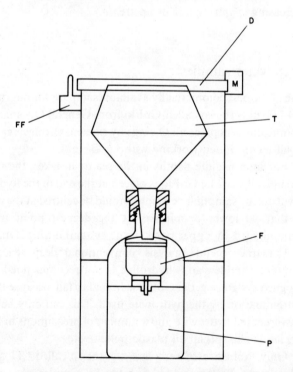

Fig. 3.4. Automated rain and snow sampler. D, cover with electromotor M; FF, humidity sensor; T, polyethylene funnel; F, filtration device; P, polyethylene sampling flask for filtrate.

Opening and closing of the cover is automatically controlled by a humidity sensor that triggers an electromotor. Typical opening and closing time spans are adjustable between 0.5 and 3 minutes. Inside the container constructed from aluminium and plastic, the sampler contains a $2\,dm^3$ polyethylene collection flask, which has a filtration device in its upper part, e.g. SM 16511, Sartorius, Göttingen, with a polyethylene funnel on top. In this manner suspended matter is filtered off. The sampler is heat controlled by the outside temperature so that snow melts and operation in winter is ensured down to $-30°C$. The sampling flask is protected from light and the outside temperature to reduce biological activity in summer. To collect small samples within preset intervals of a rain- or snow-fall, the usual $2\,dm^3$ collection flask can be substituted by a carousel or band with an appropriate number of small polyethylene flasks. All components of the sampler coming into contact with precipitates can be subjected to the cleaning procedures described in Section 3.4.

The sampler needs a normal electric power supply for its operation. At the end of a collection period, usually one week in wet deposition monitoring, the collection flask is taken out, the funnel and filtration device are separated and the water collected is poured under contamination-free conditions into a precleaned polyethylene transport flask.

The sampling station has additionally to be equipped with instruments to measure the usual meteorological parameters (wind direction, wind strength and precipitation volume per unit area).

For deposited polar snow with extremely low trace metal levels (0.0001 to 0.01 $\mu g\,kg^{-1}$), a special manual collection technique is described elsewhere [15].

3.4   Additional Operations

3.4.1   Cleaning of sampling flasks and other labware [3]

3.4.1.1   Polyethylene bottles

First, polyethylene bottles have to be treated in a laboratory washer using conventional detergents in order to remove grease. The temperature can rise to 95°C and subsequently the bottles must be rinsed with distilled water.

Polyethylene bottles for sample storage, usually $250\,cm^3$ are cleaned only with dilute HCl. For the different acid cleaning steps, large polyethylene containers are used in order to clean at least 50 to 80 bottles in one run. Such convenient containers can be obtained by cutting off the top of commercially available $60\,dm^3$ polyethylene containers of moderate price. The small bottles are filled with tenfold diluted analytical reagent grade HCl and are loaded into the container, and filled subsequently with the same acid. Manipulations must be carried out using polyethylene gloves, which have been rinsed before use. The caps of the bottles are added separately. In order to increase the leaching rate, a temperature of 70°C is attained by partially immersing a 2 dm³ Erlenmeyer silica flask filled with water heated to ~75°C by a temperature controlled electric immersion heater. The build-up of a temperature gradient is avoided by bubbling a current of filtered nitrogen through polyethylene tubing which reaches to the bottom of the container. The container, placed under a fume hood, has to be carefully covered by plastic sheets against dustfall or condensing water from the water heating device with its metallic immersion heater and electric connections.

After ~4 days treatment with this hot dilute acid in order to release heavy metals from the bottle walls, the acid is poured out and the container is filled with fresh analytical reagent grade acid, diluted 1:10, and is heated again for ~4 days.

The procedure is repeated once again for another 4 days, but this time the acid must be high purity HCl, e.g. Suprapur E. Merck, Darmstadt, or equivalent. The acid concentration should be ~2%. The water for dilution should be pure, as described below. Manipulations must be performed with care. The operator must wear sleeve protectors made of polyethylene and the usual polyethylene gloves.

To allow for possible low contamination during cleansing of the inner and outer surfaces of bottles by manipulations in dustpoor laboratory ambiance, filling the bottles with 1% ultrapure HCl for the last cleaning step is performed at a clean bench. This time, the bottles are sealed before heating in similarly diluted pure HCl, as described before, taking into account all possible contamination sources during handling and heating. After a heating period of ~4 days the bottles are emptied at a clean bench, filled with triply distilled water, acidified by adding 1 cm³ ultrapure HCl per 1 dm³ water. Once again, all handling must be carried out wearing polyethylene gloves. The capped bottles are put into polyethylene bags to protect them from dust during storage and transport. The upper part of the first precleaned polyethylene bag is twisted together and the bag is wrapped in a second bag closed by a plastic fastener. For storage at room temperature or for transport, the wrapped bottles are packed in groups of 20 in large polyethylene bags. The whole cleaning procedure with the three leaching periods, each lasting 4 days, applying HCl of progressively higher purity grade, requires about a fortnight to perform altogether.

Polyethylene bags are first rinsed with normal tap water, then they are partially filled with analytical reagent grade HCl, diluted 1:10 and immersed in the same acid in a polyethylene container for half a day. After rinsing with pure water, the procedure is repeated, this time with 1:10 ultrapure HCl for 1 day. The bags are rinsed with triply distilled water and dried at a clean bench.

3.4.1.2 Teflon flasks and other items

Teflon flasks and machined teflon cells and cell coverings generally made of TFE Teflon, need a stronger cleaning process, in order to avoid long lasting contamination during use. TFE is a sintered material and may contain metal contamination particularly iron resulting from manufacturing, which is difficult to remove. In addition to the cleaning procedure described previously for polyethylene ware, Teflon should be treated in a preliminary step with 50% analytical reagent grade HNO_3. For this purpose, Teflon cells and other small items are heated to 80 or 90°C in large silica beakers over a period of 1 week. Large numbers of bottles can be filled with HNO_3, but they must be capped tight before heating in the usual diluted HCl acid bath to avoid hot HNO_3 coming into contact with the polyethylene container.

3.4.1.3 Silica ware

Silica ware can be cleaned just like Teflon, beginning with 50% analytical reagent grade acids, HNO_3 or HCl.

3.4.1.4 Filters and filtration apparatus

Cellulose acetate membrane filters such as those from Sartorius, 0.45 μm medium pore size, 47 mm diameter, available under the specification SM 11306 are recommended. Such filters cannot be applied unprepared, as heavy metals, e.g. Cd and mainly Pb, are leached even after rinsing with 1 dm^3 of water. For cleaning purposes, filters are soaked in cold 50% HCl, analytical reagent grade, for 2 days. The acid is changed and the filters are left in the fresh solution for 2 further days. After this, they are rinsed several times with pure water and soaked in tenfold diluted ultrapure HCl for about 1 week. After repeated careful rinsing with triply distilled water, filters are kept in 1% HCl, ultrapure (or equivalent) for longer storage. Some days before use they are rinsed with pure water. For use in sea water they are conditioned with sea water, which usually has a very low level of heavy metals, during their storage until use in a mission. The sea water used must correspond to the cleanest sample that can be expected during the filtration. The sea water should be changed frequently during the continuous removal of the filters.

Blanks for filters treated in this manner, as determined by differential pulse anodic stripping voltammetry (DPASV) subsequent to low temperature ashing in an oxygen plasma, are ~3 ng of Pb and 1 ng of Cd for filters of 47 mm size.

The filtration equipment is separated into its component parts before cleaning, including all O-rings. The parts can be cleaned as described for polyethylene flasks.

3.4.2 **Reduction of blanks in water and chemicals** [3]

3.4.2.1 Ultrapure water

For cleaning purposes and dilution, pure water with extremely low heavy metal blanks is required. The recommended water supply system, based on special ion-exchange columns, is commercially available from Millipore, Bedford, Mass., USA. The Milli-Q-system is fed by normal ion-exchange water from a central supply and consists of a prefilter and a charcoal absorption column followed by 2 mixed-bed ion-exchange columns. After deionization, water passes through a microfilter (0.22 μm) to remove particulates and can be drawn off into 10 dm^3 polyethylene containers at a rate of ~ 1 dm^3 per minute. When not in use, water in the ion-exchangers is circulated periodically for 1 minute each hour. The resistivity measurement taken after use of the ion-exchanger cannot be relied upon as an indication of water quality concerning heavy metals. Thus, their blanks have to be monitored weekly by the appropriate voltammetric method.

After a flow-through of at least 300 dm^3 over a period of 3 months, a typical analysis of water sampled at the microfilter outlet yields the following results:

Cd: <0.1 ng kg^{-1}

Pb: 1 ng kg^{-1}

Cu: 5 ng kg^{-1}

Zn: 50 ng kg^{-1}.

This deionized water has lower heavy metal blanks than waters from a silica distillation apparatus and is, therefore, superior for trace analysis of heavy metals.

3.4.2.2 Acids

In Europe, amongst the easiest available ultrapure chemicals are those of E. Merck, Darmstadt, Suprapur quality. Levels of trace metals in HCl, mainly Pb, as important contaminants, are certified to be below 5000 ng kg^{-1}. Usually, however, the actual levels are only 500 ng kg^{-1}. Taking into account the dilution 1 to 1000 in acidification of a sample (1 cm^3 of 10 mol dm^{-3} HCl added to 1000 cm^3 of water sample in order to adjust to pH $\leqslant$3) this ultrapure acid will usually be good enough for acidification of water samples, as the resulting trace metal blank will be below 0.001 μg kg^{-1}. This blank will be noticeable, however, with extremely pure water samples, e.g. deep sea water in the case of Pb, as well as in polar snow samples. There is a convenient way to reduce blanks in acids by subboiling distillation. The apparatus used should be all silica, e.g. from H. Kürner, Rosenheim, FRG. The trace metal levels in the acid after subboiling distillation are usually below 0.02 μg kg^{-1}.

3.4.3 Requirements for laboratory atmosphere [3,4]

3.4.3.1 Home (base) laboratory

Generally an existing laboratory has to be converted into a clean room. This has usually to be achieved with a minimum of alterations and at reasonable cost. The high cost of a class 100 laminar-flow room, for maintaining this standard during operation, can be avoided pragmatically by adopting a less stringent approach. The laboratory is flushed with a laminar flow of filtered air, and within this 'clean' area, several laminar-flow boxes (clean benches), class 100, provide dust-free conditions for all critical manipulations. Further precautions are required, such as elimination of corroding metal components and unnecessary instrumentation, and also covering walls and ceilings with a non-shedding paint.

The effectiveness of upgrading the ambient laboratory air with respect to the dust level can be checked by monitoring dust particles with a particle-counting device, such as the Partoscope R (Kratel KG, Stuttgart, Germany). Measurements in ordinary laboratories without these provisions generally yield around 2×10^5 particles (with a diameter above 0.5 μm) per cubic foot (28 dm^3). This corresponds to ~ 6 μg of dust per cubic foot. Background levels (per cubic foot) during the night usually range up to 1×10^5 particles with a diameter above 0.5 μm, with a few particles > 5 μm. Unrestricted coming and going, and in particular smoking, increases the particle numbers to levels that can no longer be counted correctly by the Partoscope as they are beyond the limit of the measurement range (2×10^5). In a dust-controlled laboratory, however, the particle count decreases to about 800 particles (diameter 0.5–5 μm) per cubic foot, while no particles above 5 μm in diameter are present. The level generally becomes higher by a factor of about 10 if two staff members are working in the dust-controlled laboratory. At the laminar-flow clean bench, the level is normally below 10 particles (0.5–5 μm) per cubic foot and can increase to 100–200 particles per cubic foot during manipulations, but only for a few seconds.

Of course, such low levels of dust can only be maintained, if the staff wear protective clean-room overalls and arm sheets, together with polyethylene gloves for manipulations at the clean bench. Normal clothing increases the dust particle numbers considerably.

The next compromise possible between use of clean benches and a completely clean laboratory, is extension of the laminar flow zone to dimensions of 2–3 m side length, so that even space-consuming manipulations can be effected in a dust-free area (K. Bleymehl, Jülich, or Babcock-BSH, Krefeld). Snow samples can be worked up in the restricted area of a clean bench with a laminar flow of cooled air ($-5°$C).

3.4.3.2   Field laboratory

Reduction of blanks, a persistent problem in trace metal analysis, is far more difficult during field work, e.g. during sampling missions on a research vessel. Laboratory air on board is severely polluted with trace metals such as Pb, Ni and Zn. It is well known from numerous observations that during a journey there will frequently be work on paint scrubbing and maintenance with Pb-based anti-corrosion paints. This of course creates serious problems of heavy contamination of ambient air with Pb. One of the best ways of arranging to work in a clean area is to install a clean room container. This approach is limited only by the cost of installation and of transport to the port of departure of the research vessel. Once again, alterations may have to be restricted to an existing laboratory. Creation of a controlled atmosphere will involve reducing airborne contaminants in the form of dust and soot. The normal fresh air supply should, if possible, be cut off. Dirt particles introduced by shoes can be captured with a sticky floor covering, such as the plastic mats made by Dycem Ltd., Bristol (The Dycem Control Screen). A transportable clean-bench with horizontal laminar air-flow corresponding to US Fed. Stand. 209 (K. Bleymehl, Reinraumtechnik, Jülich, FRG) provides a clean working area of about 1×0.6 m at a height of 0.8 m. Horizontal air-flow is preferable for clean working with deep-sea sampling units.

In mobile laboratories in vans for field missions on land, the installation of a clean-bench again creates satisfactory working conditions for ultratrace determination of heavy metals.

3.5   General Remarks

The essential elements of the analytical procedure are summarized by a flow chart in Fig. 3.5.

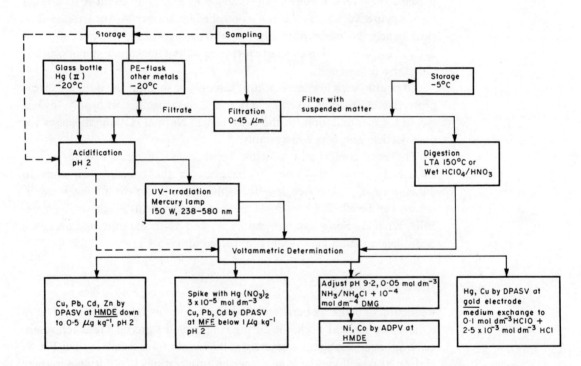

Fig. 3.5. Flow chart of analytical procedure with voltammetric determination stage. Full line, Class 4 water samples; dotted line, simplified procedure for Class 1 and also certain Class 2 and Class 3 water samples.

Table 3.3 provides a survey of the attainable detection limits and the precision of the described voltammetric methods which usually increase with concentration. For DPASV/MFE, the precision remains practically unaltered up to the reasonable upper limit of its application range around $1~\mu g~kg^{-1}$.

Table 3.3. Detection limits and precision in natural waters in $\mu g\,kg^{-1}$

Method	Detection limit at RSD±20%							Concentration determinable with RSD±10%						
	Cu	Cd	Pb	Zn	Ni	Co	Hg	Cu	Cd	Pb	Zn	Ni	Co	Hg
DPASV/HMDE	0.05	0.01	0.01	0.02	—	—	—	1.0	0.1	0.3	0.5	—	—	—
DPASV/MFE	0.007	0.001	0.001	—	—	—	—	0.03	0.005	0.005	—	—	—	—
DPASV/gold	0.02	—	—	—	—	—	0.08	0.1	—	—	—	—	—	0.2
ADPV/HMDE	—	—	—	—	0.001	0.001	—	—	—	—	—	0.02	0.02	

These detection limits also refer to the dissolved heavy metal concentrations in natural waters and molten snow. They apply additionally to the heavy metal contents of aqueous analytes resulting from digestions. With respect to digested material, however, these detection limits are not attainable, due to inevitable additional contamination in the digestion step. Nevertheless, the detection sensitivity remains more than sufficient with respect to the higher heavy metal levels in suspended matter.

With an RSD of ±40% the detection limits for Cd and Pb have even been improved to 0.0001 and 0.0005 $\mu g\,kg^{-1}$, respectively.

The order of the costs for instrumentation given in Table 3.4 emphasize, that despite its extraordinary detection sensitivity, high accuracy and remarkable convenience, advanced voltammetry represents one of the cheapest instrumental methods available for aquatic trace metal chemistry.

Table 3.4. Costs for equipment (1984–85)

Device*	US $
Normal polarograph with manual control	8 000
Polarograph with certain automated controls	10 000
Voltammetric devices with automated control and information storage	20 000
Fully automated voltammetric devices with microprocessor control and information storage	25 000

* All devices contain the differential pulse mode. In the newer fully automated class, even certain sample preparation steps and automated standard additions are included [13]. All costs contain 2500 $ for the usual accessories (electrodes, cell, etc.).

3.6   References

1   H.W. Nürnberg, *Pure Appl. Chem.*, **54**, 853 (1982); *Sci. Tot. Environ.*, **37**, 9 (1984).
2   H.W. Nürnberg, L. Mart, H. Rützel and L. Sipos, *Chem. Geology*, **40**, 97, (1983).
3   L. Mart, *Z. anal. Chem.*, **296**, 350, (1979).

4 L. Mart, *Talanta*, **29**, 1035, (1982).

5 P. Ostapczuk, M. Stoeppler, P. Valenta and H.W. Nürnberg, *Z. anal. Chem.*, **317**, 252, (1983).

6 L. Sipos, J. Golimowski, P. Valenta and H.W. Nürnberg, *Z. anal. Chem.*, **298**, 1, (1979).

7 L. Mart, H.W. Nürnberg and P. Valenta, *Z. anal. Chem.*, **366**, 350, (1980).

8 B. Pilhar, P. Valenta and H.W. Nürnberg, *Z. anal. Chem.*, **307**, 337, (1981).

9 L. Sipos, H.W. Nürnberg, P. Valenta and M. Branica, *Anal. Chim. Acta*, **115**, 25, (1986).

10 L. Mart *Z. anal. Chem.*, **299**, 97, (1979).

11 B. Schaule and C.C. Patterson. In *Lead in the Marine Environment* (M. Branica, Z. Konrad, Eds), Pergamon Press, Oxford pp. 31–43, (1986).

12 L. Mart, H.W. Nürnberg and D. Dyrssen. In *Trace Metals in Sea Water* (C.S. Wong, E. Boyle, K.W. Bruland, J.D. Burton, E.D. Goldberg, Eds), Plenum Press, New York-London pp. 113–130, (1983).

13 P. Valenta, L. Sipos, I. Kramer, P. Krumpen and H. Rützel, *Z. anal. Chem.*, **312**, 101, (1982).

14 H.W. Nürnberg, P. Valenta and V.D. Nguyen. In *Deposition of Atmospheric Pollutants* (H.W. Georgii and J. Pankrath, Eds), D. Reidel, Dordrecht-Boston pp. 143–157, (1982).

15 L. Mart, *Tellus*, **35B**, 131, (1983).

16 W. Dorten, P. Valenta and H.W. Nürnberg, *Z. anal. Chem.*, **317**, 264, (1984).

Section 4 Neutron Activation Analysis

By

Commission V.7
IUPAC

E. STEINNES
Department of Chemistry
University of Trondheim
7055 Dragvoll
Norway

4.0   Preface

Neutron activation analysis (NAA), based on irradiation in a nuclear reactor, shows high sensitivity for many elements, possibilities of systematic errors are few, and contamination problems are restricted to the pre-irradiation handling. Purely instrumental NAA is a convenient multielement technique applicable to fresh waters, where 15–20 trace elements can be determined, and to particulate phases separated from natural waters. In order to facilitate trace element determinations in sea water and to approach the ultimate determination limits in fresh-water analysis, a post-irradiation radiochemical separation normally has to be introduced before the radioactivity measurement. After pre-concentration by freeze-drying, about 50 trace elements may be determined in natural fresh waters at ambient levels. In the case of sea water, pre-irradiation separations must be employed in order to determine some of these elements. NAA is particularly useful in various research investigations and in analytical quality control work with respect to trace elements in water.

4.1   Introduction

Neutron activation analysis (NAA) based on irradiation in a nuclear reactor is the only technique within radio-analytical chemistry that is of general use for the determination of trace elements in natural waters. Two somewhat different approaches are of interest in this respect. The purely instrumental activation analysis (INAA) allows the simultaneous determination of a considerable number of trace elements at the concentration levels of occurrence in natural fresh waters. In studies more specifically aimed at one particular element or a small group of elements present in very low concentrations, it may be necessary to employ a radiochemical separation in order to remove interfering radionuclides and approach the ultimate sensitivity of NAA. This version (RCNAA) is, in general, mandatory in work involving sea water and other waters of high salt content. If not, a pre-irradiation separation procedure is employed to remove the major salt components.

Advantages commonly listed for NAA are the following:
- High sensitivity for many elements;
- High specificity;
- Few methodological sources of systematic errors (in particular for water samples);
- No reagent blank (for post-irradiation operations).

Some less favourable features inherent in this technique should also be mentioned:
- Access to a nuclear reactor is necessary for irradiation;
- The turnover time may be long (in order to allow sufficient decay of short-lived radioactivity before the handling/measurement);

- RCNAA is often laborious and time-consuming;
- Relatively high operator skill is required.

The role of NAA in water analysis is, therefore, generally confined to research applications and quality control rather than large-scale routine activity.

A flow chart indicating the main steps in NAA is shown in Fig. 4.1.

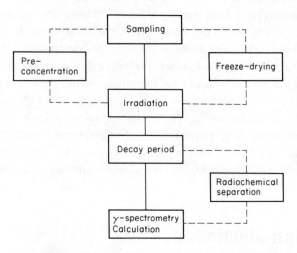

Fig. 4.1. Flow sheet of NAA applied on neutral waters.

4.2 Sample Preparation and Irradiation

In general, the pre-irradiation handling of samples of NAA is limited as far as possible in order to avoid contamination. When dealing with trace elements in water, however, it may be necessary to deviate from this principle mainly for the following reasons:

(a) Sea water cannot be irradiated as such for any length of time in a closed vessel because of radiolytic reactions leading to the build-up of high gas pressure.

(b) Fresh water is less subject to radiolysis, but in most cases the sample volume allowed for irradiation is only a few cm^3, restricting seriously the determination limits that can be achieved.

Freeze-drying is often used to remove the water prior to irradiation, and is particularly useful in the case of sea water [1]. In the case of fresh water where it may be difficult to recover the small amount of freeze-dried salt, a small amount of a high-purity substance, e.g. 25 mg of Na_2CO_3 [2], may be dissolved in the water in advance. Volatilization losses of trace elements during freeze-drying appear to be negligible except for volatile elements such as I and Hg [3]. A starting volume of 100 cm^3 is typically employed.

Irradiations are carried out in sealed ampoules made of silica or polyethylene. The ampoules must be thoroughly cleaned with strong acid

(HCl, HNO$_3$, p.a.) and deionized water before use. The selection of high-purity silica for ampoule material may be rather critical [4]. The use of polyethylene for the irradiation of water samples may be restricted to a total neutron exposure not exceeding $\sim 5 \times 10^{17}$ n cm^{-2}. Using silica ampoules, considerably higher integrated neutron doses may be acceptable by irradiation of fresh-water samples in low- and medium-flux reactors. Salbu et al. [5], for example, irradiated 5 cm^3 samples in 10 cm^3 silica ampoules to an integrated flux of 4×10^{18} n cm^{-2}. In certain high-flux reactors, however, liquid samples may not be acceptable for irradiation. In some reactors, on the other hand, water samples of 50–100 cm^3 may be irradiated in polyethylene containers in the frozen state [6]. The sample preparation and irradiation procedures to be used, thus do not only depend on factors associated with the problem and sample investigated, but also considerably on the irradiation facilities available.

The use of freeze-drying may substantially increase the integrated flux to which a given sample can be exposed. In the procedure by Lieser and Neitzert [4], for example, the constituents of a 200 cm^3 water sample were exposed to an integrated flux of 9×10^{18} n cm^{-2} after freeze-drying, which represents a potential improvement in sensitivity by a factor of nearly 100 compared to direct irradiation of a small volume of water.

In cases where the sensitivity achievable by more straightforward procedures is not sufficiently high, or where the radiation level from activated sea-salt components (mainly 37.3 min ^{38}Cl and 15.0 h ^{24}Na) is considered too high for safe work, pre-concentration from larger volumes of waters may have to be used, in connection with NAA, in the same way as used for other trace analytical techniques, with similar risks of contamination and other problems associated with the separation procedures employed. Separation techniques that have been successfully used for pre-irradiation separations of trace elements from natural waters include coprecipitation on inorganic or organic precipitates, ion exchange, use of chelating resins, and adsorption on carbon.

4.3 Post-Irradiation Chemistry

As indicated above, radiochemical separation may be necessary in order to remove interfering radioactivity formed during the irradiation. In the case of sea water samples, such treatment is necessary in almost all cases. These operations are carried out without the risk of blank contributions from reagents and equipment used. Moreover the separation of the desired radionuclide need not be quantitative if a 'carrier' technique involving chemical yield determination is properly used. Sometimes a very simple separation step, e.g. based on ion exchange [6] or electrolysis [7], may be sufficient to remove the major disturbing activities in natural waters, leaving a fraction suitable for multi-element assay by γ-spectrometry.

4.4 Instrumental Determination

The radioactivity measurements are carried out with a γ-spectrometer usually consisting of a solid-state photon detector with associated electronics, a multichannel analyser, and a small digital computer. A satisfactory system of this kind can be obtained at a cost of 30 000–40 000 US $. For a few elements the radionuclide in question does not emit γ-rays, leaving the measurements to be based on β-counting. This requires simpler equipment, but more extensive radiochemical separation work prior to the measurement.

The quantitative analysis is usually carried out relative to a standard containing the elements to be determined in known amounts, irradiated together with the samples. This relative method is now increasingly being replaced by a 'monostandard' approach, where only a neutron flux monitor is irradiated along with each batch of samples, and the element concentrations are subsequently calculated on the basis of measured peak areas, efficiency factors of the detector system used, and nuclear data pertinent to the activation and decay processes. This requires knowledge of the thermal/epithermal neutron flux ratio in the irradiation position and frequent checks on the calibration of the detector.

In most cases rather simple numerical operations are sufficient for the calculations involved with γ-spectrometric data from water analysis. For more advanced procedures the data may have to be transferred to an external computer.

4.5 Applicability of NAA to Trace Elements in Natural Waters

INAA is well suited for types of fresh water (rain, snow, river water, lake water, ground water, etc.), but is of limited use for sea water [8]. RCNAA, on the other hand, has been shown to be very useful in sea water investigations ever since the pioneer work of Smales and Pate [9].

The sensitivity of NAA shows large variation among the elements of the Periodic Table. Whereas an element such as Pb shows a very poor performance, there are certain elements that can still hardly be determined by any other analytical technique down to the concentration levels at which they occur in natural waters. A proper example is the determination of the lanthanide elements in sea water [10].

The practical determination limits in INAA depend on a number of factors including sample composition, neutron flux, irradiation time, decay time, counting time, and detector efficiency. For RCNAA a somewhat more generally applicable set of values may be defined. In Table 4.1 some representative values of determination limits are presented for different experimental conditions. Elements present as major constituents in natural water or showing little promise for NAA have been omitted. For comparison,

Table 4.1. Examples of determination limits in NAA

Element	Half-life of radio-nuclide	Determination limits fresh water (μg dm^{-3})			Concentration level (μg dm^{-3}) in natural waters[11]	
		INAA [5]	Freeze-drying, INAA [4]	RCNAA†**	Fresh water, median	Sea water, mean
Al	2.25 min	1			300	2
P*	14.3 d			0.05	20	60
Sc	84 d	0.005	0.005	0.002	0.01	0.0006
Ti	5.8 min	8			5	1
V	3.75 min	0.05			0.5	2.5
Cr	27.7 d	0.5	0.5	0.5	1	0.3
Mn	2.58 h	0.5		0.001	8	0.2
Fe	45 d	40	10	5	500	2
Co	5.3 y	0.05	0.05	0.05	0.2	0.02
Ni	2.6 h		100	0.1	0.5	0.56
Cu	12.7 h	5‡	100	0.01	3	0.25
Zn	244 d	3	0.5	0.5	15	5
Ga	14.1 h	0.5		0.02	0.09	0.03
Ge	83 min			0.2		0.05
As	26.4 h	0.2	0.1	0.005	0.5	3.7
Se	120 d	0.5	0.1	0.1	0.2	0.2
Br	35.3 h	0.2	0.02	0.002	14	6730
Rb	18.7 d	1		0.1	1	120
Sr	2.81 h			0.2	70	7900
Y*	64 h			0.005		0.013
Nb	6.3 min			2		0.01
Mo	66 h	0.5	0.2	0.05	0.5	10
Ru	39 d			0.05		0.0007
Rh	4.4 min			1		
Pd	13.5 h			0.01		
Ag	250 d	0.1		0.05	0.3	0.04
Cd	2.3 d	0.5	0.5	0.1	0.1	0.11
In	54 min			0.0005		0.00011
Sb	2.7 d	0.2	0.01	0.01	0.2	0.24
Te	8.0 d			0.05		
I	25.0 min	0.2		0.02	2	60
Cs	2.06 y	0.03		0.02	0.02	0.3
Ba	11.5 d	8		1	10	13
La	40.2 h	0.05	0.1	0.002	0.1	0.0034
Ce	32.5 d	0.05		0.01	0.2	0.0012
Pr*	19.2 h			0.05		0.0006
Nd	11.0 d			0.1	0.15	0.0028
Sm	47 h	0.02		0.001	0.06	0.00045
Eu	12.4 y	0.005	0.002	0.002	0.006	0.00013
Gd	18.6 h			0.05		0.0007
Tb	72 d			0.002	0.003	0.00014
Dy	2.35 h	0.01		0.0005		0.0009
Ho	26.7 h			0.001		0.0002
Er	7.5 h			0.05		0.0009
Tm	129 d			0.002		0.0002
Yb	4.2 d	0.1		0.01	0.01	0.0008
Lu	6.7 d			0.0005	0.003	0.00015
Hf	42 d	0.03		0.01	0.01	0.007
Ta	115 d			0.002	<0.002	0.002
W	23.8 h	0.2		0.005	0.03	0.1

Table 4.1. (*continued*)

Element	Half-life of radio-nuclide	Determination limits fresh water (μg dm^{-3})			Concentration level (μg dm^{-3}) in natural waters[11]	
		INAA [5]	Freeze-drying, INAA [4]	RCNAA†**	Fresh water, median	Sea water, mean
Re	17.0 h			0.002		0.004
Os*	15.4 d			0.02		
Ir	74 d			0.0002		
Pt	3.13 d			0.2		
Au	2.70 d	0.002	0.001	0.01	0.002	0.004
Hg	2.7 d	0.08	0.05	0.02	0.1	0.03
Tl*	3.8 y			0.5		0.02
Th	27.0 d	0.02		0.005	0.03	0.001
U	2.36 d	0.3	0.05	0.01	0.4	3.2

* Pure β-emitter, requires extensive radiochemical separation.
† Conditions: 5 cm^3 sample; irradiation time max. 3 d (days); neutron flux 2 $\times 10^{13}$ n cm^{-2} s^{-1}; induced activity 50 Bq for short half-lives ($t_{\frac{1}{2}} < 10$ h), 10 Bq for long half-lives; 15 min decay for short half-lives, 1 d for long half-lives.
** Values are also applicable to sea water in cases of $t_{\frac{1}{2}} > 3d$ and irradiation of freeze-dried samples.
‡ Depends on Na level.

typical trace element concentration values for fresh water and sea water [11] are included in the table.

It appears that 15–20 elements can be determined quantitatively in natural fresh waters at normal levels by straightforward INAA [5]. RCNAA based on the same irradiation conditions may increase the number by an additional 15–20 trace elements. The introduction of freeze-drying [4] does not appreciably increase the number of elements determined by INAA, but it brings about the necessary concentration factor for the determination of an additional 10–20 elements by RCNAA.

In sea water, Br, Rb, Sr, and Cs seem to be the only trace elements generally capable of determination by INAA. RCNAA based on a freeze-dried sample is applicable to an additional 15–20 trace elements in sea water. If, however, quantitative separation from a 1 dm^3 sample prior to activation can be achieved, RCNAA has sufficient determination power for all the 60 elements listed in Table 4.1, for sea water as well as fresh water, provided a sufficiently low reagent blank is achieved.

INAA exhibits an excellent multi-element capability for samples such as silicate rocks, coal and coal ash, and air particulates. Surprisingly little use has been made of INAA for the analysis of suspended material separated from natural waters by filtration, although the possibilities appear similarly favourable in this case [12]. Analysis of the particulate phase along with the filtered sample is likely to become increasingly important in future trace element studies in natural waters. This opens an interesting possibility of combination of INAA with techniques such as AAS and anodic stripping voltammetry (ASV) which are more suitable for the determination of some of the high-priority trace elements in the soluble phase.

Since NAA does not distinguish between different chemical forms of an element, any applications of this analytical technique to *speciation studies* must rely on separations carried out *before* the irradiation step. INAA has proved to be quite useful in speciation studies particularly aimed at distinguishing between truly soluble species in natural waters and those associated with colloidal and suspended matter [12–14]. In this work based on dialysis, ultra-filtration, centrifugation, etc., INAA provided useful information on the speciation of elements such as Al, Sc, Cr, Mn, Fe, Co, Zn, and Ba in fresh waters.

Even though NAA can provide excellent data for many of the elements presently in the focus of attention, it is only in occasional cases likely to become the selected technique for routine investigations of these elements in natural waters, because more straightforward techniques such as AAS or ASV can do the job. NAA is however a very powerful tool in many research investigations involving elements present in ultra-trace concentrations. Furthermore, being based on a unique physical principle compared to other relevant trace element techniques and being less susceptible to systematic errors than most of the alternative techniques, NAA should be extremely useful in analytical quality control related to trace elements in natural waters.

4.6   References

1   D.F. Schutz and K.K. Turekian, *Geochim. Cosmochim. Acta,* **29,** 259 (1965).
2   D.P. Kharkar, K.K. Turekian and K.K. Bertine, *Geochim. Cosmochim. Acta,* **32,** 285 (1968).
3   S.H. Harrison, P.D. LaFleur and W.H. Zoller, *Anal. Chem.,* **47,** 1685 (1975).
4   K.H. Lieser and V. Neitzert, *J. Radioanal. Chem.,* **31,** 397 (1976).
5   B. Salbu, E. Steinnes and A.C. Pappas, *Anal. Chem.,* **47,** 1011 (1975).
6   T.M. Tanner, L.A. Rancitelli and W.A. Haller, *Water, Air, Soil Pollut,* **1,** 132 (1972).
7   K. Jørstad and B. Salbu, *Anal. Chem.,* **52,** 672 (1980).
8   D.Z. Piper and G.G. Goles, *Anal. Chem. Acta,* **47,** 560 (1969).
9   A.A. Smales and B.D. Pate, *Analyst,* **77,** 188 (1952).
10   O.T. Høgdahl, S. Melsom and V.T. Bowen, *ACS Advances in Chemistry Series,* **73,** 308 (1968).
11   H.J.M. Bowen, *Environmental Chemistry of the Elements,* Academic Press, London (1979).
12   M.T. Ganzerli Valentini, N. Genova, L. Maggi, S. Meloni and R. Stella, *Nuclear Methods in Environmental and Energy Research,* USDOE CONF-800433, 106 (1980).
13   P. Beneš and E. Steinnes, *Water Res.,* **8,** 947 (1974).
14   P. Beneš and E. Steinnes, *Water Res.,* **9,** 741 (1975).

Section 5   Measurement of pH in Natural Waters

Commission V.5
IUPAC

By
A. K. COVINGTON AND P. D. WHALLEY
Electrochemistry Research Laboratories
Department of Physical Chemistry
University of Newcastle
Newcastle upon Tyne NE1 7PU, UK

W. DAVIDSON
Freshwater Biological Association
The Ferry House, Ambleside
Cumbria LA22 0LP, UK

M. WHITFIELD
Marine Biological Association
The Laboratory, Citadel Hill
Plymouth PL1 2PB, UK

5.1 Introduction

pH is probably the most commonly measured quantity in environmental research and water quality control. Its determination is so routine that it is often taken for granted, little attention being paid to the adequacy of the measurement procedure and there being little or no need to consider the meaning of the resulting number. Yet, if we regard it as an analytical technique for measuring hydrogen ions, the vast concentration range and extreme sensitivity immediately commands respect (pH 3–10 is equivalent to 1 mmol dm^{-3} to 0.1 nmol dm^{-3}). No other measurement is performed over such a range of concentration using simple equipment and scant regard for contamination problems, while expecting rapid, reliable results and applicability to all types of water.

Most applications require only a knowledge of the values of pH relative to previous measurements. Thus, although internal consistency must be maintained, it is not usually necessary to relate the pH values obtained to those of other workers. Such measurements can often be obtained quite readily with a minimum of attention. However, many studies of the natural environment involve comparisons between different sites, both in time and space, and so there is a need for comparability of pH measurements made in different laboratories. The precise measurement of the pH of sea water has received a great deal of attention over the past ten years [1]. Although much time and effort have been devoted to the development of procedures which are specific to this reasonably constant composition natural water, problems of comparability still exist [1]. In contrast, relatively little work has been done until recently on the measurement of the pH of fresh waters and estuaries, where the composition of the medium is highly variable. This deficiency has been highlighted recently by workers concerned with the phenomenon of acid rain. The reproducibility of measurements made at low pH and low ionic strength has been questioned [2,3] and the inadequacies of past data which have been used to study trends in acidification have been exposed [4,5].

When pH is used to determine defined environmental parameters such as stability constants or kinetic rate constants, it not only needs to be reproducible between laboratories, but there is also a need for its interpretation in terms of hydrogen ion activity or concentration. This is particularly so when the chemical processes operate in both fresh and saline systems, requiring the validity of the 'constant' to be checked for waters of wide-ranging composition.

This Section discusses the problems, both theoretical and practical, which beset the measurement of pH in natural waters. It identifies the liquid junction, formed between reference electrode bridge solution and test or standard solutions, as the chief source of error, and considers how this problem may be minimized. It makes the recommendation that renewable, free diffusion junctions should be employed when high reproducibility and repeatability are required.

5.2 Fundamental Considerations

5.2.1 Definitions

The definition of pH in terms of the hydrogen ion relative activity [6]:

$$pH = -\lg a_H \tag{5.1}$$

is purely notional since single ion activities cannot be measured and a practical or operational approach is required. The operational definition is based on sequential measurements in the test solution and a series of standard buffers [7] using cells of the form:

Reference electrode	KCl > 3.5 mol dm^{-3}	¦ ¦ ¦ ¦ ¦ ¦	Test (X) or Standard (S)	H$^+$-responsive electrode

where the hatched vertical lines indicate the presence of a liquid junction. The value of pH(X) is given by

$$pH(X) = pH(S) + \{E(S) - E(X)\}/gT - \{E(JS) - E(JX)\}/gT \tag{5.2}$$

where $g = (R/F)\ln 10$ and $E(S)$ and $E(X)$ are the cell potential differences in the standard and test solutions respectively, and $E(JS)$ and $E(JX)$ are the liquid junction potentials with standard and test solutions. The two cells have the same temperature throughout, and the same glass and reference electrodes. The operational pH is defined by setting the term involving the liquid junction potential equal to zero.

The definition of pH is completed by assigning values of pH at 5° intervals of temperature to one or more chosen standard reference solutions [7]. Potassium hydrogen phthalate is the reference value standard and six other buffer solutions are primary standards with values assigned from measurements on cells without liquid junction. This involves making an assumption about the single ion activity coefficient. In addition to the primary standards there are alternatively numerous operational standard solutions to which pH values have been assigned on the basis of measurements on cells with liquid junctions formed in a highly reproducible fashion. The difference between using primary or operational standards for measurements in natural waters is negligible. The problems associated with measuring the pH of natural waters arise from attempting to put the operational definition into practice. The various factors concerned will be reviewed.

5.2.2 Glass electrodes

Most pH measurements are done with glass electrodes rather than with the platinum–hydrogen gas electrode. For this procedure to be valid, these two electrodes must respond identically to hydrogen ions. Although there are errors associated with glass electrodes at very high and very low pH, the response is very good for pH 2–10, which encompasses the pH of most natural waters. There have been reports of relevant errors, either attributable to very low buffer capacity [8] or to colloidal interactions [9], but for measurements of most natural waters these effects should be very small and the glass electrode can be assumed to give the theoretical response or close to it. Even so, careful selection and checking of glass electrodes is necessary to ensure theoretical or near-theoretical response (see Section 5.3.1).

5.2.3 Reference electrodes

A reference electrode has to provide a stable potential in potassium chloride solution between successive measurements of standard and test solution. Properly prepared mercury–mercury(I) chloride (calomel) or silver–silver chloride electrodes give highly reproducible potentials [6] and should not be a source of error in high concentration, neutral potassium chloride solutions.

5.2.4 Liquid junction potentials

The liquid junctions which appear in the operational cells have to be carefully considered in all practical pH measurements. The concentrated potassium chloride solution which actually forms part of the reference electrode assembly is in contact with either the test solution(X) or the standard solution(S). Junctions between two different ionic solutions always give rise to a potential difference, which is included as part of the measured cell potential difference. However, pH defined operationally ignores the last term in eqn (5.2); that is, it is assumed that the liquid junctions are identical in the two cells. Therefore, if the liquid junction potentials differ, then this term will be a source of error in the measured pH(X). The liquid junction potential depends upon composition and ionic strength of the solution in contact with the potassium chloride. So, the liquid junction potential will not usually be the same in the two cells. The difference is known as the residual liquid junction potential [6,7]. This residual liquid junction potential error is present to a greater or lesser extent in all pH measurements and constitutes the greatest obstacle to interpreting pH in terms of hydrogen ion activity, and has led to attempts to eliminate its effect by development of special scales of pH for sea water. Its irreproducibility is the cause of the greatest practical problems associated with the measurement of pH. The compromises which are adopted to circumvent these fundamental and practical difficulties are outlined below.

5.2.5 Low ionic strength solutions (<0.1 mol dm^{-3})

The assignment of pH values to the reference value standard and primary standard buffer solutions is made by making an assumption that the single ion activity coefficient for the chloride ion is given by an appropriate version of the Debye–Hückel equation [4] known as the Bates–Guggenheim convention [10]:

$$-\lg\gamma_{Cl} = AI^{1/2}/(1+1.5I^{1/2}) \tag{5.3}$$

165   FUNDAMENTAL CONSIDERATIONS

where A, the Debye–Hückel constant is temperature-dependent. Various tests [6, 11] show that this is a reasonable convention to adopt, but its limit of validity is $I = 0.1$ mol dm^{-3}. Since liquid junction potentials cannot be determined without making assumptions about single ion activities and vice versa, the adoption of a convention for a single ion activity coefficient leads to conventional values for the residual liquid junction potentials, by comparing the assigned primary standard and operational pH standard values for the same buffer solutions [7]. For those solutions selected as primary standards the conventional residual liquid junction potentials are in fact small. This is also true for the residual liquid junction potential between the IUPAC-recommended concentration and diluted (1 : 10) standard reference solutions [12]. Further, if the test solution matches the selected standard reference solution in ionic strength and composition then the residual liquid junction potential may be expected to be small. Thus, if the liquid junction is properly formed (see later), standard reference buffers should be suitable for calibrating glass-reference electrode pairs, which are to be used for measuring the pH of all low ionic strength solutions such as fresh water and rain water.

5.2.6 Solutions of high ionic strength (> 0.1 mol dm^{-3})

Sea water of 35‰ salinity has an ionic strength of 0.72 mol dm^{-3}. At this ionic strength the simple Debye–Hückel equation [4] no longer holds and the residual liquid junction potential contribution is large. This complicates both measurement and interpretation of pH in sea water and has led to the development of new pH scales for sea water [1,13]. Whilst it is possible to make measurements on sea water using the standard reference buffers of ionic strength < 0.1 mol dm^{-3}, the interpretation of the measured values in terms of hydrogen ion activity or concentration is fraught with difficulty. A pH scale with a clearer conceptual significance can be defined [14] in terms of the total hydrogen ion concentration of the test solution by

$$\text{pH(SWS)} = -\lg {}^{T}m_{H} \tag{5.4}$$

where ${}^{T}m_{H}$ is the sum of the free and complexed hydrogen ion concentrations in the sea water test sample. The pH(SWS) scale is based on synthetic sea water containing sulfate and buffered with Tris. Owing to the formation of the hydrogen sulfate ion, the total hydrogen ion concentration differs from the concentration of free, uncomplexed hydrogen ion. The pH(SWS) of the Tris reference buffers is determined by titration with hydrochloric acid to a final pH of 3. By matching the ionic strength of the saline standards to sea water, the residual liquid junction potential is minimized. If sea water rather than pure water is considered as the solvent, then the activity coefficients of components contributing 1% of the total ionic strength of the medium will be close to unity. The pH(SWS) scale is therefore equivalent to a hydrogen ion activity scale in the sea water medium. For estuarine samples the concept has been extended [14] to cover the salinity range 10–40‰ in 5‰ intervals and the assigned values for the Tris reference buffers in fluoride-free sea water diluted with distilled water to the appropriate salinity can be summarized [15] by eqn (5.5)

$$\text{pH(SWS)} = (2559.7 + 4.5\ S)/T - 0.5523 - 0.01391\ S \tag{5.5}$$

where S is the salinity in ‰ and T is in Kelvin.

Millero [16] has recently re-evaluated the data on the dissociation of Tris buffers with special reference to the provision of reliable pK^* values at low salinities. He suggests that the pH on the total hydrogen ion concentration scale can be calculated from eqn (5.6):

$$\text{pH(SWS)} = pK_T^* - (9.73 \times 10^{-5}S - 6.988 \times 10^{-5}S^2)m_{\text{Tris}} \tag{5.6}$$

where pK_T^* is given by

$$\begin{aligned} pK_T^* = {} & -22.5575 + 3477.5496/T + 3.328\,67\ \ln T \\ & - (2.3755 \times 10^{-2}S - 6.165 \times 10^{-5}S^2)/T + 6.313S \end{aligned} \tag{5.7}$$

and m_{Tris} is the molality of the Tris buffer. Eqn (5.6) is reckoned to be more reliable than eqn (5.5) at salinities below 20‰.

A third scale for sea water has been devised based on the free hydrogen ion concentration in synthetic, sulfate-free sea water buffers [17,18] but this cannot be determined by simple titration and is evaluated instead from cells without liquid junction. We shall not consider this scale further here.

The relation between the IUPAC pH and pH(SWS) scales is obtained through the quantity $f'_H(X)$, which is a function of the activity coefficient of the hydrogen ion in X and the residual liquid junction potential [1,19] and is capable of direct measurement for particular glass-reference electrode pairs (see Section 5.5).

5.3   Practical Considerations

5.3.1   Sources of errors

So far we have considered the fundamentals of measurement of pH and practical problems have hardly been mentioned. The glass electrode is almost universally used as the hydrogen-ion responsive electrode and, as already mentioned, this has been shown to give a theoretical or near theoretical response [6,20,21]. However, there can be differences between commercial products. There is a wide range of glass compositions which show a pH function, and each manufacturer usually markets several. The detailed composition and fabrication are often unique to the manufacturer, and indeed electrodes from different batches may show different performance characteristics of response time and scale range. Experience shows that if samples are carefully selected and subjected to frequent performance checks [20,21], no significant error should originate from this source, with the exception of low ionic strength freshwaters which have very poor buffer capacity. They are, therefore, prone to contamination, which can come from the glass electrode itself by desorption of ions previously adsorbed during earlier measurements or standardization, and from dissolution of the glass. Some form of solution flow, such as gentle stirring, is necessary to avoid errors due to this effect.

As previously discussed, reference electrodes are capable, if properly prepared, of providing time-independent potentials in potassium chloride solution and there is no inherent problem from this source, except hysteresis, if the temperature is changed. Unfortunately, the same is not true of the liquid junction formed between the potassium chloride salt bridge and the standard or test solution. Almost all the problems of practical pH measurements are associated with these junctions and so these will now be discussed in detail.

5.3.2   Liquid junctions

The liquid junction potential depends on the geometry of the junction except for junctions between two different concentrations of the same electrolyte (homo-ionic junctions). Guggenheim in 1930 [22] described the characteristics of five different types of junction comparing their merits and noting that their limitations were sometimes severe. For the successful, precise measurement of pH (± 0.01), it is essential that the liquid junction is formed reproducibly and renewed for each experimental measurement. The free diffusion junction, which meets these requirements, will be described later in detail. First, we shall consider the inadequacies of commercially available junctions.

5.3.3   Commercial liquid junctions

Although there is a wide range of commercial junctions available, the commonest employ a porous ceramic plug to separate the potassium chloride from the test or standard solution. These suffer from the following disadvantages:

(i) there is no standard design; a variety of materials and geometries is employed.

(ii) the junction potential of supposedly identical electrodes can differ markedly [1]. Deviations of 0.2 in pH have been observed and discrepancies between different types of electrode can be much greater.

(iii) although new electrodes can exhibit the above errors, there is additional evidence [23] that the potential can be influenced by contamination. The porous material can exhibit a memory effect from the previous solution, or from storage in media other than potassium chloride.

(iv) the potential depends on the outward flow rate of the potassium chloride solution. The cell potential difference in a stirred solution is not the same as in an unstirred solution, probably because the geometry of the junction is different. The effect is neither repeatable nor reproducible [24]. There will also be an inward diffusion of water tending to dilute the concentrated potassium chloride solution.

In recognition of the above problems, alternative commercial designs have become available. Sleeve electrodes, which form a junction in a ground glass or plastic joint have gained in popularity because the junction is renewable and contaminants can be flushed out. However, their characteristically high leakage rate can cause contamination of poorly buffered waters. Moreover, experiments have shown that they are liable to give irreproducible results [25, 26]. This may be due to their ill-defined geometry, the leakage path not always being the same. Some manufacturers have introduced renewable porous plugs to combat contamination or ageing effects. Although these changes can help to alleviate obvious deteriorations in performance, the inherent problems of using restrained junctions remain. The problems associated with such restrained junctions are so great that their use should be avoided whenever work of the highest reproducibility and repeatability is required. Some workers have devised procedures for obtaining reasonable performance from commercial reference electrodes, but it is advisable to check for any bias against free diffusion junctions.

5.3.4   Free diffusion junctions

We believe, as does Culberson [1], that reproducible pH measurements require not only standard buffers, but also the use of a standard form of reference electrode. We propose that the free diffusion junction should be adopted as this standard form. Guggenheim [22] stated that in this form of junction 'the transition layer should be initially short compared with the distance between the electrodes, an absolutely sharp boundary being, of course, impossible and that unconstrained diffusion should be allowed to take place'. Under these conditions, the length of the transition layer is always increasing, but if the junction has cylindrical symmetry, such that the gradients of concentration and potential are parallel to a line perpendicular to the initial boundary, the potential difference is essentially time independent. This type of junction is easy to realize experimentally by various methods using, for example, capillaries, open tubes, or glass taps.

For use as a practical standard, the junction requires further specification so that it is reproducible and repeatable:
(i) it should be formed in a capillary tube which has a bore of 0.5–2 mm.
(ii) the tube should be arranged vertically so that the solution being measured overlies the denser, concentrated potassium chloride, thus preventing gravity-induced convection.
(iii) the junction should be readily renewable.
Figures 5.1 and 5.2 show two designs which incorporate free diffusion junctions meeting these criteria.

The flow cell which is illustrated in Fig. 5.1, is that used by Covington *et al.* [12] after the original design of Culberson [1]. A sharp T-junction was found to be critical to the formation of a reproducible liquid junction and to the fast response of the cell. This was made by using a tungsten wire to form a pin-hole in the wall of a glass capillary tube. A second capillary tube was sealed round the hole to form the T-junction. Release of potassium chloride from the reservoir (Fig. 5.1) enabled a fresh liquid junction to be formed. After some of the test solution had been flowed through the cell, the liquid junction was

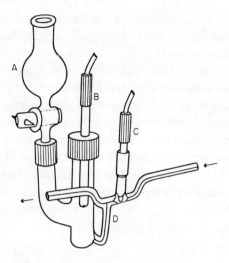

Fig. 5.1. Modified flow cell after Culberson [1]. A, KCl reservoir; B, reference electrode; C, glass electrode; D, T-junction.

clearly visible as a sharp boundary a few mm below the T-junction. Thus, the essential requirements of cylindrical symmetry are met.

The dip electrode [26] illustrated in Fig. 5.2 uses similar principles. It consists of a liquid junction formed at a pin-hole T-piece, similar to that used in the flow cell. The bridge solution is overlain by the less dense test solution. Thus the lower capillary carries the sample solution to the T-piece and the

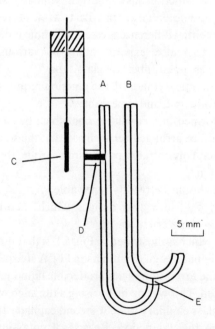

Fig. 5.2. Renewable free-diffusion liquid junction reference electrode. A, from KCl reservoir via syringe; B, to waste via syringe; C, Ag/AgCl electrode in 3.5 mol dm^{-3} KCl (sat. with AgCl); D, ceramic plug; E, free diffusion liquid junction.

excess sample and bridge solution to waste. A fresh junction is formed by using reciprocating syringes connected to the two capillaries. Simultaneous action allows the bridge solution to be extruded from the T-junction as the test solution is drawn into the top capillary. To ensure a well-defined junction, the ratio of the displacements of the syringes should be at least 2:1 so that the volume of the sample withdrawn into the capillary is in excess of the injected potassium chloride. This method not only ensures a fresh junction, but successfully eliminates contamination problems of the sample by the bridge solution. Some sample has to be removed to form the junction, but if the capillaries have a bore of 0.5 mm, the volume of this is very small (typically $< 10 \ \mu dm^3$), so such a design can be used for titrimetry. The choice of reference electrode to be used with either of these designs of free diffusion junction is irrelevant because it is maintained in concentrated potassium chloride. Even a commercial reference electrode with ceramic plug can be used provided it is ensured that the concentration of potassium chloride filling solution is the same as that used for the bridge solution.

The dip design of free diffusion junction was shown to have the following advantages over commercial designs of restrained junction:

(a) It gives very reproducible potentials (± 0.2 mV).

(b) It is not prone to contamination or storage effects because the junction can be readily renewed.

(c) Once formed the potential remains stable for at least several hours.

(d) In solutions of low ionic strength ($I \leqslant 0.1 \ mol \ dm^{-3}$) with $4 < pH < 10$, the conventional residual liquid junction potential between standard buffers and 10 times diluted standard buffers has been shown to be effectively zero.

(e) The potential is not affected by convection; provided that the junction is 2 cm away from the open solution, it is the same in quiescent or stirred solution and electrical noise does not increase with stirring.

It is clear from the foregoing that the use of this design of junction will provide pH data which are capable of comparison between laboratories. Covington et al. [26] used it to measure the second apparent dissociation constant of phosphoric acid titrimetrically in solutions of different ionic strengths. The pK values obtained using the free diffusion junction agreed to within 0.01 whereas commercial electrodes gave results differing by up to 0.3.

5.4 Procedures for Measuring pH of Freshwaters

In the preceding sections it has been shown that it is essential to use a free diffusion junction, for only then is the measurement of sufficiently good quality for interpretation in terms of hydrogen ion activity or concentration to be worthy of consideration. However, the use of such a procedure is only warranted if adequate attention has been paid to sampling and handling procedures.

The pH of poorly buffered freshwater samples is not stable but can change with time [27]. These changes may be due to the following factors: (i) the ionic equilibria in the solution are temperature dependent; (ii) the concentration of dissolved gases present in solution may be perturbed by re-equilibration with the atmosphere, photosynthesis, respiration or microbiological degradation processes; (iii) the water may not be at chemical equilibrium so that reactions with suspended solids are possible. Thus, for natural waters, *in-situ* pH measurements have been recommended [27]. Where this is not possible, alternative practical procedures have to be adopted. Samples should be collected in well-washed, darkened, borosilicate glass bottles and not in plastic, metal or soda glass containers. The bottles should be pre-rinsed with the sample and then completely filled so that, when the stopper is replaced, it displaces the sample in the neck, preventing the entrainment of any air. Ideally, the bottle should be maintained at the *in-situ* temperature of the natural water and pH measured at that same temperature. When this is not possible, the measurement should be made at some other selected temperature and the *in-situ* pH calculated from the temperature dependence of the carbonate stability constants [28]. The fact that the bottle is sealed preventing gaseous exchange at the air–water interface makes this calculation possible [1]. Poorly buffered waters of high biological productivity should be measured as quickly as possible. Photosynthesis in sealed bottles has been shown to change the pH from 7.8 to 9.3 in 2 h [29]. Darkening the bottle encourages respiration, which changes the pH from 7.8 to 7.3 in the same time. These represent extreme examples, but generally pH measurements on natural water samples should be carried out within a few hours.

Acid waters are neither buffered by the carbonate system nor are they biologically productive. Therefore they are much less prone to pH changes caused by atmospheric exchange or biological processes and so storage prior to measurement is less detrimental. Rain water samples can be affected by microbiological processes which modify the nitrogen species, but this is unlikely to affect the pH significantly. Notwithstanding these better storage characteristics, it is still prudent to perform the pH measurement as soon as possible. Because of problems associated with glass electrode carry-over, procedures which advocate performing measurements on quiescent solutions [2,3] should be avoided. Unlike the procedure recommended for buffered waters at near neutral pH, it is not possible to predict the temperature dependence, and so every effort should be made to preserve samples at their original temperature. Alternatively, the temperature dependence for a particular water may be estimated by experiment. The standardization procedure for near neutral waters is not applicable. Potassium hydrogen phthalate buffer can be used for standardization and 1:1 phosphate for slope checking [7]. Some authors [2,3] have advocated using dilute mineral acids; this is acceptable providing there is no danger of carry-over contamination changing the pH of such poorly buffered samples.

Some handling problems can be overcome by using a flow cell such as that shown in Fig. 5.1. The sample can be arranged to flow directly into the cell by gravity feed. Exposure to the atmosphere is minimized, as is exchange of carbon dioxide. When such an arrangement was used to perform measurements on fresh waters [12], it proved necessary to maintain a flowing solution.

Although the pH was constant over a range of flow rates ($1-4 \text{ cm}^3 \text{ min}^{-1}$), at low flows ($< 1 \text{ cm}^3 \text{ min}^{-1}$) and in stationary solutions, the pH tended to drift. This instability was attributed to poor solution flushing at the glass electrode. The continual flow of solution allows the liquid junction to be positioned close to the glass electrode ($\sim 1 \text{ cm}$) without risk of contamination from potassium chloride. The cell resistance, an important consideration in very dilute solutions, is consequently minimized. A further advantage of a flow system results from the high impedance electrode circuit not being broken between samples, and so delay associated with re-establishing electrical equilibrium is avoided [30]. The time lag between the introduction of a new solution and the initial electrode response was 15–30 s, and it usually took between 1–3 minutes to reach a pH within 0.02 of the equilibrium value. For buffers, equilibrium was achieved in less than 2 minutes. Fresh water samples, preceded by a standard buffer, required 2–10 minutes. A cautionary note regarding the use of flow cells with peristaltic pumps is necessary. The use of electrical pumps inevitably causes interference to the cell potential difference by transmission of electrical 'noise' along the tubing particularly with dilute solutions. Also, peristaltic pumps produce additional problems associated with their pulsed action. There are indications that this can physically disturb the liquid junction, irrespective of whether the pump is upstream or downstream of the cell. All these 'noise' problems are eliminated when the solution is supplied by gravity feed.

Although a flow cell simplifies the handling of air-sensitive samples, there are some situations (e.g. titrations) which preclude its use. A dip cell using a reference electrode with free diffusion junction should then be used (Fig. 5.2). It is essential to stir the solution to prevent 'contamination' problems associated with the glass electrode. Because the free diffusion junction is not affected by stirring, the discrepancies which arise with commercial restrained junctions are avoided. Restrained junctions are subject to stirring noise and perform best in quiescent solutions [2,3]. Thermostatting is an important consideration for any precise measurement of pH. It not only defines the system, but eliminates some apparent noise problems.

Electrode pairs should be standardized using the IUPAC primary reference solutions [7]. The difference in using the IUPAC operational standard reference solutions [7] is negligible for the present purposes. Covington *et al.* [12] have determined lg a_H for four diluted standard buffers ($I \sim 0.01 \text{ mol dm}^{-3}$) using cells without liquid junction. The buffer capacity of these solutions is obviously inferior to the standard buffers, but pH should still be stable to ± 0.01 for freshly prepared solutions. Their use for standardization of electrode pairs may be advantageous in reducing glass electrode carry-over problems.

5.5 Procedures for Measuring pH of Sea Water and Estuarine Waters

pH measurements in sea water are most commonly made using electrode pairs standardized in standard reference buffers. These buffers retain their popularity largely because certified materials are available. In contrast, saline buffers required for the standardization of electrode pairs on the pH (SWS) scale have to be prepared individually using a rather tedious procedure [31]. However, measurements relative to standard reference buffers can be subject to systematic errors approaching 0.1 in pH [1,19] because of variations in the residual liquid junction potential [see eqn (5.2)] caused by differences in the reference electrode design. To ensure that pH measurements made relative to standard reference buffers are strictly comparable, it is necessary to estimate the variable contribution made by the liquid junction potential. This can be done experimentally by determination of the parameter f'_H which may be defined by the equations:

$$\lg f'_H = \lg {}^T\gamma_H^T + [E(JS) - E(JX)]/gT \tag{5.8}$$

$$= -\lg {}^T m_H^T - pH(X) \tag{5.9}$$

The conventional hydrogen ion activity coefficient ($^T\gamma_H$) is characteristic of the sea water medium only, so that differences between f'_H values determined for different electrode pairs will provide a direct indication of the systematic errors in pH associated with their use. The f'_H value can be determined experimentally by two methods:

(a) In the acid titration method [32], a sea water sample of the appropriate salinity is first adjusted to pH 3–4 by addition of strong acid, and the carbon dioxide released is purged. Subsequent aliquots of acid will then provide known increments of $^T m_H$, so that eqn (5.9) can be written as

$$10^{[-pH(X)]} = f'_H \cdot {}^T m_H \tag{5.10}$$

A plot of the left hand side of eqn (5.10) against $^T m_H$ gives a straight line of slope f'_H.

(b) In the buffer method, [1,19] an electrode pair standardized in standard reference buffers is used to determine the pH of a saline buffer of the appropriate salinity. The measured pH of the saline buffer [pH(X)] and the pH assigned to the buffer on the total hydrogen scale [pH(SWS), eqn (5.5)] are then related to give

$$\lg f'_H = pH(SWS) - pH(X) \tag{5.11}$$

This is clearly the simplest procedure experimentally, but it does require the preparation of saline buffers according to the procedure outlined by Hansson [14]. Since the salinity of sea water rarely varies by more than ±1‰, the direct use of saline buffers to measure pH(SWS) is not subject to significant systematic errors due to variations in the liquid junction potential. In addition, the electrode response is more rapid when the electrodes are transferred between the buffer and saline samples if saline buffers rather than standard reference buffers are used. The Joint Panel on Oceanographic Tables and Standards (JPOTS) Subcommittee on Carbon Dioxide has consequently recommended that the saline buffer procedure be adopted for sea water pH measurements, and that the use of IUPAC standard reference buffers for this purpose should be discontinued. For this recommendation to become effective, certified reference materials must be made available. In the meantime, all measurements made relative to standard reference buffers should be accompanied by measurements of the f'_H values at the appropriate temperature and salinity for the electrode pair employed.

When estuarine waters are measured, the residual liquid junction potential of the electrode pair will contribute significant systematic errors to the measurement whether the electrodes are standardized in standard reference buffers or saline buffers. These errors could in principle be avoided by using a series of saline buffers to cover the salinity range encountered, but in practice this would be unacceptably complicated. For electrodes standardized in standard reference buffers, the systematic errors can be assessed by measuring f'_H values over the appropriate salinity range as described above.

If a saline buffer is to be used, a salinity in the middle of the estuarine range (between, say, 15 and 20‰) would be appropriate. The systematic errors incurred by the use of this single buffer could then be estimated by measuring the potential difference of the electrode pair in a series of saline buffers spanning the appropriate salinity range. The difference between the pH assigned to a particular buffer by eqn (5.5) [pH(SWS)] and the pH measured by the electrode pair in that buffer relative to the selected standard [pH(SWS)*] is given in ref. [19].

$$\Delta pH = pH(SWS) - pH(SWS)* \tag{5.12}$$

$$= -\lg \left[{}^T\gamma_H(S)/{}^T\gamma_H(X) \right] - [E(JS) - E(JX)]/gT \tag{5.13}$$

The first term only depends on the composition of the saline buffer and is independent of the electrode pair used in the measurement. The ΔpH values obtained can, therefore, be used for assessing the systematic errors in the use of a single saline buffer over the whole estuarine salinity range. The measurement of f'_H and ΔpH values for pH cells containing a range of conventional reference electrodes [19] indicates that, although there is good internal consistency in measurements made with individual electrode pairs (standard deviation < 0.02 in pH), systematic errors approaching 0.1 in pH can be observed between different electrode pairs irrespective of the standard buffers used. Clear differences were observed between various types of commercial electrode, and the f'_H and ΔpH values varied with time and between nominally identical commercial electrodes of the same design. The differences in behaviour between various electrode pairs were, in general, more pronounced if the electrodes were standardized in 1:1 phosphate buffer rather than a saline buffer, and the time for a steady reading to be reached was often longer in the standard reference buffer. However, the differences between the performance of the electrode pairs in IUPAC standard reference and in saline buffers were small in comparison with the large differences observed between different reference electrodes on both pH scales.

It is clear that thermodynamically useful pH measurements could be obtained with considerably less effort and enhanced reliability if the stability and reproducibility of liquid junction potentials could be increased.

Culberson's flow cell for sea water measurements [1], incorporated a renewable free diffusion liquid junction, formed within a vertical capillary tube. This cell permits measurements to be made on small, static sample volumes ($5-20 \, cm^3$) injected into the cell. Lack of solution agitation does not introduce appreciable errors in these concentrated solutions. Culberson's cell was based on a Beckman micro-blood pH assembly. It initially used a palladium annulus liquid junction [33], but this caused long equilibration times in sea water samples after calibration in standard buffers and would be unacceptable for anoxic samples, where redox reactions may occur at the metal junction causing spurious potentials. Subsequent versions of this cell have incorporated a renewable free diffusion liquid junction formed in a capillary tube. Culberson's cell was modified [19] for estuarine pH measurements at varying ionic strengths (see Fig. 5.1). In addition to a significant improvement in reproducibility and a considerable reduction in the systematic errors associated with pH measurements across the salinity range, the flow cell has the additional practical advantages of rapid and reproducible pH response under field conditions, where the salinity varies from sample to sample. The cell retains its standardization for several days to within ± 0.002 in pH.

The practical design and use of the flow cell differs in some details from its use in fresh waters. It is not so important in a high ionic strength medium to keep the junction close to the glass electrode. Neither has it been found necessary to use a flowing solution. The buffer capacity of sea water is much greater than fresh water and so contamination problems associated with the glass electrode are reduced. A technique found useful [19] was to take samples for flow cell measurements in $25 \, cm^3$ disposable plastic syringes taking care to avoid cavitation and undue gas exchange. Any air above the sample is dispelled from the syringe before storage and the syringe closed with a close fitting luer cap. The sample was applied to the flow cell directly through a three-way tap to assist venting any air bubbles to waste. The sample may be left stationary or slowly flowed.

The flow cell was used [19] to measure the pH of saline buffers on the pH(SWS) scale and differences from the values calculated, with reference to the 20‰ saline buffer, using the true values from eqn (5.5). Much smaller systematic errors were found than when commercial reference electrodes were used. Only at 5‰ and 35‰ were the errors approaching 0.02 in pH, thus the use of a free diffusion junction significantly reduces the variability of the measurements. The flow cell was found to be capable of producing good quality estuarine water data under field conditions [19].

5.6   Summary of Recommendations

5.6.1   Fresh water samples

If a precision of ± 0.01 in pH measurements in freshwater samples is required, it is recommended that:

5.6.1.1   The sample is at controlled temperature and collected in well-washed, darkened borosilicate glass bottles filled to the neck. Poorly buffered water samples of high biological productivity should be analysed as quickly as possible.

5.6.1.2   Electrode pairs should be standardized with IUPAC primary standard reference buffers, e.g. for samples of pH 5–8, use 1:1 phosphate buffer with 1:3.5 phosphate for slope checking, but for acidic samples of pH 3–5, use phthalate buffer with 1:1 phosphate for slope checking.

5.6.1.3   Use a reference electrode with renewable free diffusion junction such as one of the designs of Fig. 5.1 and 5.2. Only by this means is a precision of ± 0.01 in pH obtainable on a routine basis.

5.6.1.4   Report details of the experimental procedure including standard reference buffers and values, design of reference electrode, temperature of samples and nature of any corrections made.

5.6.2   Sea and estuarine water samples

If precision of ± 0.01 in pH measurements in sea and estuarine water samples is required, it is recommended that:

5.6.2.1   The sample is at a constant, controlled temperature.

5.6.2.2   Electrode pairs should be standardized on either the IUPAC pH or pH(SWS) scales, but the commercial availability of standard reference materials for the former makes this more convenient in practice for the time being.

5.6.2.3   The electrode pair should be characterized over the appropriate salinity and temperature range to assess the systematic errors associated with variations of liquid junction potential and hydrogen ion activity coefficient. This requires the determination of f'_H or ΔpH values as appropriate, and the conversion factors should be published along with the measured pH(X) or pH(SWS) values.

5.6.2.4   Considerable advantages ensue from using a reference electrode design with renewable liquid junction such as the flow cell of Fig. 5.1. By this means a precision of ± 0.01 in pH is obtainable on a routine basis and standardization is only necessary at the beginning and end of a working day.

5.7 References

1 C. Culberson, in *Marine Electrochemistry*, (Eds M. Whitfield and D. Jagner), Wiley, New York, (1981).
2 J.N. Galloway, B.J. Crosby and G.E. Likens, *Limnol. Oceanogr.*, **24**, 1161, (1979).
3 N.R. McQuaker, P.D. Kluckner and D.K. Sandberg, *Environ. Sci. Technol.*, **17**, 431, (1983).
4 J. Kremer and A. Tessier, *Environ. Sci. Technol.*, **16**, 606A, (1982).
5 G. Howells, *Trans. Am. Fish. Soc.*, **111**, 779 (1982).
6 R.G. Bates, *Determination of pH, theory and practice*, Wiley, New York (1973).
7 A.K. Covington, R.G. Bates and R.A. Durst, *Pure Appl. Chem.*, **55**, 1476, (1983).
8 W.H. Beck, A.E. Bottom and A.K. Covington, *Anal. Chem.*, **40**, 505, (1968).
9 D.P. Brezinski, *Talanta*, **30**, 347, (1983).
10 R.G. Bates and E.A. Guggenheim, *Pure Appl. Chem.*, **1**, 163, (1960).
11 R.G. Bates, *CRC Crit. Rev. Anal. Chem.*, **10**, 247, (1981).
12 A.K. Covington, P.D. Whalley and W. Davison, *Analyst*, **108**, 1528, (1983).
13 R.G. Bates, *Pure Appl. Chem.*, **54**, 229, (1982).
14 I. Hansson, *Deep Sea Res.*, **20**, 479, (1973).
15 T. Almgren, D. Dyrssen and M. Strandberg, *Deep Sea Res.*, **22**, 635, (1975).
16 F.J. Millero, *Liminol. Oceanogr.*, in press (1986).
17 R.G. Bates and J.B. Macaskill, *Adv. Chem. Ser.*, 147, (1975).
18 R.W. Ramette, C.H. Culberson and R.G. Bates, *Anal. Chem.*, **49**, 833, (1977).
19 R.A. Butler, A.K. Covington and M. Whitfield, to be published.
20 M.F.G.F.C. Camoes and A.K. Covington, *Anal. Chem.*, **46**, 1547, (1974).
21 A.K. Covington and M.I.A. Ferra, *Anal. Chem.*, **49**, 1363, (1977).
22 E.A. Guggenheim, *J. Am. Chem. Soc.*, **52**, 1315, (1930).
23 J.A. Illingworth, *Biochem. J.*, **195**, 259, (1981).
24 D.P. Brezinski, *Analyst*, **108**, 425, (1983).
25 D. Midgley and K. Torrance, *Analyst*, **101**, 833, (1976).
26 A.K. Covington, P.D. Whalley and W. Davison, *Anal. Chim. Acta.*, in press.
27 *The Measurement of Electrical Conductivity and the Laboratory Determination of the pH of natural, treated and Waste Waters.* Standing Committee of Analysts, Department of the Environment, HMSO, London (1978).
28 W.F. Langelier, *J. Am. Waterworks Assn*, **38**, 179, (1946).
29 J.F. Talling, *J. Ecol.*, **64**, 79, (1976).
30 E. Pelletier and J. Lebel, *J. Fish Aquatic Sci.*, **37**, 703, (1980).
31 K.S. Johnson, R. Voll, C.A. Curtis, R.M. Pytkowicz, *Deep Sea Res.*, **24**, 915 (1977).
32 T. Almgren and S.H. Fonselius, in *Methods of Sea Water Analysis*, (Ed K. Grasshoff), Verlag Chemie, Weinheim pp. 97–115, (1976).
33 T. Takahashi, R.F. Weiss, C. Culberson, J.M. Edmond, D.E. Hammond, C.S. Wong, L. Yuan-Hui, and A.E. Bainbridge, *J. Geophys. Res.*, **75**, 7648, (1970).

Section 6 Electroanalytical Measurement of Trace Metals Complexation

By

Commission V.5 **J. BUFFLE**

IUPAC *Department of Inorganic, Analytical and*
Applied Chemistry, University of Geneva
Sciences II
30 quai E. Ansermet
1211 Genève 4
Switzerland

6.1 General Considerations

6.1.1 Objectives and stages of complexation measurements

Electrochemical techniques are powerful tools for the measurement of total concentrations of a number of metals (Cr, Mn, Fe, Co, Ni, Cu, Zn, Ag, Hg, Pb, Cd, Tl, Bi) in natural waters, which are either toxic or essential for life [1,2]. Obviously, the behaviour of a given metal, in the aquatic environment, (such as its uptake by organisms, adsorption/desorption on solid particles, or precipitation/dissolution) is not only dependent on its total concentration, but also, very much on the nature and properties of its various chemical species [2–6].

Figure 6.1 gives the order of magnitude of the concentrations of trace metals and major organic and inorganic ligands in fresh and sea water. Solid or colloidal particles are included in this list as adsorption reactions of trace metals on solid surfaces can be considered, conceptually, as a complexation reaction of the metal ion with a complexing site of the particle [5,7]. Furthermore, most of the naturally occurring complexing agents (proteins, fulvic compounds, polysaccharides, metallic oxides or hydroxides and silicates) have many complexing sites. They may even consist of many different molecules, since it is not generally possible to isolate all the ligands of the aquatic mixture (see below). Hence, in the following, the term 'complexing agent' will refer to a group of complexing molecules or particles, 'ligand' will be used for a specific complexing molecule, whereas 'complexing site' will refer to a particular coordinating group in a molecule.

From Fig. 6.1 (see also ref. [8]) it may be inferred that the variety of complexes of a given metal ion may be extremely large and that the task of determining their concentration and properties is very difficult. It is not therefore surprising that, at present, no general procedure can be given and only a few important guide-lines can be recommended to facilitate the development of methods and the interpretation of results in each particular case. Three basic approaches, hereafter called objectives, can be adopted for this purpose, as shown in Fig. 6.2. The role and relative importance of steps I to V, particularly the electroanalytical steps (III and IV), vary from one objective to the other.

Objective A (Fig. 6.2A): Measurement of the distribution of inert species

'Inert' species refer here to those species whose lifetime is large enough compared to the total duration of sampling, separation and analysis procedures (steps I to III). The concentration, $|M|_i$, of each ith inert species can be measured, after separation and possibly decomposition of the complexing agent (e.g. wet digestion), by any method enabling the determination of the total concentration of M at the desired level.

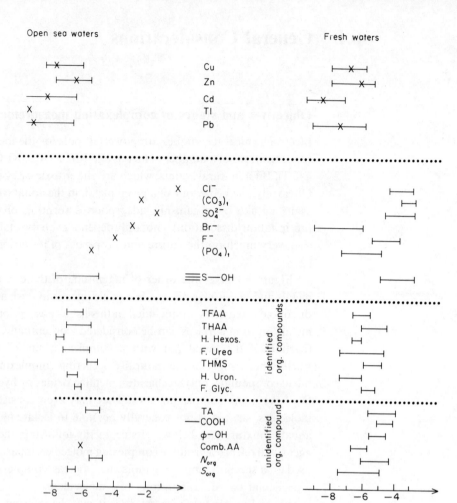

Fig. 6.1. Ranges (——) and average values (X) of reported concentrations of the electro-analytically measurable cations and of the major organic and inorganic complexing agents (from ref. [8] and [114]). The higher levels of metal concentrations are generally due to pollution. TFAA (THAA), total free (hydrolysable) aminoacids; H. Hexos, Hydrolysable Hexosamines; F. Urea, free urea; THMS, total hydrolysable monosaccharides; H. Uron, hydrolysable uronic acids; F. Glyc., free glycollic acid; TA, total acidity; –COOH, COOH groups; ϕ–OH, phenolic groups; comb. AA, combined amino acids; N_{org}, total org. nitrogen; S_{org}, total organic sulfur; 70–80% of unidentified compounds are classified as fulvic. $\equiv$S–OH, –OH site of metallic oxides.

Two types of separation processes can be distinguished:

(i) *the isolation* of well-characterized species (e.g. organo-mercurials [9], organo-tins [10]) and

(ii) *the fractionation* of groups of non-identical but homologous complexes with similar properties [11–14]:

because, in natural waters, many of the complexing agents are very complex mixtures (Fig. 6.1), most of the inert complexes cannot be isolated individually, but only fractionated into groups. In general the term 'fractionation' will be

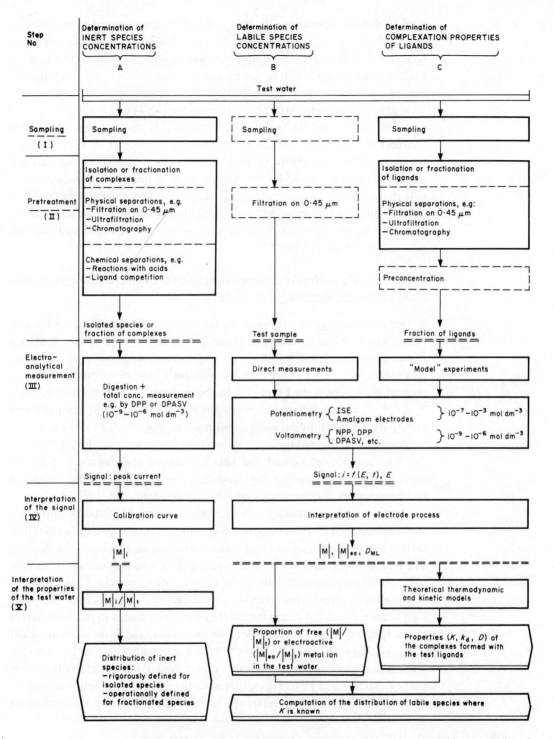

Fig. 6.2. The most important steps involved during the measurement of concentration of inert complexes (A), of free metal ion concentration in unperturbed sample (B), and of complexation properties of ligands (C) |M|, concentration of the free metal ion; $|M|_i$, concentration of the ith complex species of M or concentration of M in the ith fraction of complex species; $|M|_t$, total concentration of M in the sample. Other symbols: see list of symbols. ⌐ ⌐ ⌐ ⌐ = steps to be avoided as far as possible.

183   GENERAL CONSIDERATIONS

used to denote separation of groups of homologous species (complexes or ligands) from a mixture too complex to enable full isolation of each individual component. Fractionation methods for inert complexes utilize ultrafiltration [11], chelating resins [12], or ion exchange resins [13,14]. UV irradiation is also used [13,14] to decompose the organic complexes selectively and enable their anlysis by anodic, stripping voltammetry (ASV) techniques.

In all these cases, the measured parameter $|M|_i$ corresponds to the total concentration of M combined in the isolated species or in the separated fraction, and the conditions of application of electroanalytical methods are similar to those described for the measurement of total metal concentration, $|M|_t$ in the global water sample [1]. Undoubtedly voltammetric methods are very useful for measuring $|M|_i$ [11]–[15] thanks to their high sensitivities. Nevertheless, for Objective A, the key step is not the electrochemical one (III or IV), but the separation (II).

Objective B (Fig. 6.2B): Direct measurement of the free metal ion concentration in the test water; distribution of the labile species

In this approach, great care must be taken to avoid, as far as possible any perturbation of the medium, in order not to modify the labile equilibria between metal ions and ligands, or the interactions (coagulation, aggregation, adsorption on surfaces) between the various kinds of complexing agents themselves. Hence steps I and II must be minimized. Nevertheless, because direct *in situ* measurements are rarely possible, a few steps (like sampling and filtration) are generally necessary, and must be done with extreme care.

A typical example of this approach is the measurement of $|Fe^{2+}|$, $|Mn^{2+}|$, $|S^{2-}|$ in anoxic waters and the computation of the distribution of the corresponding labile species, by combining the experimental data with the literature values of the relevant equilibrium constants [16–18].

Another type of application is the measurement of the total complexing site concentration of waters (the so-called complexation capacity), by titrating them with the test metal ion [generally $Cu(\text{II})$], to saturation of the ligands. The measured signal is interpreted in terms of residual free metal ion and the equivalence point corresponds to the complexation capacity [19–23].

Although potentiometric methods are the simplest ones for measuring free ion concentrations, for heavy metals they can be used only in polluted waters due to their relatively low sensitivity, so that voltammetric methods are often preferred. However, in these cases, the intricate composition of the medium may make difficult the interpretation of their signals in terms of free metal ion concentrations (step IV in Fig. 6.2). Hence, any further interpretation (step V) in terms of distribution of labile species or complexation capacity must be done with great care.

Objective C (Fig. 6.2C): Measurement of the complexation properties of the complexing agents

A complementary approach to the previous one is to: (i) separate the various types of complexing agents, (ii) study their complexing properties separately, and (iii) use the corresponding parameters to recompute the distribution of the species in the initial test water.

In such an approach, great care must be used in the sampling and fractionation procedures (step I and II), in particular to avoid, as much as possible, the denaturation of the ligands, e.g. break-up of tertiary structures, induction of aggregation or coagulation. In this respect, preconcentration steps should be avoided as far as possible.

Nevertheless, compared to the previous approach, some sacrifice has to be made regarding the possible slight perturbations in the ligand's structure, in order to enable the separation of the various complexing agents and to improve the final rigour of the electroanalytical interpretation (step IV). As in the case of separation of inert complexes (Objective A) complete isolation of a pure ligand is seldom possible, but the complexing agents can be fractionated into groups of compounds having similar properties. Those can be studied in detail by means of 'titration experiments' (change in $|M|_t/|L|_t$, in pH, etc.).

Approach C may result in a stronger perturbation in the test ligands than Objective B, but in addition to an easier interpretation of electroanalytical signals, it enables one to:
- test and develop theoretical models of interactions between metal ions and natural complexing agents;
- determine the corresponding characteristic parameters (Fig. 6.3: equilibrium constants (K), kinetic constants (k_f, k_d), diffusion coefficients (D);
- use these parameters to compute the distribution of labile species in a particular water, as well as to make predictions on the biogeochemical behaviour of trace metals.

Approach C has been applied for instance, for the study of carbonato complexes of Pb(II) [24, 28], copper complexes of aquatic proteins and peptides [25], or complexes of Pb(II) with aquatic fulvic acids originating from soils [26, 27].

6.1.2 Interpretation of complexation properties by means of electroanalytical methods

The above discussion shows that electroanalytical methods can be used in two different ways for complexation studies in natural waters: (i) as sensitive detection methods after a separation procedure, or (ii) as sensitive and selective techniques enabling direct discrimination of free metal ion from its complexes. While other detection methods can be used, in case (i), the electrochemical ones have the unique advantage, in case (ii), of being capable of directly measuring the free metal ion concentration $|M|$, or better still its activity $[M]$.

This is of utmost importance as $[M]$ is often the key factor indicating the state of kinetic or thermodynamic processes regulating the trace element activity in waters [29]. This is shown schematically in Fig 6.3A, where L is a dissolved ligand and ML is the corresponding complex formed with M. This figure sums up how $[M]$ is related to important parameters in the various processes: adsorption equilibrium on solid particles (all the K_a values), adsorption equilibrium on cell surfaces (K_M^c), kinetics of biological uptake

(K_M^c), equilibrium (K) and kinetic $(k_d$ and $k_f)$ complexation constants for reactions in solution, and all diffusion coefficients of the dissolved species.

It is worth noting the analogy between processes occurring at:
- solid surfaces (Fig. 6.3A) and potentiometric electrodes (Fig. 6.3B)—in both cases the flux of M is zero;
- cell surfaces (Fig. 6.3A) and voltammetric electrodes (Fig. 6.3B)—in both cases a flux of M passes through the interface.

Because of these similarities to reactions in solution, the electroanalytical signals are dependent on the same basic parameters as those of environmental processes, namely:

$$|M|, \quad K, \quad k_d, \quad D \tag{6.1}$$

This unique capability of electroanalytical methods makes them a very powerful tool for understanding and predicting the biogeochemical role of trace metals in waters.

In the following pages the optimal conditions to be used in this type of

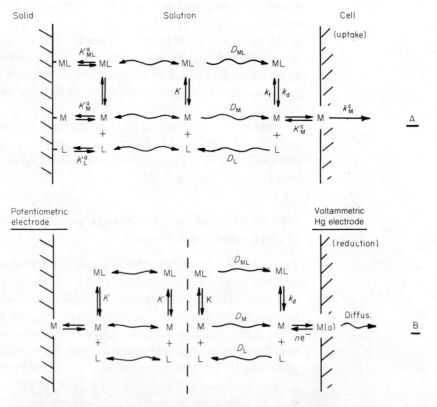

Fig. 6.3. Schematic representation of the basic reaction parameters of M, L and ML with natural surfaces (A) and electrodes (B). Scheme B represents ISE measurements (potentiometry) and reduction of M to metallic state (M(o)) on mercury (voltammetry). Similar schemes can be drawn for other potentiometric and voltammetric electrodes. K_M^a, K_L^a, K_{ML}^a, adsorption equilibrium constants on solid particles; K_M^c, adsorption equilibrium constant of M on the membrane of organism; k_M^c, rate constant of uptake of M by organism. Other symbols: see list of symbols.

application are discussed at length. The problems to be tackled can be grouped into four categories (Fig. 6.2):

- contamination and adsorption effects, as for any trace analysis measurement (Fig. 6.2, steps I, II, III);
- perturbation of labile equilibria during sampling and pretreatment (Fig. 6.2, steps I and II);
- interpretation of electrode processes (Fig. 6.2, steps III and IV);
- interpretation of reactions occurring in the bulk of test samples (step V).

These aspects together cover a wide scientific domain: trace analysis, environmental chemistry and biology, electrochemistry, colloid chemistry, physical and inorganic chemistry. Obviously all these cannot be described in detail here.

This report focusses particularly on the interpretation of the electro-analytical data (Fig. 6.2, step III–IV). Contamination and adsorption problems, as well as general methodological aspects, are similar to those discussed in detail in refs. [1], [30] and [113] for the analysis of total concentrations of metals. Only the most important aspects of sampling and pretreatment, in relation to perturbation of chemical composition, are discussed here. Similarly, a few basic concepts of the properties of polyfunctional and polyelectrolytic naturally occurring ligands are given to clarify the behaviour of the components which may affect the electroanalytical signal. For more detail, the reader should refer to specific reviews [8, 31, 36, 44, 45, 49].

The chief aim of this report is to provide environmental scientists who may not have specialized in electrochemistry, with some practical and simple guidelines to enable them to interpret experimental data. For that purpose, the basis of the relevant electroanalytical processes are briefly given, but emphasis is laid on their use in environmental conditions. Indeed, extrapolation from synthetic solution to natural water conditions is often not straightforward, as several assumptions used in the former case may not be valid in the latter. Reference to basic monographs [32–34] is recommended for detailed electroanalytical principles. A previous IUPAC report [35] should be consulted as well, for the determination of stability constants in synthetic solutions.

6.2 Basic Properties of Aquatic Complexing Agents

6.2.1 Differences between synthetic solutions and natural water conditions

In pure solution containing only one metal ion M and one ligand L, the complexation reaction [35] may be written as:

$$m\mathrm{M} + h\mathrm{H} + l\mathrm{L} \leftrightarrows \mathrm{M}_m\mathrm{H}_h\mathrm{L}_l \qquad (6.2)$$

(e.g. $\mathrm{Ca} + 2\mathrm{H} + \mathrm{Cit} \leftrightarrows \mathrm{CaH}_2\mathrm{Cit}$)

where charges are omitted, M, H and L are the hydrated species of the metal ion, proton and ligand respectively, and Cit = citrate. The thermodynamic stability constant corresponding to eqn (6.2) is defined by:

$$\beta_{m,h,l} = \frac{[M_m H_h L_l]}{[M]^m [H]^h [L]^l} \tag{6.3}$$

For a rigorous determination of stability constants one needs:
- mass balance equations [35] for the total concentration of M, $(|M|_t)$ and L, $(|L|_t)$, where

$$|M|_t = |M| + \sum_{m,h,l} m |M_m H_h L_l| \tag{6.4}$$

$$|L|_t = |L| + \sum_{m,h,l} l |M_m H_h L_l| \tag{6.4'}$$

- a charge balance equation
- measured values of $|M|$, $|H|$, and/or $|L|$
- computed values of activity coefficients to convert the concentrations ($||$) into activities ($[\ ]$).

These determinations are relatively simple, particularly when conditions are so chosen that only a few complexes are simultaneously formed in the solution, and $|L|_t \gg |M|_t$. In natural water conditions, the situation is much more difficult due to the presence of large numbers of ligands (Fig. 6.1). Three cases can be considered, however:

(a) *Waters containing a mixture of well-characterized ligands only*, the total concentrations of all of which can be measured. An example of this is mineral waters in which the only ligands present are the major inorganic anions. For this type of multiligand system, the degree of complexation, α_w, is obtained by combining eqns (6.3)–(6.4') for all ligands L_i (e.g. see ref. [51]). If there is no formation of mixed complexes one obtains:

$$\alpha_w = \frac{|M|_t}{|M|} = 1 + \sum_{i=1}^{n} \sum_{m,h,l} \left[m \frac{{}^i\beta_{m,h,l}}{{}^i\gamma_{m,h,l}} [M]^m [H]^h [L_i]^l \right] \tag{6.5}$$

where the index w refers to well-characterized ligands and ${}^i\gamma_{m,h,l}$ is the activity coefficient of the complexed species. Superscript i refers to complexes with ligand L_i. Since all the ligands are known, the values of ${}^i\beta_{m,h,l}$ can be determined as before, although the calculations are more tedious and necessitate the use of curve fitting procedures and computer programs. Furthermore, the statistical errors incurred in ${}^i\beta_{m,h,l}$ may become appreciable for a large number of ligands.

(b) *Fractionated water samples containing a group of ill-characterized but homologous compounds* including a large number of similar but non identical complexing sites L_i (Objective C in Fig. 6.2). This is the case for most organic complexing agents and many inorganic colloids. In contrast to the previous case, the following conditions are different:
(i) The total concentration of each type of complexing site $(|L_i|_t)$ is unknown.

It is generally much lower than the overall concentration of complexing sites ($|L|_t$).

(ii) In many cases (Fig. 6.1): $10 < |L|_t/|M|_t < 100$ and consequently $|L_i|_t$ is not much larger than $|M|_t$.

(iii) h [eqn (6.2)] is unknown for each complexing site.

(iv) Often the charge and nature of the complexing site is unknown, so that first, a charge balance equation cannot be written, and second, activity coefficients cannot be computed. Hence, with these ill-characterized complexing agents, only the free metal ion concentration, $|M|$, or its degree of complexation α_i, can be determined rigorously.

(c) *Unfractionated waters containing both well- and ill-characterized complexing agents* (Objective B in Fig. 6.2). In this case, the total degree of complexation, α, can be written:

$$\alpha = \alpha_w + \alpha_i - 1$$

As before, only α or $|M|$ can be rigorously determined.

In the last two cases, the interpretation of $|M|$ or α in terms of physico-chemical parameters (K, k_d), it is often useful to obtain more insight into the nature and properties of the complexing agents or to make predictions about the behaviour of M in aquatic media. However, all such interpretations are necessarily based on important assumptions so that the physical meaning of the parameters thus obtained must be taken with caution. These assumptions are briefly discussed below. (For further details see refs. [8, 31, 36–39]).

6.2.2 Interpretation of physico-chemical properties of fractionated groups of ligands

6.2.2.1 Formation of 1/1 complexes

Since the complexity of the test medium often hinders more detailed interpretation, 1/1 ML_i are often assumed to be formed between M and each site L_i. Therefore $m = l = 1$ [eqn (6.2, 6.3)]. Furthermore, because of conditions (iii)–(iv) (section 6.2.1.6), only apparent equilibrium constants:

$$K_i = \frac{|ML_i|}{|M||L_i|} \tag{6.6}$$

valid for constant ionic strength and pH, are generally computed [ML_i is the equivalent of $M_mH_hL_l$ in eqn (6.3) and will be used instead hereafter]. It must be mentioned that attempts have been made to take into account other types of reactions, for instance (i) the dependence of K_i on pH for metallic oxides [39,40] or fulvic and humic compounds [31,37,38], (ii) the formation of ML_2 complexes with sites of metallic oxides [39] or fulvic compounds [41,42], or (iii) mixed complexes [43].

6.2.2.2 Distribution spectrum of complexation properties

Gamble *et al.* [31, 44, 45] were the first to suggest, for fulvic compounds, that they should be considered as a mixture including an infinity of sites with different binding energy. Each site being present at an infinitely small concentration: $|L_i| = d|L| \rightarrow 0$ and, consequently $|ML_i| = d|ML| \rightarrow 0$. This concept also is applicable to other aquatic groups of ligands, in particular polysaccharides, proteins or solid particles [40], all of which are characterized by a polyfunctional and a polyelectrolytic nature. In such a case, when the complexing agent is titrated with M, the equilibrium constant does not take a finite number of discontinuous values, but transforms to a continuous so-called 'differential equilibrium function'

$$K = d|ML|/|M|d|L| \tag{6.7}$$

Obviously K decreases when $|M|_t/|L|_t$ increases, particularly due to (i) the large number of different sites, the strongest being first saturated, and (ii) the high electric (generally negative) charge which decreases with saturation of L. This change in K with $|M|_t/|L|_t$ is of paramount importance because:
(a) it is a key factor in the complexing properties of natural macromolecular and colloidal ligands (FA, HA, polysaccharides, proteins, metallic oxides), and
(b) it constitutes a basic difference between these ligands and 'simple' small ligands such as Cl^-, CO_3^{2-}, amino-acids and NTA. The mode of computation of the function K is beyond the scope of this report (see ref. [8, 31, 45]).

The change in K with $|M|_t/|L|_t$ can also be represented by a 'distribution spectrum' [36] (Fig. 6.4):

$$dN_{ML}/d\log K = f(\log K) \tag{6.8}$$

where dN_{ML} is the number of moles of M complexed to the sites, the binding constants of which are between $\log K$ and $(\log K + d\log K)$. Several theoretical models have been proposed to explain the observed distribution function [37, 48], but the important point is that the value of K may vary over

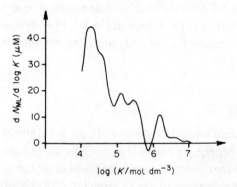

Fig. 6.4. Distribution spectrum of equilibrium complexation properties of Cu(II)–fulvic acid systems (adapted from ref. [36]): dN_{ML}, number of moles of ML whose the value of equilibrium constant K is between K and $K + d\log K$. Fulvic acid from Ogeechee Estuary. Molecular size 1.3–3.1 nm; pH = 6.0.

several orders of magnitude with $|M|_t/|L|_t$ (Fig. 6.4) [8, 47], even for fractionated groups of ligands.

6.2.2.3 Distribution spectrum of other properties

The above concept is not restricted to equilibrium complexation reactions. For instance it has been widely used to represent the size distribution of suspended particles [49]. Two properties are particularly relevant for the application of electroanalytical methods:

(i) the size and/or diffusion coefficient of complexing agents. Their size ranges between a few Å (Cl^-, CO_3^{2-}) and several microns (oxide and silicate particles). The distribution of several important aquatic complexing agents is given in Fig. 6.6.

(ii) the dissociation rate of complexes. An example of distribution is shown in Fig. 6.5, for the complexation of Cu(II) by fulvic compounds.

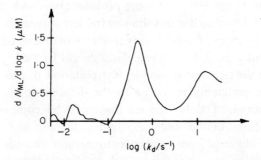

Fig. 6.5. Distribution spectrum of kinetic complexation properties of Cu(II)–fulvic acid system (adapted from ref. [36]). k_d, first order dissociation rate constant of the complexes. Fulvic acid from Ogeechee Estuary. Molecular size 1.3–3.1 nm; pH = 7.5. The integral under the curve represents 23% of the total number of sites.

Due to the scarcity of information presently available, the above figures must only be considered as illustrative of the range variation of D and k_d values in aquatic media. It is seen that they are very wide and it is to be expected that these distribution spectra may have an important influence on the measured electroanalytical fluxes, i.e. on the measured current.

6.2.3 Experimental applications

From the above discussion it is obvious that, with aquatic complexing agents, the averaging of K or k_d values measured at various $|M|_t/|L|_t$ ratios will be less meaningful for the more inefficient of the fractionation procedures. In practice two kinds of average equilibrium parameters, $\bar{K}$ and $\tilde{K}$, are often used. They are based on the assumption that all the complexing sites in the medium are equivalent, their total concentration being $|L|_t$.

The average equilibrium function, $\bar{K}$, is directly derived from eqn (6.6):

$$\bar{K} = \frac{|ML|}{|M| \cdot |L|} = \frac{\sum_i |ML_i|}{|M| \sum_i |L_i|} = \frac{|M|_t - |M|}{|M| \cdot (|L|_t - |M|_t + |M|)} \tag{6.9}$$

For an infinite number of non-equivalent sites, K changes with $|M|_t/|L|_t$, so that $\bar{K}$ also decreases with increase in this ratio.

The average equilibrium quotient, $\tilde{K}$, is defined as the average value of $\bar{K}$ between the two experimental limits of $|M|_t/|L|_t$. For a group of ligands, it is a weighted average value which depends primarily on the choice of the above limits, and in particular on the sensitivity of the method used. For instance, the log $\tilde{K}$ values computed from the data in Fig. 6.4 could vary between 4 and 7 depending on the chosen limits. Hence $\bar{K}$ and $\tilde{K}$ should be used with great care. It is beyond the scope of this paper to discuss the exact relationships between K, $\bar{K}$ and $\tilde{K}$ [8, 31, 47], but it may be shown that the values of $\bar{K}$ and $\tilde{K}$ are close to K only when $|L|_t \sim 2(|M|_t - |M|)$.

Obviously the same averaging problem arises with the other properties (k_d and D). In particular in voltammetric techniques, the measured flux is a weighted average of fluxes of many species with different diffusion coefficients. Consequently the following guidelines should be used:

- On the unfractionated sample, interpretation of complexation measurements are preferably restricted to the determination of $|M|$ and to the concentrations of those species that may be computed from $|M|$, the concentrations of the well-characterized ligands and their complexation constants obtained from independent measurements [50, 51].
- On fractionated samples K and k_d distribution functions should be determined instead of average parameters. It must be noted that, for application of voltammetric methods to these measurements, size fractionation of the sample is particularly appropriate, as it enables us to work with groups of compounds having similar diffusion coefficient values. Simultaneously, Fig. 6.6 suggests that a certain homogeneity in chemical nature can also be obtained, inside a given fraction, by choosing the corresponding size cut-off limits correctly.
- As far as possible, complexation data should be checked by at least two independent methods [52] and/or by 'titration experiments'. For well-characterized ligands, the latter consist of a step-by-step reconstitution of the test medium, by monitoring the influence of successively added components on the complexation parameters [53, 54]. For ill-characterized complexing agents, this is done by following the change in complexation parameters which occur by varying the experimental conditions (such as pH, $|L|_t$, ionic strength, competing cations [41, 55].)

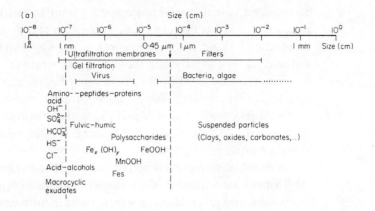

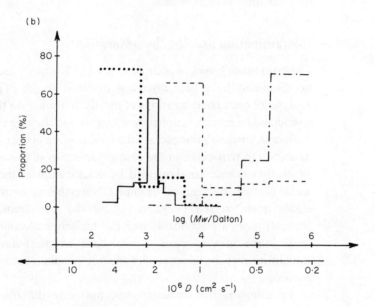

Fig. 6.6. Molecular weight (Mw) distribution of the most important aquatic organic ligands. The values on the diffusion coefficient scale are only order of magnitudes.
(a) General scheme adapted from ref. [5; chap. 6, and 29], (b) Distribution of macromolecular "dissolved" compounds (for detailed references see ref. [8]): . . . . , sea water fulvic acids; —, soil water fulvic acids; ----, fresh water peptides+proteins; —·—, fresh water polysaccharides.

6.3 Sampling and Pretreatment

Sampling and handling must be tailored to the particular problem under study. For Objective B (Fig. 6.2) it is very important to use all the conditions which help to preserve the sample in its natural unperturbed state. In particular, whenever possible, *in situ* measurements are preferable. For Objective C, the nature of the water, the location and period of sampling are particularly important in relation to further interpretation of complexation data. Indeed, environmental processes may result in a natural fractionation of aquatic components that can be used profitably to collect samples with fairly simple composition. This is sometimes a neater way than laboratory methods and may at least greatly ease the subsequent treatment of the sample. For instance, organic matter from soils, in water bodies, can be more selectively sampled in autumn or winter when primary productivity is minimal. Similarly, selective sampling of particulate and dissolved complexing agents, in an estuary, may be based on the change in their relative concentrations along the estuary.

An excellent review on sampling and handling has been written by Riley [58], for the measurement of total concentrations of organic and inorganic compounds and gases. Many aspects covered in his review are also useful for complexation measurements.

6.3.1 · Contamination and loss by adsorption

The pronounced risks of poor accuracy due to contamination or adsorption on containers have been discussed in detail in refs [1] and [30], for the analysis of total concentrations of metals. Not only do they also have to be considered for complexation measurements, but they are even more difficult to overcome, since, in principle, as little modification of the medium as possible is desirable. In particular, in the case of Objective B, losses of trace metals by adsorption cannot be eliminated by acidification of the sample, since this would perturb the chemical equilibria. Nevertheless, most of the recommendations made in refs [1, 30, 56, 57] for the measurement of total metal concentrations are applicable here. Particularly, collection of the samples has to be done in very clean polyethylene or plexi-glass flasks, and pre-equilibration of the labware with the test sample is recommended whenever possible.

In complexation measurements, not only the trace metals but also dissolved organic carbon (DOC) contamination must be avoided. This is essential, as many of the ligands under investigation are organic compounds, and DOC is generally used as the analytical parameter for measuring their concentration. Any DOC contamination may cause an error in the measurement of $|L|_t$, and in turn in the analysis of complexation data, even if these contaminations are non-complexing in nature. Sampling in such cases should be done in carefully washed polyethylene flasks whose release of DOC is generally below the detection limit (< 0.1 mg dm^{-3}). Other possible sources of trace metals or DOC contamination are the water used for preparing standards or electrolyte solutions. Suprapur, or equivalent, electrolytes should be used, but sometimes, even these require further purification, e.g. by electrolysis using a mercury cathode, for removing trace metal contamination and UV irradiation with a mercury lamp [1], for minimizing DOC contamination.

6.3.2 Collection of samples

The choice of the sampling procedure strongly depends on the nature of the problem to be solved. For Objective C, the methods described in ref. [1] for oxygenated waters, in surface or at low depth, are generally applicable [58]. The use of a peristaltic pump is not recommended unless contamination from the tubing has been carefully tested. Silicone tubing produces high DOC contaminations compared to Tygon tubing. It is also necessary to ensure that the pump system does not damage the particulate matter or plankton. To minimize contamination from the container walls it is best to collect a large sample (1 dm^3 or preferably more).

Very great caution must be taken when the water is sampled for Objective B, particularly in deep and/or in anoxic waters, to avoid contamination by oxygen or degassing of H_2S or CO_2, due to pressure changes. Applications of electroanalytical methods have been studied in detail, for such cases, e.g. by Davison [17, 59] and Zali [16]. In anoxic waters containing Fe(II) or Mn(II), their oxidation by traces of oxygen can produce enough very reactive colloidal FeOOH or MnO_2 to modify the speciation of trace elements strongly. A change in CO_2 concentration may, of course, influence all acid–base and complexation equilibria. The problem of CO_2 is not restricted to deep waters: surface oxic waters can be either depleted or supersaturated with respect to CO_2, due to photosynthesis or respiration. Sample collection in sealed borosilicate glass or perspex systems is essential in such cases. Immediate field filtration on 0.45 or 0.2 μm filters is also necessary to eliminate most living organisms which, otherwise, would continue to produce or consume CO_2 inside the sampling bottle.

6.3.3 Storage

Storage of the sample must be reduced to the minimum time, particularly for Objective B. The sample filtered on 0.45 or 0.2 μm filters should be kept in the dark, at the same temperature as that prevailing *in situ*. Changes in temperature may facilitate the above mentioned precipitation processes. Samples should not be frozen, because of the resulting modifications in the structure and properties of macromolecular and colloidal ligands. Storage of anoxic samples must be done in air-tight, carefully sealed systems, due to rapid oxidation of H_2S and Fe^{2+} by even traces of oxygen . Glass or plastic bottles must be used in such cases. It is also well known that *ortho*-phosphate is easily lost from samples if particulate matter is not removed and when samples are stored in polyethylene or polyvinylchloride containers [58].

6.3.4 Pretreatment and fractionation

As mentioned before, all pretreatment and complexation determinations must be performed as soon as possible after sampling. This is particularly important in fresh waters where many strong ligands are present in a metastable colloidal state, or where supersaturation with respect to precipitation is prevalent, e.g.

ferrous sulfide in anoxic waters or calcium carbonate in eutrophic surface waters.

6.3.4.1 Separation of particulate from soluble components

Because of the scavenging role of living and non-living particulate matter in unmodified waters [60], its separation from soluble components should be performed immediately after sampling. Membranes with well defined 0.45 μm pore size are usually adopted for this purpose, but 0.2 μm can also be used. Before filtration of large volumes of samples on these membranes, it is recommended, particularly for waters with large content of particulate matter, that the samples be filtered on a more porous membrane, e.g. 8 μm, in order to minimize the risks of clogging. This last effect can be checked by controlling the flow-rate of the filtrate which should not drop below that of pure solvent by $> 20\%$. Detailed comparison of the actual and nominal pore sizes of various filters is recommended [58].

Before use, membranes should be carefully washed to avoid leaching of trace metals and DOC [24, 61, 62], by filtering acid solutions, e.g. 5×10^{-3} mol dm^{-3} HCl. A minimum volume of 20 cm^3 cm^{-2} is necessary to bring contamination of DOC and Fe down to < 0.2 mg dm^{-3} and 10^{-7} mol dm^{-3} respectively. Simply soaking the membrane in deionized water is highly inadequate. The cleaning operation can also be advantageously used to measure the flow-rate of the solvent through the membrane. Whenever possible, after washing, preconditioning of the membrane is recommended in order to decrease as much as possible the risks of adsorption of trace elements. This can be done by filtering a small volume of prefiltered sample or of synthetic solution which has a similar composition of major ions to the sample. Furthermore, the first fraction of filtrate of the test sample (~ 1 cm^3 cm^{-2} of membrane) must be discarded.

The filtrate thus obtained can be used either for direct determination of $|M|$ (Objective B) or further fractionation (Objective C). Separation of particulate matter can also be performed by centrifugation, for avoiding contamination from the membrane or adsorption on it. However, centrifugation may produce a scavenging of colloidal organic matter by sedimenting particles. Above all, for the interpretation of complexation properties, a clear-cut fractionation is important (Section 6.2) and is more easily obtained by filtration.

6.3.4.2 Fractionation of soluble components

The choice of fractionation procedure largely depends on the nature of the sample and the purpose of complexation measurements, but any condition minimizing changes in physical (e.g. temperature) or chemical conditions (e.g. introduction of reagents) must be preferred. In this respect, physical methods of fractionation, e.g., dialysis [61], gel filtration [63, 64], adsorption on resins [4, 65] and membrane ultrafiltration (organic compounds [11, 66]; metallic oxides [115]) are generally less perturbing than chemical ones. Ultrafiltration is often favourable, particularly because adsorption problems are minimized, due to the fairly small ratio of surface of membrane/volume of solution, and to the fact that chemical composition inside the ultrafiltration cell can be carefully controlled. As mentioned in Section 6.2.3, it is particularly useful for further voltammetric measurements.

The chemical fractionation of complexes is generally based on their ease of dissociation. Various procedures have been proposed [12–15, 20, 67] for stepwise dissociation of complexes of increasing stability followed by the determination of the concentrations of the metal ions released. This type of approach gives a general idea of the reactivity of the metal ion initially present in the test sample, although the fractions obtained do not necessarily correspond to representative groups in natural waters.

6.3.5 Concentration techniques

Preconcentration is only performed on fractionated complexing agents, i.e. for Objective C, and must be avoided for Objective B, particularly because precipitation reactions, e.g. that of $CaCO_3$ or coagulation of colloidal components, may occur due to increase in salt concentration [68]. Even for a given fraction, the concentration step must be limited as much as possible, as it may cause non equilibrium or slowly reversible processes such as aggregation of macromolecules. Amongst concentration methods, evaporation under reduced pressure at room temperature, or freezing concentration techniques are the least perturbing and applicable to large initial volumes (10–100 dm^3). For < 10 dm^3, ultrafiltration can be used for simultaneous fractionation and concentration of the sample. Extraction procedures may also be useful in particular cases, although many aquatic ligands have hydrophilic properties and are difficult to extract with non-polar solvents.

6.4 Complexation Measurements by Potentiometric Techniques at Zero Current

The potential E of ion-selective (ISE) or amalgam electrodes may serve as direct measurements of the free metal ion activity, [M] of the cation to which the electrode is sensitive. The theoretical background of the relation between E and [M] and its analytical applications is discussed in detail elsewhere [69]–[71]. Only a few general considerations and some specific applications to complexation in natural waters are discussed here.

6.4.1 Principle of complexation measurements

For both ISE and amalgam electrodes, E is related to [M] by the Nernst equation:

$$E = E_0 + s \log [M] \tag{6.10}$$

where E_0 and s are constants characteristic of the potentiometric system used. For ideal Nernstian behaviour: $s = 2.303\,RT/nF$ where n is the charge of M for ISE or the number of exchanged electrons for amalgam electrodes. For the latter, E_0 is a function of the standard redox potential of the M/M(o) couple [M(o)=metal atom] and of the concentration of M(o) in the amalgam. [M] can be written as:

$$[M] = |M|_t\,\gamma_M/\alpha \tag{6.11}$$

where γ_M = activity coefficient of M and $\alpha = |M|_t/|M|$. Hence, combining eqn (6.10) and (6.11), α or $|M|$ can be directly determined from:

$$E = E'_0 + s\,\log|M| \tag{6.12}$$

or

$$E = E'_0 + s\,\log|M|_t - s\,\log\alpha \tag{6.12'}$$

where $E'_0 = E_0 + s\,\log\gamma_M$ is constant provided that γ_M is kept constant during the experiment. Although this last condition is not essential, its fulfilment enables more reliable data to be obtained in practical cases. Figure 6.7 shows a calibration curve obtained in non-complexing medium [eqn (6.12) or (6.12') with $\alpha = 1$; curve no. 1] and a 'complexation' curve, obtained in complexing solution [eqn (6.12'); curve no. 2]. This last case is an example of the titration of a sample at constant ligand concentration ($|L|_t$) and varying $|M|_t$. At any point of the titration curve, α can be computed from eqn (6.12') and from the difference in potential, $\Delta E = E_1 - E_2$, between the calibration and the

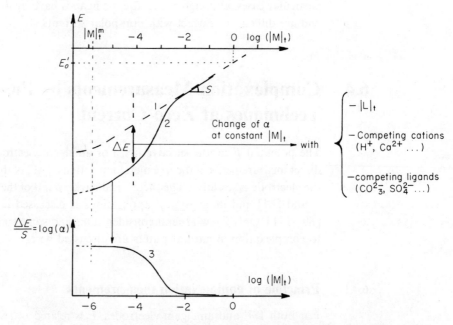

Fig. 6.7. Schematic representation of potentiometric measurement of $\alpha = |M|_t/|M|$ and of the titration experiments which are commonly used to determine the equilibrium complexation properties of aquatic complexing agents (see text for details). $|M|_t$ in $mol\,dm^{-3}$.

complexation curves:

$$\log \alpha = \Delta E/s = f(|M|_t). \tag{6.13}$$

This gives curve no. 3 (Fig. 6.7): when $|M|_t$ increases, more and more binding sites are occupied, so that $\log \alpha \to 0$ and $\Delta E \to 0$.

Obviously, $\log \alpha$ can also be measured by a single point determination in the unperturbed sample, provided the total concentration, $|M|_t$, is greater than the sensitivity limit, $|M|_t^m$ of the electrode used. For most metal ISEs, $|M|_t^m$ is not lower than 10^{-6} mol dm^{-3} (for exception see ref. [72]). A slightly lower level can be attained with amalgam electrodes, e.g., Pb, Cd, Zn [71]. Not only metals, but also some inorganic ligands can be measured by means of ISE, the sensitivity limits of which are as follows:

$F^-:|F|_t^m = 10^{-7}$ mol dm^{-3}; $I^-:|I|_t^m = 10^{-7}$ mol dm^{-3};

$Br^-:|Br|_t^m = 10^{-5}$ mol dm^{-3}; $Cl^-:|Cl|_t^m = 10^{-4}$ mol dm^{-3};

$CN^-:|CN|_t^m = 10^{-6}$ mol dm^{-3};

$S^{2-}:|S|_t^m = 10^{-7}$ mol dm^{-3}; $OH^-:0 < pH < 14$.

It must be emphasized that the above mentioned sensitivity limits refer to total concentrations, but that much lower free metal ion (or ligand) concentrations can be measured accurately, provided that $|M|_t > |M|_t^m$. Nevertheless, this last condition mostly limits the use of potentiometric electrodes to polluted waters or Objective C studies, particularly in cases where preconcentration is possible. In these latter, potentiometric techniques enable an easy recording of titration data, not only of L by M, but also for other complementary experiments (Fig. 6.7), where $|M|_t$ or sometimes $|M|$, is kept constant. For instance [41]:

● varying $|L|_t$ gives indications of the M/L stoichiometry
● titration with competing cations (in particular H^+) at constant $|L|_t$ and $|M|_t$ is important to assess their exact role in natural waters
● titration by other ligands may give information on the possible formation of mixed complexes.

6.4.2 Optimum conditions for potentiometric measurements

Compared with voltammetric signals, fewer factors affect potentiometric measurements so that their interpretation is easier, particularly in systems containing adsorbable compounds or where the ligands are not in excess. The two most important factors influencing the accuracy and precision of α or $|M|$ are: the precision of the analytical measurement of E and the constancy and correctness of the values of E'_0 and s used in eqn (6.12) or (6.12′).

(a) Statistical errors in $|M|$ and α

From eqn (6.12), it may be shown that, for given values of E'_0 and s, the error incurred in $|M|$, $\delta|M|$, is related to that in E, δE, by:

$$\delta|M|/|M| = 2.303 \, \delta E/s.$$

For divalent metal ions ($s = 29.5\,\text{mV}$ at $T = 25°C$), $\delta|M|/|M| = 7.8\%$ for $\delta E = 1\,\text{mV}$. The relative error in α is identical to that in $|M|$. Hence the accuracy of the measured mV must be ascertained by using instruments capable of measuring potentials to 0.1 mV. The apparent stability of the value displayed by the potentiometer may be an important source of error, when the potential is drifting very slowly over a long period of time. Therefore, the stability of the potential must be monitored by recording potential–time curves and the steady potential is noted when the drift is $< 0.02\,\text{mV}\,\text{min}^{-1}$. Furthermore, a minimum electrolyte concentration of $10^{-3}\,\text{mol}\,\text{dm}^{-3}$ is generally necessary for sufficient stability of potentiometric measurements.

(b) Constancies of s and E_0'

These are best obtained in the following way. The titration cell must be carefully thermostatted ($\pm 0.1°C$). Either saturated silver chloride or calomel electrodes can be used as reference electrodes, but a salt bridge with the same electrolyte as that of the test solution must be used, to avoid leakage of Cl^-. The stability of the junction potential may be checked, during the course of the experiment, by recording the E versus time curves (see above), and by calibrating the electrode before *and* after the titration. In the latter case, E_0' and s values must remain unchanged. Stable values of the junction potential, γ_M and hence E_0' are also favoured by the use of a medium of constant ionic strength. In some cases the natural electrolyte concentration of the sample may be sufficient, otherwise an electrolyte is added to the solution. $NaClO_4$ is preferred to KNO_3 to avoid any formation of ion pairs of M with NO_3^- [73].

(c) Determination of s and E_0' values

This is preferably done as follows [41, 70, 71, 108]:
• directly from the complexation curve, when L is titrated with M. The portion of the curve corresponding to complete saturation of L ($50 < |M|_t/|L|_t < 500$) is used to compute s and E_0'
• from calibration curves recorded just before and after complexation measurements for other types of determination. Changes in E_0' and s should be < 1.0 and 0.1 mV, respectively. Mean values are then introduced in eqn (6.13).

(d) Nernstian behaviour

The validity of complexation data can be ensured only when the response of the electrode is Nernstian. Hence eqn (6.10) must be obeyed with $s = 29.5 \pm 1.5\,\text{mV}$ at 25°C, for divalent cations. The value of s should be checked from eqn (6.12), in two ways;
• from a calibration curve in uncomplexing medium (Fig. 6.7) and
• by keeping $|M|_t$ constant and varying $|M|$ by means of known complexing agents or metal ion buffers [74].

(e) Interferences

Conditions should be chosen in such a way that the electrode does not respond significantly to other cations. Sufficiently accurate corrections for interference effects by using selectivity coefficients are often not possible due to their changes with experimental conditions. Malfunctioning of the electrode, in particular adsorption problems, is often indicated by rather low values of s, too high values of $|M|_t^m$ and sluggish response time. Cl^- may strongly affect the response of the Cu–ISE [75, 76].

6.5 Complexation Measurements by Voltammetric Methods

6.5.1 Principles of application of voltammetric techniques

6.5.1.1 General considerations

For the sake of simplicity the term voltammetric will be used synonymously for both polarographic and voltammetric methods. This type of electrochemical technique must be chosen for unconcentrated and unpolluted samples because of the generally low levels of electroactive metals $(10^{-10} < |M|_t < 10^{-6}\ mol\ dm^{-3})$ and ligands $(10^{-8} < |L|_t < 10^{-5}\ mol\ dm^{-3}$ with exception of the major anions). The general principles of complexation studies by voltammetric methods are described in several monographs [32–34] and the reader should refer to these for detailed theoretical discussion, and to the review by Nancollas and Tomson [35] for guidelines on complexation measurements in pure solutions of synthetic ligands forming labile complexes with reversibly reduced metal ions.

(a) Schematic classification of electrode processes

Figure 6.8a summarizes the most important phenomena that may influence electrochemical signals in environmental samples:

1 The diffusion of ML, M and L (D_{ML}, D_M, D_L).

Since many aquatic ligands are large molecules (Fig. 6.6), frequently $D_{ML} \sim D_L \ll D_M$. Furthermore, complexing agents consisting of a group of homologous compounds are not characterized by a well-defined value of D_L, but by a corresponding distribution spectrum (Section 6.2.2). Consequently, the overall diffusion rate of the complex species is a weighted average value which depends on the $|M|_t/|L|_t$ ratio. Only for size-fractionated samples can an average value of D_{ML} be considered as meaningful.

2 Thermodynamics of complexation reaction (K) and kinetics of ML dissociation (k_d).

As in the above case, groups of homologous ligands cannot be characterized by unique values of K and k_d but by the corresponding distribution spectra.

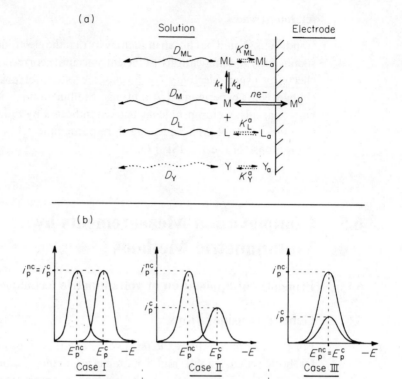

Fig. 6.8. Schematic classification of voltammetric electrode processes in natural waters.
(a) General scheme of the most important reactions at the electrode surface. Only the full line reactions are assumed to be present in discussion of sections 6.5.1.2–6.5.1.3.
(b) Schematic classification of DPP or DPASV results for: chemically labile complexes with $D_{ML} = D_M$(I) and $D_{ML} < D_M$ (including $D_M = 0$) (IIa) chemically inert complexes (III), and quasi-labile complexes (IIb). Assumptions: reversible charge transfer, no adsorption effects; $|L|_{t,0}/|M|_{t,0} > 10$. Superscript c and nc refer to complexing and non-complexing media, respectively.

3 Kinetics of reduction of M (and/or ML)

4 Adsorption of L and formation of adsorbed complexes, ML.

This is found to occur with most organic ligands and several inorganic ones, such as colloidal iron hydroxyde and sulfide.

5 Adsorption of non complexing compounds, Y.

This can occur with possible modification of the redox process of M.

Owing to these complexities, in general, only systems where M is reversibly reduced and ML is electroinactive or reduced at much more negative potentials (point **3**) are accessible for interpretation. Adsorption effects (points **4** and **5**) are often undesirable and hence conditions must be chosen to minimize them. Hence any interpretation of voltammetric data should be based on preliminary tests to discriminate between the relative importance of processes **1** to **5**. Some of the most useful ones are described in Section 6.5.2.

The range of the possible distribution of K, k_d or D values, particularly the latter, must be minimized as far as possible by appropriate fractionation and must be taken into account in the models used to: (i) transform the electrochemical signals (i, E) into $|M|$ or α values, and then (ii) interpret $|M|$ and α in terms of K or k_d. Aspect (ii) was briefly discussed in Section 6.2. The importance of point (i), which is considered below, has been stressed by recent controversies [77]–[80] which show the necessity to specify clearly the basic assumptions made in the various interpretation models.

(b) Basic assumptions for voltammetric interpretation of data

According to the above discussion, the following conditions must generally be fulfilled for a simple interpretation of the data (Fig. 6.8a).
 (i) M is reversibly reduced,
 (ii) ML is not reduced nor adsorbed in the studied potential range,
(iii) L and Y are not adsorbed,
(iv) $\alpha = |M|_t/|M|$, in the bulk of the solution, does not change during the time scale of the voltammetry. This is obviously the case when a fast equilibrium exists between M and ML, but it may also be true, even in the opposite case, provided the reaction is very slow.

Figure 6.8b represents a set of typical DPP curves (I to III) that may be obtained with systems satisfying the above conditions. The change in peak parameters (E_p, i_p) would be similar for other techniques such as ASV, DPASV, cyclic voltammetry and ac polarography. These cases have been discussed in detail [32–34] for well-characterized ligands. The corresponding basic equations for dc polarography and formation of a well-characterized 1/1 ML complex are given below as an introduction to facilitate further discussion of the application of voltammetric techniques to aquatic conditions.

When conditions (i)–(iii) above are valid, the degree of complexation, $\alpha_0 = |M|_{t,0}/|M|_0$, at the electrode surface (subscript 0), is related to the electrode potential, E, and the measured current, i, through several alternatives.

The first is the Nernst equation for the M/M(0) couple:

$$E = E_0 + \frac{RT}{nF} \ln \left[\frac{|M|_0}{|M(o)|_0} \right] = E_0 + \frac{RT}{nF} \ln \left[\frac{|M|_{t,o}/\alpha_0}{|M(o)|_0} \right] \tag{6.14}$$

where E_0 is the standard redox potential, n is the number of exchanged electrons, R, T and F have their usual meaning, and activities of M and M(o) are replaced by the corresponding concentrations. Note that α_0 is defined here only as an analytical parameter: it does not imply that equilibrium is achieved between M and ML at the electrode surface.

The second is the Ilkovič equation for the diffusion of M(o) inside the electrode:

$$i = K_1 \sqrt{D_{M(o)}} \, |M(o)|_0 \tag{6.15}$$

where $K_1 = 0.73 \, nF \cdot m^{2/3} \cdot t_d^{1/6}$, m = flow rate of mercury (mg s^{-1}), t_d = drop time.

The last alternative is the equation for the diffusion of M and ML, from the bulk of the solution to the solution side of the electrode surface. The form of this equation depends on the dissociation rate of the ML complexes (see Section 6.5.1.2–6.5.1.4).

In non-complexing media (superscript nc) containing neither L or ML, the total concentration of M, $|M|_t$, is equal to that of the free ion, $|M|$, and the Ilkovič equation for the diffusion in solution is:

$$i = K_1 \sqrt{D_M} \, (|M| - |M|_0) \tag{6.16}$$

Combination of eqn (6.14) to (6.16) gives:

$$E = E_{1/2} + \frac{RT}{nF} \ln \left[\frac{i_1^{nc} - i}{i} \right] \tag{6.17}$$

where $\quad E_{1/2}^{nc} = E_0 + \frac{RT}{nF} \ln \sqrt{D_{M(O)}/D_M} \tag{6.18}$

and $\quad i_1^{nc} = K_1 \sqrt{D_M} \, |M|_t = K_1 \sqrt{D_M} \, |M| \tag{6.19}$

$E_{1/2}$ and i_1 are the half-wave potential and the limiting current of the dc polarographic wave, respectively.

6.5.1.2   Chemically labile complexes (Cases I and IIa in Fig. 6.8b)

(a) Well-characterized ligand L in pure solution

Labile complexes are defined here as those for which the formation or dissociation reactions of ML are so fast ($k_d \to \infty$, Fig. 6.8a) that the equilibrium condition is always fulfilled at the electrode surface, despite the flux due to the reduction of M. It may be shown [32, 46, 85] that such a condition applies when

$$k_d \gg K \, |L|/t_d \tag{6.20}$$

This condition is applicable to other voltammetric techniques, by replacing t_d by the relevant time scale of the technique, e.g. pulse duration in DPP or NPP.

When a labile complex is formed, the diffusion of the free and complexed forms are not independent, because of the fast equilibrium between them. Instead, it may be considered that a given ion M passes part of its diffusion time in the ML form, and the other part in the free form. Hence, the overall diffusion of M + ML is governed by an average diffusion coefficient [32]:

$$\bar{D} = D_{ML} \frac{|ML|}{|M|_t} + D_M \frac{|M|}{|M|_t} \tag{6.21}$$

$$= D_{ML} \frac{\alpha - 1}{\alpha} + D_M \frac{1}{\alpha} \tag{6.22}$$

where α must be replaced by α_0 at the electrode surface.

The Ilkovič equation for the diffusion in solution is:

$$i = K_1 \sqrt{\bar{D}} \left(|M|_t - |M|_{t,0} \right) \tag{6.23}$$

The equation of the current–potential curve is obtained by combining eqn (6.14)–(6.15) with eqn (6.23). But a simple treatment is possible only when the following condition is fulfilled:

$$|L|_t > 10 |M|_t \tag{6.24}$$

Under this condition, the amount of ligand liberated at the electrode surface by the reduction of the 1/1 ML complex is negligible compared to $|L|_t$, and $\alpha \simeq \alpha_0$. Condition (6.24) is even more severe for ML_n complexes [32]. Eqn (6.22) can then be introduced into eqn (6.23). Combining it with eqn (6.14)–(6.15) and (6.19) gives:

$$E = E_{1/2}^c + \frac{RT}{nF} \ln \left[\frac{i_1^c - i}{i} \right] \tag{6.25}$$

with:

$$i_1^c = K_1 \sqrt{D_{ML} \frac{\alpha - 1}{\alpha} + D_M \frac{1}{\alpha}} \, |M_t|$$

or

$$\frac{i_1^c}{i_1^{nc}} = \sqrt{\frac{1}{\alpha} + \frac{D_{ML}}{D_M} \cdot \frac{\alpha - 1}{\alpha}} \tag{6.26}$$

and

$$E_{1/2}^c = E_0 + \frac{RT}{nF} \ln \sqrt{\frac{\alpha D_{M(O)}}{D_M + D_{ML}(\alpha - 1)}} - \frac{RT}{nF} \ln \alpha \tag{6.27}$$

Finally, by combining eqns (6.18), (6.26) and (6.27), one obtains the DeFord and Hume equation ([82]; Fig. 6.9):

$$\ln \alpha = \frac{nF}{RT} (E_{1/2}^{nc} - E_{1/2}^c) + \ln(i_1^{nc}/i_1^c) \tag{6.28}$$

Equation (6.28) is applicable to other voltammetric methods (in particular DPP and DPASV), $E_{1/2}$ and i_1 (Fig. 6.9) being replaced by E_p and i_p, respectively (Fig. 6.8b I, IIa). Eqn (6.28) is also directly applicable to 'stripping polarography'. In this method ([23, 83, 84, 111, 112]; see case study 1), polarograms similar to those of Fig. 6.9 are reconstructed by plotting the peak current, i_p^c, of ASV or DPASV curves versus the corresponding deposition potential.

Since, in general, $D_{ML} \leqslant D_M$, a smaller value of i_1^c compared to i_1^{nc} is observed for many labile complexes. This result must not be confused with the decrease in current observed with chemically inert complexes (Section 6.5.1.3, Fig. 6.8b, case III). In contrast to the latter, for labile complexes:

• $i_1^c/i_1^{nc} \neq |M|/|M|_t$, even for $D_{ML} = 0$ [eqn (6.26)].
• $E_{1/2}$ (or E_p) is always shifted to negative values, even when $D_{ML} = 0$ [eqn (6.28) and (6.26)].

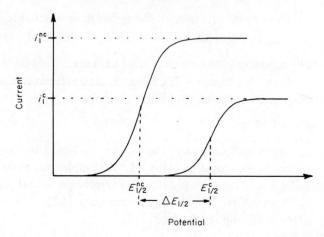

Fig. 6.9. Schematic change in dc polarogram or stripping polarograms from non-complexing medium (nc) to complexing medium (c). Same assumptions as in Fig. 6.8b II.

Three limiting cases are of interest:

$D_{ML} = D_M$

Then eqn (6.26) becomes:

$$i_l^c/i_l^{nc} = 1 \qquad (6.26a)$$

or $i_p^c = i_p^{nc}$ (case I). Hence in this case, the complexation reaction results only in a change in $E_{1/2}$ or E_p compared to the reduction of the uncomplexed ion, but the current is unchanged. The condition $D_M \simeq D_{ML}$ is often valid for complexes with 'simple' small ligands (Cl$^-$, amino-acids, and so on).

$\alpha - 1 \gg D_M/D_{ML}$

This is the case of very strong complexation ($\alpha \gg 1$), in particular with not too large ligands (e.g. peptides, fulvics). Then

$$i_l^c/i_l^{nc} \cong \sqrt{D_{ML}/D_M} \qquad (6.26b)$$

As before, α is measurable only from the change in $E_{1/2}$ [eqn 6.28], but now additional information is obtained about the size of ML, from the change in i_l (Fig. 6.8, case IIa).

$D_M/D_{ML} \gg \alpha - 1$

$$ \qquad (6.29)$$

In this case, the complex may be called physically inert, since the transport of free M is predominant compared to that of ML. This case applies to relatively weak complexation with very large macromolecules, such as DNA or polysaccharides. Eqn (6.26) becomes:

$$i_l^{nc}/i_l^c = \sqrt{\alpha} = \sqrt{|M|_t/|M|} \qquad (6.26c)$$

In this particular case, α may be obtained both from the decrease in limiting current or the change in $E_{1/2}$ (case IIa), but no information is gained about the size of ML.

(b) Application to aquatic conditions

Two aspects are particularly important in natural samples:

(i) *Fulfillment of condition (6.24)* ($|L|_t > 10 |M|_t$)

Figure 6.1 shows that this condition is valid for major inorganic ligands (Cl^-, CO_3^{2-}, etc) but it is not always the case with other ligands, particularly organic ones. Hence the validity of this condition must be checked carefully in each particular case. Complexation measurements without excess of ligands (unbuffered media) are discussed in ref. [32].

(ii) *Mixture of ligands*

In solutions containing many complexes, eqn (6.21) should be generalized to include a sum of terms corresponding to all the diffusing species.

$$\bar{D} = D_M \frac{|M|}{|M|_t} + \sum_i D_{ML_i} \frac{|ML_i|}{|M|_t} \tag{6.30}$$

When all the diffusion coefficients of the complexes have the same value ($D_{ML_i} = D_{ML}$), eqns (6.25)–(6.28) are still valid. This is often the case in mineral waters where ligands are only inorganic anions. In such a case α can be determined by eqn (6.28) and the β values of each particular species can be obtained from eqn (6.5) with $\alpha_w = \alpha$.

However, many surface waters also include macromolecular organic and inorganic complexing agents with widely different diffusion coefficients. In such cases, eqn (6.30) and (6.23) show that the resulting current is a weighted average value of all individual contributions. There is, to date, no rigorous solution to interpret complexation data for mixtures of ligands with widely varying D_{ML} values. Hence, in studying the formation of labile complexes with aquatic complexing agents, it is essential to size-fractionate them in such a way that the diffusion coefficients of all the complexes in the studied fraction should be close enough to enable a meaningful average value to be used. The interpretation is simplified, however, when M is combined only with very large ligands. If condition (6.29) applies for each of them, i.e. $D_M/D_{ML} \gg |ML_i|/|M|$ then eqn (6.26c) and (6.28) are valid even when the D_{ML} values are widely different.

6.5.1.3 Chemically inert complexes (Case III, Fig. 6.8b)

(a) Well-characterized ligands

Inert complexes are defined here as being those which do not dissociate significantly ($k_d \to 0$, Fig. 6.8a) during the time scale of the voltammetric measurement, e.g. drop time in dc polarography or pulse duration in DPP and NPP. It may be shown [32, 46, 85] that, for dc polarography, this condition is met when:

$$k_d \ll K|L|/t_d \tag{6.31}$$

where $K = |ML|/|M| |L|, |L|$ is the free ligand concentration and t_d = drop time.

Under such conditions, M and ML behave as two independent species. At potentials where ML is not electroactive, only M is reduced, so that eqn (6.16) can be applied here. Hence equations similar to eqn (6.18)–(6.19) are obtained for $E_{1/2}^c$ and i_l^c:

$$E_{1/2}^c = E_{1/2}^{nc} \tag{6.32}$$

and

$$i_l^c = K_l \sqrt{D_M} |M| \tag{6.33}$$

but now $|M| < |M|_t$, so that, by combining eqn (6.19) and (6.33):

$$\alpha = i_l^{nc}/i_l^c = |M|_t/|M|. \tag{6.34}$$

Also note that, in contrast to the case of labile complexes, no excess ligand is necessary for chemically inert complexes.

As mentioned above, eqn (6.31) is valid for other voltammetric techniques, by replacing t_d with the relevant time scale of the method. Hence, the definition of inert complexes is an operational one which depends on the nature of the technique used. This has two important consequences:

(i) complexes which behave in an inert manner with a fast technique (small time scale) may be labile or quasi-labile with a slower one (large time scale). This is examplified in Section 6.6.2.2.

(ii) The fact that a given voltammetric technique measures inert complexes, does not imply that the whole solution is kinetically stable. For instance, complexes whose dissociation times are e.g. a few hours behave in an inert manner during the time scale of any voltammetric technique. But when a set of voltammograms is recorded, e.g. at 15 minute intervals, a slow change in current is then observed. This approach can be used to measure the dissociation rate of electrochemically inert complexes [32].

(b) Aquatic samples

Equation (6.34) is applicable to samples containing mixtures of ligands, regardless of the nature of the various complexes, provided that condition (6.31) is fulfilled for all of them. As mentioned above, it is possible to discriminate between chemically inert complexes and chemically labile, but physically inert ones, by testing:

- the change in i_l^c with α [eqn (6.26c) and (6.34)]
- the change in $E_{1/2}^c$ with α [eqn (6.26c), (6.28) and (6.32)].

This discrimination is very useful for interpreting the biogeochemical role of M.

6.5.1.4   Quasi-labile complexes (Case IIb, Fig. 6.8b)

These are complexes for which k_d values are intermediate between conditions (6.20) and (6.31). Such cases are rather complicated to interpret, even for well-characterized ligands, and will not be discussed here in detail. It should be mentioned that the observed change in $E_{1/2}$ or E_p is less than for labile complexes [eqn (6.28)] but more than for inert ones [eqn (6.32)]. Similarly, the decrease in i_l or i_p is less than for inert complexes [eqn (6.34)], but more than for labile ones [eqn (6.26)]. The mathematical treatment for a mixture of well-characterized ligands forming complexes with the same diffusion coefficients was first given by Koutecky et al. [101]. Turner et al. [102,103] applied it to aquatic conditions.

Interpretation of such systems in the case of a mixture of aquatic ligands with different diffusion coefficients is very difficult since the measured weighted average flux depends simultaneously on distribution spectra of both D and k_d (Fig. 6.5 and 6.6). For similar reasons, it is also very difficult to interpret correctly the changes in current measured during the titration of the water sample with M. Indeed, in such a case, one must take into account not only the variation of K with $|M|_t/|L|_t$ as in potentiometric measurements, but also those of k_d and D_{ML}. In particular, ML may change gradually from fully inert to fully labile during the course of the titration (Fig. 6.5). The change in D_{ML} strongly depends on the composition of the sample.

6.5.2   Discriminating tests

In order to choose the correct interpretation of the data, the nature of the most important electrode processes must be tested. A set of tests is given in Table 6.1b and c for typical cases most often encountered in aquatic systems as defined in Table 6.1a. Tests in Table 6.1 are valid for reversible charge transfer

Table 6.1. Some tests for discriminating between different typical electrochemical behaviour in DPP and DPASV.

Table 6.1a. Definitions of typical behaviours considered in Tables 6.1b and 6.1c

| | | Kinetic classification (see also Fig. 6.8) | | Supplementary conditions | | |
| | | | | | $|L|_t/|M|_t$ in | |
Type of system under study	Case no	Dissociation rate of ML	Diffusion of ML	Adsorption	DPP	DPASV
A	I	non limiting	$D_{ML}=D_M$	no	>10	>1000
B	IIa	$(k_d \simeq \infty)$	$D_{ML}=0$	no	>10	>1000
C	I		$D_{ML}=D_M$	no	<10	<1000
D	IIa		$0<D_{ML}<D_M$	yes	>10	<1000
E	III	$k_d=0$	non limiting $(0 \leqslant D_{ML} \leqslant D_M)$	no	non-limiting factor	

$\Delta E_p^D = RT/2F \ln(i_p^c/i_p^{nc}) + RT/2F \ln\alpha = \Delta E_p$ for a diffusion controlled system; $\Delta E_p^{max} = RT/2F \ln\alpha$: In Table 6.1b and 6.1c, variations of the electrochemical parameters (ΔE_p, $W_{\frac{1}{2}}$ or i_p^c/i_p^{nc}) are given for decrease or increase in $t_d(t_d^\downarrow$ or $t_d^\uparrow)$ or increase in $|M|_t(|M|_t^\uparrow)$. Criteria given under headings $|M|_t$ are valid provided the nature of complexing sites does not change with $|M|_t/|L|_t$ (see Section 6.2.2). S–c=S-shaped curve: b–c=broken curve; cte=constant; p=proportionality constant; $\downarrow$=decrease; $\downarrow\rightarrow$X=tends to X by decreasing values. (cte$<1)\rightarrow1$=tends to 1 from a constant value, <1. Other symbols: see Table of symbols.

Table 6.1b. Some electroanalytical tests in DPP (t_d = drop time)

System	$\Delta E_p = E_p^{nc} - E_p^c$ Value	$t_d\downarrow$	$\|M\|_t\uparrow$	$W_{\frac{1}{2}}$ Value (mV)	$t_d\downarrow$	$\|M\|_t\uparrow$	i_p^c/i_p^{nc} Value	$t_d\downarrow$	$i_p^c = f(\|M\|_t)$
A	$=\Delta E_p^{max}$	cte	cte	45.2	cte	cte	1	cte	$i_p^c = p\cdot\|M\|_t$
B	$=\Delta E_p^D$	cte	cte	45.2	cte	cte	<1	cte	b c{break for $\|M\|_t=\|L\|_t$}
C	$\geqslant\Delta E_p^{max}$	cte	$\downarrow\to 0$	>45.2	cte	pass through maximum	<1	—	s-c $(i_p^c \neq p\|M\|)$
D	$\gg\Delta E_p^{max}$	$\downarrow\to\Delta E_p^D\downarrow$		$\geqslant 45.2$	$\downarrow\to 45,2$	—	$0\text{-}(>1)$	$\downarrow\to(\text{cte}<1)$	$i_p^c = p\|M\|_t$ {break for $\|M\|_t=\|L\|_t$; $i_p^c = p\cdot\|M\|$}
E	0	cte	cte	45.2	cte	cte	<1	cte	b c {break for $\|M\|_t=\|L\|_t$; $i_p^c = p\cdot\|M\|$}

Table 6.1c. Some electroanalytical tests in DPASV (for detail see also ref. [6]). (t_d = deposition time)

System	$\Delta E_p = E_p^{nc} - E_p^c$ Value	$t_d\uparrow$	$\|M\|_t\uparrow$	i_p^c Value	$f(t_d)$	$f(\|M\|_t)$
A	$=\Delta E_p^{max}$	$\simeq$cte	$\simeq$cte	$=i_p^{nc}$	$i_p^c = p\cdot t_d$	$i_p^c = p(\|M\|_t)$
B	$=\Delta E_p^D$	cte	cte	$<i_p^{nc}$	$i_p^c = p\cdot t_d$	b c{break for $\|M\|_t=\|L\|_t$}
C	$0\leqslant\Delta E_p\leqslant\Delta E_p^{max}$	$\downarrow\to 0$	$\downarrow\to 0$	$<i_p^{nc}$	s–c	s–c{$i_p^c \neq p\cdot\|M\|$}
D	$\gg\Delta E_p^{max}$	$\downarrow\to 0$	$\downarrow\to 0$	$<i_p^{nc}$	b–c	b c {two breaks; 2nd break at $\|M\|_t=\|L\|_t$}
E	0	cte	cte	$<i_p^{nc}$	$i_p^c = p\cdot t_d$	b c {break at $\|M\|_t=\|L\|_t$; $i_p^c = p\|M\|$}

systems and are helpful to characterize:
- diffusion versus kinetically controlled current
- adsorption phenomena on the electrodes
- effects due to non excess of ligand and insufficient pH buffering at the electrode surface.

These tests are discussed for DPP and DPASV on DME and HMDE, respectively, which are the most frequent methods used in practice. However, they also apply to other methods.

From the discussion in Section 6.5.1, clearly the more useful tests are: the change in $\Delta E_p = E_p^{nc} - E_p^c$ and i_p^c or i_p^c/i_p^{nc} with $\|M\|_t$ and/or t_d (drop time on DME; deposition time on HMDE). In some cases, the width of the peak, $w_{1/2}$, at $i = i_p/2$, is also a useful criterion. It must be emphasized that, the data in Table 6.1b and c apply to well-characterized ligands, or at least well-fractionated complexing agents.

6.5.2.1 Reversibility of charge transfer process

The fulfillment of tests listed in Table 6.1b and c serve as criteria for reversibility. In particular, for a reversible system involving fully labile or inert complexes, $w_{1/2} = 45.2$ mV in DPP and DPASV, when the pulse height, ΔE, is < 10 mV. Furthermore, in DPP, the absolute value of i_p must be independent of the sign of the applied pulse ($+\Delta E$ or $-\Delta E$), and the corresponding peak potentials must be separated by ΔE. Non-fulfillment of these conditions may indicate either a slow electrochemical or slow chemical process, or that the ligands are not present in excess.

6.5.2.2 Kinetic versus diffusion-controlled current

It is fairly easy to discriminate between chemically inert complexes on the one hand (case E in Table 6.1) and chemically labile (cases A,B) or quasi-labile complexes on the other, particularly by inspecting the value of ΔE_p. $\Delta E_p = 0$ for inert complexes.

It is more difficult, but very important for interpretation of the data, to discriminate between chemically sluggish (case IIb in Fig. 6.8b) and physically sluggish or inert (case IIa in Fig. 6.8b) complexes. This entails a detailed investigation which is beyond the scope of this report. The rotating electrode is very useful in this respect [33]. The effect of variation in the sign of the pulse potential or the pulse duration in DPP, or in the frequency in ac polarography, may also be used as good tests [32]–[34].

6.5.2.3 Adsorption effects

In water analysis, adsorption effects have been studied in detail mostly for organic compounds [26, 27, 87–94], but similar effects may be observed with inorganic compounds such as colloidal iron hydroxides [95]. Natural organic matter was found [88, 93, 96] to be significantly adsorbed on mercury, even at low concentrations (down to fractions of mg dm^{-3}) and for small times of contact with the electrode solution (down to a fraction of a second). Comparison of the behaviour of the various types of electrode seems to indicate that adsorption is weak or non-existent on glassy carbon or wax-impregnated graphite electrodes [90] but strong on Pt [90,92] and Hg [88,89,91,93,94,96].

The best characterized consequence of adsorption of organic matter on complexation measurements is the formation of adsorbed complexes at the electrode surface [26, 27, 89, 97]. Both the current and, above all, the negative shift in E_p may be greatly increased by this effect (case D in Table 6.1; for details see Section 6.6.3). Although there is no report dealing specifically with the effects of adsorption processes on the kinetics of charge transfer in natural systems, these effects are well-known with well-characterized solutions [32]. In water analysis they might play a role in the formation of double peaks in the ASV determination of Cu(II) [98] or in the stronger adsorption effects observed in DPASV compared to ASV [99].

The following tests (Table 6.1, case D) are particularly useful in verifying the possible influence of adsorption effects.

Comparison of ΔE_p with $\Delta E_p^{max} = (RT/2F) \ln \alpha$, where the maximum possible value is given to α for this test. $\Delta E_p > \Delta E_p^{max}$ often indicates the adsorption of ligands and formation of adsorbed complexes.

In natural waters, adsorbed ligands are often macromolecules which are present at low concentrations. As a result, adsorption may be slow and this can be tested by looking at the change in ΔE_p with drop time (t_d), in DPP: adsorption results in an increase in ΔE_p with t_d (Section 6.6.3). In DPASV ΔE_p decreases when deposition time increases [47]. In both cases the independency of ΔE_p on t_d is a necessary requirement for the absence of adsorption effects. Tests making use of the change in i_p with t_d (Table 6.1b and c) are also very useful.

With NPP, variation of pulse duration can also be used as a good criterion for testing adsorption of ligands [89, 97].

6.5.2.4 Metal-to-ligand ratio and pH buffer capacity

As mentioned before, for an easy interpretation of complexation by labile complexes forming ligands, L must be in excess at the electrode surface. For reduction methods, it was seen that the condition $|L|_t/|M|_t > 10$ must be fulfilled. For ASV or DPASV methods, the condition is much more severe, owing to the large value of $|M|_{t,0}$ during the reoxidation step. It may be shown [47, 86] that in this case the condition is $|L|_t/|M|_t > 1000$. If these conditions are not fulfilled (case C in Table 6.1), one observes too high values for $w_{1/2}$, non linear i_p against $|M|_t$ curves, and non linear i_p against t_d curves [47, 86].

The pH buffer capacity must also be large enough at the electrode surface to ensure that $\alpha_0 = \alpha$. Insufficient pH buffering may produce the same type of results as those obtained without excess of ligand. In natural samples which contain HCO_3^-, purging the solution with a mixture of N_2/CO_2 both removes O_2 and buffers the solution [16, 17, 24]; see also case study 1. In some cases, the natural ligands themselves may provide some pH buffer capacity ([27], see case study 3). In other cases, weakly complexing buffers such as MES, PIPES or HEPES [100] may be used at low concentrations.

6.5.3 Optimal conditions for complexation measurements

Optimal conditions must be chosen in each particular case, on the basis of the relative importance of the processes discussed above. General guidelines are provided here (Table 6.2). Obviously compromises must usually be made. For instance, for the choice of electrode material, Hg is generally preferable to glassy carbon, despite stronger adsorption effects being found with the former (Table 6.2d and e). As far as the technique is concerned, DPP is often preferred to NPP, because of better sensitivity and selectivity, although the interpretation of NPP signals are simpler [89] (Table 6.2a and b). Similarly, DPASV is more sensitive than DPP, but the interpretation of signals may be more

Table 6.2. Important factors to be considered for speciation in natural waters and examples of the choice of the best electroanalytical conditions

Factors considered	Condition No 1	better than	Condition No 2
(a) Simple interpretation→use of a technique based on simple theory	NPP		DPP
	ASV		DPASV
(b) High sensitivity requirement	DPASV bet. th. ASV		DPP bet. th. NPP
(c) $|L|_t^0/|M|_t^0 \simeq |L|_t/|M|_t$	DPP		DPASV
	HMDE		TMFE

$$\left[\begin{array}{l} \text{Simple interpretation for} \\ \left\{\begin{array}{l} |L|_t/|M|_t > 10 \text{ with DPP} \\ |L|_t/|M|_t > 1000 \text{ with DPASV} \end{array}\right] \end{array}\right.$$

(d) Reproducible electrode surface	Hg		Solid electrode
	HMDE		TMFE
(e) Minimization of adsorption:			
Nature of electrode	Glassy C or graphite		Hg or Pt
'History' of electrode	Hg		Solid electrode
(renewability	HMDE or DMF		TMFE
of the surface)	DPP		DPASV
short t_d bet. th. large t_d			

In this table: bet. th. means 'better than'. Comparison between HMDE and TMFE is given for use with ASV techniques only.

difficult, because, during the two electrochemical steps (deposition and stripping) different reactions may occur at the electrode surface [47].

In general, DPP on DME is the best compromise for $|M|_t > 10^{-6}$ mol dm^{-3}, DPASV on HMDE is used for $|M|_t > 10^{-8}$ mol dm^{-3}, and DPASV on TFME is necessary for lower concentrations.

When a supporting electrolyte must be used, e.g. for fresh waters, its concentration should be as low as possible (e.g. 2×10^{-2} mol dm^{-3} NaClO$_4$) to minimize any change in the sample, and any possible contamination. This is also compatible with the use of pulse techniques which need less concentrated electrolytes than others. Furthermore, on Hg, the $Zn^{2+}/Zn(o)$ couple is more reversible the lower the concentration of electrolyte [32].

The previous sections showed that voltammetric methods depend on a large number of factors. Several of them can be quantitatively controlled and furthermore, many experimental conditions can be varied to choose the most appropriate ones for solving the particular problem in hand. Thanks to these capabilities, voltammetric methods serve as very powerful tools for the determination of complexation properties of metals in environmental samples [2, 22, 26, 110, 116, 117]. Obviously, because of the relatively large number of factors which must be tested, the non-specialist in electroanalytical chemistry is required to make some effort to understand the basic electroanalytical processes and the various possible ways of interpreting and controlling them.

But, as discussed in the introduction, this effort is also beneficial for the understanding of biogeochemical processes themselves.

The forgoing part of this report provided the most important general guidelines which may facilitate correct use of these techniques. In the following paragraphs, three typical cases reported in the literature have been taken as examples to demonstrate how these guidelines may be applied. They correspond to the following cases in Table 6.1a:

- case A: labile, fast diffusing complexes
- case E: chemically inert complexes
- case D: labile, slowly diffusing complexes, with adsorption of ligands.

It must be borne in mind that these descriptions are only given as examples and that their applicability, under different conditions, should be carefully checked, in particular by referring to the original papers.

6.6 Case Studies

6.6.1 Labile fast diffusing complexes (Case A, Table 6.1): study of Pb(II) carbonato complexes in sea water by stripping polarography (basic references [22,24,28,118])

6.6.1.1 Purpose and conditions

The aim of this work was to study the relevance of the literature values for the complexation constants of Pb(II) carbonato complexes, in normal sea water conditions, i.e. particularly in the presence of many other possible complexing agents, and at the ultra-trace level of the metal ion studied.

Experimental conditions

Solutions simulating sea water were used. The relative concentrations of Cl^- and HCO_3^- were varied to check their complexing properties, and the ionic strength was kept constant at 0.7 mol dm^{-3} by adding appropriate amounts of $NaClO_4$. The following solutions were studied:

1 0.7 mol dm^{-3} $NaClO_4$
2 0.02 mol dm^{-3} $NaHCO_3 + 0.1$ mol dm^{-3} $NaCl + NaClO_4$
3 0.10 mol dm^{-3} $NaHCO_3 + 0.1$ mol dm^{-3} $NaCl + NaClO_4$
4 0.40 mol dm^{-3} $NaHCO_3 + 0.1$ mol dm^{-3} $NaCl + NaClO_4$.

Solutions were thermostatted at $25\,°C$. In each case the pH, and hence CO_3^{2-}, was varied, by changing CO_2 partial pressure in the N_2/CO_2 gas mixture used to deaerate the solution. This procedure ensured a sufficient buffer capacity in the solution (Section 6.5.2.4). Only in the reference $NaClO_4$ solution, was pH kept constant at 3.0 to avoid the formation of hydroxo complexes of Pb(II).

In all the cases: $|Pb(II)|_t = 6 \times 10^{-9}$ mol dm^{-3}. This very low concentration entailed the use of ASV on TMFE (Section 6.5.3). The following electroanalytical conditions were used: rotating electrode—1500 rpm; deposition time—5 min; quiescent period between deposition and stripping—20s; positive scan rate—100 mVs^{-1}.

Nature of the sample

This system belongs to the first of the three categories listed in Section 6.2.1, i.e. a mixture of well-characterized ligands whose total concentrations are known exactly. Hence the β values of each complex can be determined from eqn (6.5) and the values of α determined experimentally.

It must be mentioned that, although the system discussed here is an artificial one, the same approach was used by the authors to study the complexation of Cd in genuine sea water [22]. In this case, the influence of ill-characterized ligands, in particular organic matter and suspended particles, was minimized by sampling the water at locations where their concentrations were low (Section 6.3.2).

6.6.1.2 Electroanalytical characteristics of the test system

In the present case, many of the simplifying assumptions discussed in Section 6.5.1 and 6.5.2 are valid.
- The $Pb^{2+}/Pb(o)$ system is known to behave reversibly at the mercury electrode.
- There is a large excess of ligand (L), even at the lowest concentration of L: $|L|_t/|M|_t = 3 \times 10^{+6} \gg 1000$ (Section 6.5.2.4).
- The buffer capacity at the electrode surface is large enough (Section 6.5.2.4). It is controlled mainly by the H_2CO_3/HCO_3^- couple and can be computed to be close to 10^{-2} mol dm^{-3}, that is $> 10^6$ times larger than $|M|_t$.
- The ligands in the solution are known not to be adsorbed on the electrode surface (Section 6.5.2.3).
- The ligands form labile complexes with Pb(II).
- Furthermore, due to the small size of the ligands, the values of the diffusion coefficients of the complexes are likely to be very close to that of Pb^{2+} (Section 6.5.1.2).

Under these conditions, eqn (6.28) is valid and can even be simplified since

$$i_1^{nc}/i_1^c = \sqrt{(D_M/D_{ML})} = 1 \text{ [eqn (6.26a)] and}$$

$$\ln \alpha = \frac{nF}{RT}(E_{1/2}^{nc} - E_{1/2}^c) \tag{6.35}$$

6.6.1.3 Description of the results

For each solution, the stripping polarograms were constructed from the heights of the ASV peaks corresponding to a series of cathodic deposition potentials, E_d, ($\sim 10\, E_d$ values) in the potential range of the hypothetic dc-polarogram (Fig. 6.10). These measurements have to be performed over an extended range of the ligand concentration (CO_3^{2-}), to obtain a ln α versus

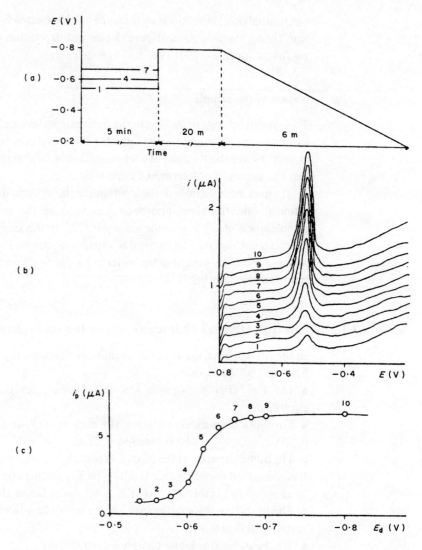

Fig. 6.10. Construction of pseudo-polarogram from automated ASV-measurements, for the Pb(II)–CO_3^{2-} system, (from ref. [22]). (a) Course and timing of polarizing voltage. (b) Family of ASV-peaks as function of cathodic deposition potential. (c) Resulting pseudo-polarogram.

$\ln|CO_3^{2-}|$ relationship (Fig. 6.11), relevant for all complexes which may exist in sea water. Hence a large number of experiments have to be done, over a range of $|CO_3^{2-}|$ covering several decades. It is, therefore, important to automate the procedure for recording ASV peaks.

6.6.1.4 Interpretation of the data

Because the solution is fully characterized, the values of α obtained in Fig. 6.11 can be used in eqn (6.5), to compute the β values of the carbonato complexes in the following way. Figure 6.11 shows that when $|CO_3^{2-}| \to 0$, α tends to a constant value, $\alpha(|CO_3^{2-}|=0)$ due to the complex formation with all ligands

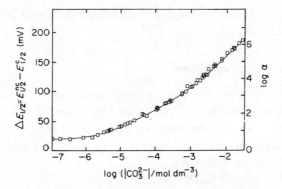

Fig. 6.11. Change in $\Delta E_{1/2} = E_{1/2}^{nc} - E_{1/2}^{c}$ or $\log \alpha$ with $|CO_3^{2-}|$ for Pb(II) reduction, at ionic strength of sea water (from ref. [118]). $|Pb(II)|_t = 6 \times 10^{-9} \, mol \, dm^{-3}$.

other than CO_3^{2-}. Any increase in α with $|CO_3^{2-}|$ is only due to formation of carbonato complexes. Hence, assuming that there is no mixed complex, eqn (6.5) becomes:

$$\alpha = \alpha(|CO_3^{2-}| = 0) + \beta_1 |CO_3^{2-}| + \beta_2 |CO_3^{2-}|^2 + \ldots \beta_n |CO_3^{2-}|^n \qquad (6.36)$$

The average value of the coordination number of Pb(II) with CO_3^{2-} is given by $\bar{n} = d \log \alpha / d \log |CO_3^{2-}|$. From Fig. 6.11 it is seen that $0 < \bar{n} < 2$ in the studied concentration range of CO_3^{2-}. Hence eqn (6.36) can be used with only the first three terms of the right hand side for evaluating β_1 and β_2. The following values were obtained in this way ($T = 25°C$; $\mu = 0.7 \, mol \, dm^{-3}$):

$\log \beta_1 = 5.62 \pm 0.22$

$\log \beta_2 = 8.80 \pm 0.35$

By using these values, together with the literature values for the stability constants of hydroxo- and chloro-complexes, the following distribution of Pb(II) species was computed for the composition of sea water (charges are omitted):

Pb: 1.8% $PbCO_3$: 43.0% $Pb(CO_3)_2$: 3.7%

PbCl: 8.6% $PbCl_2$: 12.6%

PbOH: 30.0% $Pb(OH)_2$: 0.5%

6.6.2 Chemically inert complexes (Case E, Table 6.1): chelation of Cd(II) by NTA in sea water (basic references [24,53])

6.6.2.1 Purpose and conditions

Nitrilotriacetic acid (NTA) has ecological relevance, as it may occur at low concentration levels, both in sea water and inland water, due to the substitution of polyphosphates by NTA in detergents. Complexation of metals, particularly the toxic ones, e.g. Cd(II), by non biodegraded NTA may have a strong influence on the productivity of natural waters. The aim of this work was to explore the general parameters and specific factors affecting the complexation of heavy metals with strong organic ligands, like NTA, in sea water. In particular the role of salinity components (Cl^-, Na^+, Ca^{2+}, Mg^{2+}) was tested.

Experimental conditions

Both natural sea water and solutions simulating it were used. In the latter, the concentrations of Cl^-, Ca^{2+} and Mg^{2+} were varied to test their separate effects. In all the cases except for natural sea water, $NaClO_4$ was used to adjust the ionic strength to $0.7 \, mol \, dm^{-3}$. The following solutions were used:

1 $0.7 \, mol \, dm^{-3} \, NaClO_4$
2 $0.1 \, mol \, dm^{-3} \, NaCl + NaClO_4$
3 $0.59 \, mol \, dm^{-3} \, NaCl + NaClO_4$
4 $0.01 \, mol \, dm^{-3} \, CaCl_2 + NaClO_4$
5 $0.0536 \, mol \, dm^{-3} \, MgCl_2 + NaClO_4$
6 artificial sea water prepared according to ref. [104]
7 natural Adriatic sea water (Rovijn, Istria).

Solutions were thermostatted at $25 \pm 0.5 \, °C$. For solutions (1)–(6) a borate buffer ($5 \times 10^{-3} \, H_3BO_3 + 5 \times 10^{-4} \, NaOH$) was used to adjust the pH value between 7.3–8.3. Adriatic sea water was passed through a $0.45 \, \mu m$ filter. All solutions were spiked with $10^{-7} \, mol \, dm^{-3} \, Cd(II)$ and titrated with NTA.

Since Cd(II) was present in sufficiently high concentrations, the experiments were performed with DPP and DME, using the following conditions: drop time—5s; flow rate of Hg—$1.02 \, mg \, s^{-1}$; pulse amplitude—50 mV; pulse duration—40 ms; scan rate—$0.5 \, mV \, s^{-1}$; reference electrode—saturated calomel electrode for solutions (6)–(7), Ag/AgCl/sat.KCl/$0.7 \, mol \, dm^{-3}$ NaCl// for solutions (1)–(5). All solutions were deaerated for 20 min with pure N_2 before the measurements.

Nature of the sample

Solutions (1)–(6) belong to the first type of water samples defined in Section 6.2.1, i.e. a mixture of well-characterized ligands whose total concentrations are known exactly. Hence the β value of each complex, in particular Cd-NTA, can be determined from eqn (6.5) and from the values of α determined experimentally. Natural sea water (solution 7) is also likely to fall under this classification owing to the low level of colloidal particles and organic matter found in the particular collected sample. This prediction can be tested by comparing the results obtained with solutions (6) and (7).

The concentration of $10^{-7} \, mol \, dm^{-3}$ used here for Cd(II) corresponds to that of polluted waters and is 100–300 times larger than in unpolluted samples. The same study was performed with DPASV at a lower concentration [22] ($3 \times 10^{-9} \, mol \, dm^{-3}$) and with other metals (Zn(II) [105]; Pb(II): [106]).

6.6.2.2   **Electroanalytical characteristics of the test system**

- Cd(II) is well known to be reversibly reduced.
- The pH buffer capacity is much larger (~ 5000 times) than the total concentration of Cd(II), $|Cd|_t$, (Section 6.5.2.4).

• The ligands are known not to be adsorbed at the electrode surface.

• The ligands forming labile complexes (see below) are in large excess (Section 6.5.2.4) compared with $|Cd|_t$, at least in solutions (2)–(6) ($> 10^6$ times). In addition, in these solutions, all cations except Cd(II) are in large excess (> 100 times) compared with NTA.

• All the test ligands are known to form labile complexes with Cd(II) (Cl^-, CO_3^{2-}, OH^-), with the exception of NTA.

Therefore $|Cd|_t$ can be split into two terms:

$$|Cd|_t = |Cd|_{t,l} + |CdNTA| \qquad (6.37)$$

where $|Cd|_{t,l}$ is the concentration of Cd(II) present as Cd^{2+} plus that bound to labile complexes. The overall degree of complexation, α, [eqn(6.5)] becomes:

$$\alpha = \frac{|Cd|_t}{|Cd|} = \frac{|Cd|_{t,l}}{|Cd|} + \frac{|CdNTA|}{|Cd|} \qquad (6.38)$$

$$= 1 + \sum_i \sum_l {}^i\beta_l |L_i|^l + K|NTA| \qquad (6.39)$$

$$= \alpha_l + K|NTA| \qquad (6.40)$$

where L_i is the ith labile complex forming ligand, l is the number of coordinated ligands in the complex, K is the equilibrium formation constant of Cd-NTA, and all ligand concentrations are those of free ligands. α_l is the degree of complexation of Cd(II) for all labile complex forming ligands.

• Since all are small inorganic anions, the diffusion coefficients of all the corresponding complexes are most likely to be close to that of Cd^{2+}. Hence in the absence of NTA, the polarographic behaviour of Cd(II) would be expected to be that of a reversibly reduced metal ion forming fast diffusing labile complexes (Section 6.5.1.2) as in case **1**. Assuming all the diffusion coefficients to be equal, a value of 1 is obtained for the ratio

$$i_{p,l}^c / i_p^{nc} = |Cd|_{t,l} / |Cd|_t \qquad (6.41)$$

where $i_{p,l}^c$ = peak current due to reduction of labile complexes and $|Cd|_{t,l} = |Cd|_t$ in the absence of NTA. By applying eqn (6.28) to DPP peaks one obtains:

$$\ln \alpha_l = (2F/RT) \cdot (E_p^{nc} - E_p^c) \qquad (6.42)$$

(see Fig. 6.12a, curves 1, 2).

• Dissociation of Cd-NTA is known to be a slow process, and the direct reduction of the undissociated complex occurs at potentials several hundred mV more negative than Cd^{2+} [107]. Hence the addition of NTA is expected to cause a decrease in i_p^c (Fig. 6.12a, curve 3), and possibly a supplementary negative shift in E_p^c, the extent of which depends on the dissociation rate of Cd-NTA (Section 6.5.1.3–6.5.1.4) and the time scale of the method used. For example, in dc polarography (relevant time scale = drop time = few seconds) a non-zero shift in $E_{1/2}$ is observed [32, 53] between polarograms recorded in the presence and absence of NTA, which indicates that CdNTA does not behave as a fully inert complex. In contrast, with DPP, for which the relevant time scale is pulse duration (40 ms), CdNTA behaves as a fully inert complex, as no shift in E_p is observed (Fig. 6.12b; Section 6.5.1.3, eqn (6.32); Table 6.1b).

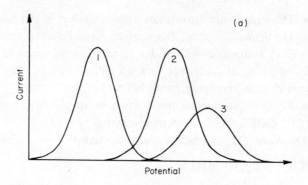

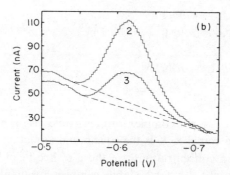

Fig. 6.12. Schematic (a) and experimental (b) change in DPP curve for Cd(II) reduction in non complexing medium (curve 1), in complexing medium forming only labile diffusing complexes with the same diffusion coefficient as Cd^{2+} (curve 2) and in medium containing labile + non labile complexes (curve 3). Curves 2b and 3b: pH = 8.0, $|Cd(II)|_t = 10^{-7}$ mol dm^{-3} (solution n° 7 in Section 6.6.2.1), curve 3b: same as curve 2b + $|NTA|_t = 10^{-4}$ mol dm^{-3}.

To ensure that the peak obtained in presence of NTA contained no contribution from the dissociation of CdNTA, its time dependency was studied. Indeed, eqn (6.26) shows that, if $i_{p,l}^c$ is diffusion controlled, the ratio $i_{p,l}^c / i_p^{nc}$ should be independent of drop time (Table Ib), a result which is found in practice [53].

Hence the fraction of Cd(II) not complexed by NTA can be computed from eqn (6.41). Note that this equation is analogous to eqn (6.34), but more general, since it is applicable to mixtures of both fully inert and fully labile complexes with equal diffusion coefficients. It must be emphasized that the peak current measured in the presence of NTA is not proportional to free Cd^{2+} concentration, but to the sum of the concentrations of all labile species.

6.6.2.3 Description of the results

Each of the solutions 1 to 7 was spiked with increasing amounts of NTA, and the corresponding DPP peak current, $i_{p,l}^c$, was recorded. The corresponding ratios of eqn (6.41) were then computed and plotted against log $|NTA|_t$ (Fig. 6.13), where $|NTA|_t$ is the total concentration of added NTA.

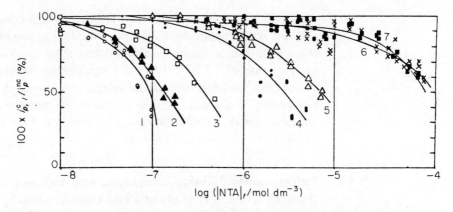

Fig. 6.13. Change in DPP peak current of Cd(II) ($i_{p,l}^c$) with NTA concentration. The ordinate gives the ratio (in %) of $i_{p,l}^c$ over the peak current which would be obtained with the same total concentration of Cd(II) ($|Cd|_t$), in non-complexing medium (i_p^{nc}). $|Cd|_t$, i_p^{nc} and ionic strength (0.7 mol dm^{-3}) are constant (see Section 6.6.2.1 for the electrolyte composition corresponding to curves 1–7). The corresponding pH values were 7.35 (1), 7.83 (3), 7.38 (4), 7.46 (5), 8.29 (6), 8.01 (7). Full curves are obtained by polynomial curve fitting. Reproducibility of experimental data: ±44% (1), ±10% (2, 3), ±8% (4), ±7% (5), ±24% (6), ±15% (7).

6.6.2.4   Interpretation of the results

Figure 6.13 shows that the value of $|NTA|_t$ necessary to decrease $|Cd|_{t,l}$ to a given fraction of $|Cd|_t$ (e.g. 50%) increases with the concentrations of (i) the inorganic anions, and (ii) the major cations (Ca^{2+} and Mg^{2+}). This is due to the competition between NTA and the labile complex-forming ligands for Cd^{2+} in case (i), and to the 'masking' of NTA by the major cations in case (ii).

The expression for the conditional stability constant, K, of CdNTA, under the studied ionic strength and temperature conditions can then be combined with eqn (6.37) and (6.41) to give:

$$K = \frac{|CdNTA|}{|Cd|\,|NTA|} = \frac{|CdNTA|}{(|Cd|_{t,l}/\alpha_l)\cdot(|NTA|_t/\alpha_{NTA})} = \frac{(i_p^{nc}/i_{p,l}^c)-1}{|NTA|_t/\alpha_l\cdot\alpha_{NTA}} \tag{6.43}$$

where

$$|NTA|_t = |NTA|\cdot\alpha_{NTA}$$

$$= |NTA|+|HNTA|+|NaNTA|+|CaNTA|+|MgNTA|. \tag{6.44}$$

In particular, eqn (6.43) shows that K is easily obtained at $i_p^{nc}/i_{p,l}^c = 2$ (50% decrease in $i_{p,l}^c$). Then

$$K = \frac{\alpha_l\alpha_{NTA}}{|NTA|_t} \tag{6.45}$$

By using the literature values for the stability constants of complexes in eqns (6.44) and (6.39), an average value of log $K = 9.54 \pm 0.2$ was found for solutions 1 and 3–7, in agreement with the literature values. This shows that the effect of major ions is indeed predictable by the α_{NTA} and α_l terms and that the stability constant of CdNTA determined at high concentration, in pure solution, is also applicable under the natural water conditions. Particularly, the coincidence of results in artificial and natural sea waters shows that complexation of Cd by NTA is not significantly influenced by any other components present in the natural sample. The approach used in this study was applied in a similar way to determine the mechanism and kinetics of formation of Cd(II), Pb(II), and Zn(II) complexes with EDTA, in the presence of Ca(II), in sea and lake waters [81].

6.6.3   Labile, slowly diffusing complexes, with adsorption of ligands (Case D, Table 6.1): study of Pb(II) complexation by pedogenic fulvic compounds of fresh waters (basic references [26,27])

6.6.3.1   Purpose and conditions

The fulvic compounds of fresh waters constitute 60–90% of the total aquatic organic matter and most of its 'unidentified' fraction (Fig. 6.1). They can be separated into two different groups of homologous compounds: the 'pedogenic', i.e. soil-originated, and the 'aquogenic' compounds, i.e. those formed in the water bodies themselves by primary productivity. Their relative importance depends on the nature of the water studied, its location and the season, but pedogenic compounds are usually present in large proportions in inland waters because of their continuous leaching by rainfall. $1–10\,mg\,dm^{-3}$ are typical values for the concentration of pedogenic fulvic compounds, but values up to $100\,mg\,dm^{-3}$ can be found. They are known to have molecular weights ranging between a few hundreds and few thousands (Fig. 6.6). Their major complexing sites are phenolate and benzene-carboxylate, and above all the chelating sites formed by their combinations (phthalate and salicylate type of sites). However, they may also include minor, but stronger sites.

    The aim of this work was to test the capabilities of voltammetric techniques for complexation studies with this important group of natural ligands, and to measure their complexation properties for Pb(II) to enable predictions about their role in the aquatic behaviour of lead. One important aspect of this work was to check how the methods used to test the nature of electrode processes in pure solution of well-characterized ligands can be applied to a group of ill-characterized but homologous natural compounds.

Experimental conditions

The concentration of pedogenic fulvic compounds (hereafter denoted by L) was varied between 10 and $160\,mg\,dm^{-3}$ ($6 \times 10^{-6}–10^{-4}\,mol\,dm^{-3}$ of complexing sites), i.e. in a range as close as possible to that of natural waters, but which also fulfilled the requirement: $|L|_t/|Pb|_t > 10$ [eqn (6.24)].

    Pb(II) was added to the solutions, and $|Pb|_t$ varied between 10^{-6} and $10^{-5}\,mol\,dm^{-3}$, depending on the method used. Due to the complexity of the test system, several methods were used for intercomparison of the data and only direct reduction methods were employed because the interpretation of the results is easier. In each case, the minimum concentration of lead required to obtain a meaningful interpretation of the above tests was used. It must be noted, however, that with this type of system, the fulfilment of the conditions required for testing the electrode process, and the adsorption properties of the complexing agent, often limit the sensitivity of the method. For instance, the sensitivity in DPP was limited by the necessity of using small pulse amplitudes and drop times (see below). The typical electroanalytical conditions used were the following (results with NPP and cyclic voltammetry are given in ref. [110]):

- dc polarography: scan rate $= 2\,mV\,s^{-1}$,
- cyclic voltammetry: scan rate $= 10-100\,mV\,s^{-1}$,
- ac polarography: frequency $= 20-1000\,Hz$; amplitude $= 10\,mV$
- DPP: scan rate $= 2\,mV$; pulse amplitude $= \Delta E = 10\,mV$; pulse duration $= 49\,ms$; sampling time at end of pulse $=$ last $5\,ms$
- NPP: pulse duration $65\,ms$; sampling time $=$ last $7\,ms$.

The electrodes used were DME (drop time $= t_d = 0.3-5\,s$; flow-rate $= 1.35\,mg\,s^{-1}$), HMDE (surface area $= 2\,mm^2$) and $Ag/AgCl/sat.KCl/0.1$ $mol\,dm^{-3}\,NaNO_3//$ reference electrode.

Solutions were thermostatted at $25 \pm 0.1°C$, ionic strength was fixed with $0.1\,mol\,dm^{-3}\,NaNO_3$, and pH was kept constant at 6.0 without using a buffer, to avoid complexation by either OH^- or the buffer itself, as well as possible interactions between the complexing agent and the buffer.

In order to check the validity of complexation measurements obtained in this work, they were compared with those obtained with Pb(II)-ISE [41, 108] and Pb(II) amalgam electrode [71]. Furthermore, the same system was also studied at lower concentrations (Pb$_t = 10^{-7}-10^{-6}\,mol\,dm^{-3}$ and $|L|_t = 5-60\,mg\,dm^{-3}$) using ASV [47].

Nature of the sample

This sample belongs to the second category in Section 6.2.1: It is formed by a group of homologous, but not fully characterized ligands. As discussed in Section 6.5 a meaningful interpretation of voltammetric results is possible in such cases, only for sufficiently homogeneous samples, particularly with respect to size distribution; on the other hand, as far as possible, the fractionation procedure used must not change the characteristics of the complexing agent. The fulfillment of the two above requirements was checked from the results of complexation measurements made on the same sample, but with different degrees of fractionation, and by detailed analysis of the test solutions:

- The water sample was collected in a pond located in Fontainebleau forest, in a non-peaty, unpolluted, granitic region (minimum amount of dissolved inorganic salts). This pond has a naturally low productivity and sampling was done at the end of autumn when the amounts of aquogenic fulvic and proteinaceous compounds are even more negligible. This ensured the original water to include mostly pedogenic fulvic compounds, so that the pretreatment could be reduced to a minimum, in order to avoid as far as possible any perturbation in the structure and nature of complexing agents. The samples were passed through a 0.2 μm filter, and the small amount of HCO_3^- was eliminated by stoichiometric quantities of $HClO_4$. Concentrated solutions were prepared by freeze drying and stored at 4°C in the dark. A concentration factor of 40 was used in this work for stock solution. Later experiments showed that similar results are obtained with a concentration factor of 6.

• The samples were characterized by thermogravimetry, UV-visible absorption and fluorescence spectroscopy [68], cascade ultrafiltration [66, 68], membrane osmometry [108], and potentiometric titrations [41, 108]. These results confirmed that the organic compounds studied resemble fulvic compounds extracted from soils, particularly with respect to their content of phenolic and carboxylic groups, their size distribution (80% between 10–20 Å; compare with Fig. 6.6b) and their average molecular weight (1700). Because of this relatively narrow size distribution, the use of an average value for the diffusion coefficient of complexes was considered meaningful (Section 6.5.1.2b).

• The sample was analyzed for trace metals, including iron and aluminium, and inorganic anions. Their concentration was negligible compared to that of the organic complexing sites. The Ca^{2+} concentration was between $1-6 \times 10^{-4}$ mol dm^{-3}, i.e. too low to compete with heavy metals [41]. The negligible role of Fe, Ca and inorganic anions of the sample was confirmed by results obtained on samples purified by cascade ultrafiltration [41].

6.6.3.2 Electroanalytical characteristics of the system

With such a system, several of the simplifying assumptions discussed in Section 6.5 are not valid, or at least must be checked carefully.

(a) As discussed before (Section 6.6.1.2), the $Pb^{2+}/Pb(o)$ couple is known to be reversible.

(b) $|L|_t$ and $|M|_t$ are chosen to fulfil the condition $|L|_t/|M|_t > 10$ [eqn (6.24)] as far as possible. Only in a few cases this ratio was smaller.

(c) There is no pH buffering agent, in the solution, other than L itself (Section 6.5.2.4). During the reduction of Pb(II), free L may be liberated at the electrode surface, due to the dissociation of PbL, and H^+ may combine with it. As a result, the pH at the surface (pH_0) may be higher than that in the bulk of the solution (pH), which may give rise to a larger value of α_0 compared with α. In such a system, where no buffer is added, but where L takes its place, the value of $\Delta pH = pH_0 - pH$ can be computed from (i) an experimental pH titration curve of L, and (ii) the values of the diffusion coefficients of the diffusing species. In the present case, it was found that for $|L|_t/|M|_t > 10$, the value of ΔpH was <0.1 unit.

(d) Natural organic compounds are often strongly adsorbed. Hence this effect (Section 6.5.2.3) was studied in detail by:

Cyclic voltammetry. The peak potentials and currents of the cathodic (E_{pc}, i_{pc}) and anodic (E_{pa}, i_{pa}) peaks were recorded. A general shift of E_{pc} and E_{pa} towards negative values was observed in complexing media, when compared with a non-complexing one (Fig. 6.14). This suggests the formation of labile complexes. However, when HMDE is left in the solution for increasing times, t_a, before beginning the scan, E_{pc} tends towards more negative values and i_{pc} increases, behaviour which cannot be explained by case A in Table 6.1. Note that in Fig. 6.14, a low value of $|L|_t/|M|_t$ was chosen because a more visible effect is observed under this condition. For this reason these results can be used only for qualitative interpretation.

DPP. A negative shift in ΔE_p was observed (Fig. 6.15) which increased with the time of electrode–solution contact (t_a = drop time in this method). Simultaneously i_p^c/i_p^{nc} increased (Fig. 6.16).

These observations could be interpreted quantitatively as resulting from slow adsorption of L and formation of adsorbable labile complexes. This produces a larger value of α_0 compared to α, hence a larger value in ΔE_p, and an increased reduction current due to the accumulation of Pb(II) at the surface. The characteristics of this adsorption process were confirmed by studies in the absence of Pb(II) [93].

(e) The dissociation rate of PbL complexes was checked in detail by using the fact that the adsorption process is fairly slow. Under this condition E_p and i_p

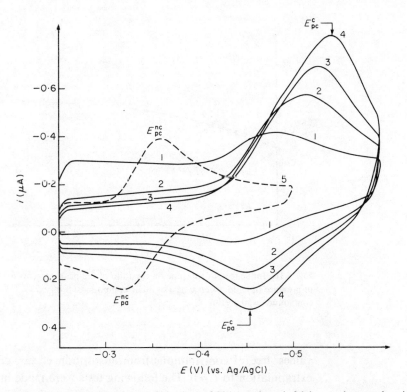

Fig. 6.14. Cyclic voltammetric behaviour of Pb(II)–pedogenic fulvic complexes on hanging mercury drop electrode. Initial potential: -0.25 V; organic matter 13 mg dm^{-3};
$|Pb(II)|_t = 2 \times 10^{-5}$ mol dm^{-3}; scan rate: 100 mV.s; electrolyte: NaNO$_3$ 0.1 mol dm^{-3}; pH = 8.0. Variation of cyclic voltammetric curves with waiting time (t_a) 1, $t_a = 3$ s; 2, $t_a = 30$ s; 3, $t_a = 60$ s; 4, $t_a = 120$ s; 5, curve of Pb^{2+} in the absence of organic matter.

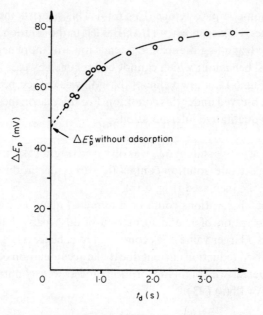

Fig. 6.15. Change in the DPP peak potential as a function of drop time, for the reduction of Pb(II)–pedogenic fulvic complexes, due to the slow adsorption of fulvic compounds. $|Pb|_t = 5 \times 10^{-6}$ mol dm^{-3}; fulvic compounds $= 100$ mg dm^{-3}.

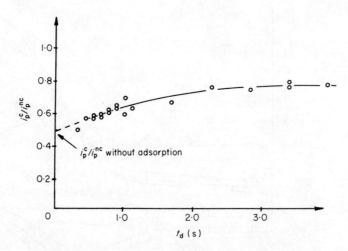

Fig. 6.16. Influence of drop time on the ratio of the peak currents for the reduction of Pb(II) in presence (i_p^c) and absence (i_p^{nc}) of fulvic compounds. $|Pb|_t = 5 \times 10^{-6}$ mol dm^{-3}, fulvic compounds $= 60$ mg dm^{-3}.

values, free of contributions from adsorption effects, can be evaluated by extrapolation to $t_a = 0$. The following tests were made in this way:

● In cyclic voltammetry, the values of $E_{pa} - E_{pc}$ and i_{pa}/i_{pc} parameters corresponded to the theoretical ones (30 mV and 1.0, respectively) for a reversible diffusion-controlled system.

● In DPP, the value of i_p^c/i_p^{nc} was 0.5 instead of 1.0, which may be due either to slowly diffusing or slowly dissociating complexes (Fig. 6.8.II a,b). Further tests were then performed to check whether or not the complexes were labile even with fast techniques:

- In cyclic voltammetry i_{pc} was found to be proportional to $\sqrt{v}$ (v=scan rate) and E_{pc} was independent of v, for $10 < v < 200\,\text{mV s}^{-1}$.
- In ac polarography, at any potential along the peak, $\cot\phi = 1$ (where ϕ = phase angle of the alternative component of the current). Furthermore, the peak current was proportional to $\sqrt{w}$ (w=frequency of the superimposed alternating potential).
- The absolute value of i_p was the same for DPP peaks with pulse amplitude $+\Delta E\,\text{mV}$ and $-\Delta E\,\text{mV}$, respectively, and the difference in E_p was equal to ΔE (Section 6.5.2.1). Furthermore, ΔE_p tends gradually to 0 when $|M|_t/|L|_t$ increases (Table 6.1b).

All these criteria indicate a reversible reduction of labile complexes, with diffusion controlled current, even for very fast techniques (for details on tests with cyclic voltammetry and ac polarography: see ref. [109]). Hence the average diffusion coefficient, D_{PbL}, of the complex can be computed from eqn (6.26) applied to DPP peaks currents: $D_{PbL} = 1\text{--}2 \times 10^{-6}\,\text{cm}^2\,\text{s}^{-1}$. This result was confirmed by the data obtained in the same way with other techniques. Furthermore, α can be computed from eqn (6.28):

$$\ln \alpha = \frac{2F}{RT}(E_p^{nc} - E_p^c) - \ln(i_p^c/i_p^{nc}) \tag{6.46}$$

where all the E_p and i_p^{nc}/i_p^c values are obtained by extrapolation to $t_d = t_a = 0$.

6.6.3.3 Complexation results

The values of $E_p^{nc} - E_p^c$ and i_p^{nc}/i_p^c were measured by DPP at constant $|Pb|_t = 5 \times 10^{-6}\,\text{mol dm}^{-3}$ and by varying $|L|_t$. For each solution, t_d was varied between 0.3 s and 3 s, and the above parameters were extrapolated to $t_d = 0$, and introduced into eqn (6.46) to compute α.

Obviously the extrapolation procedure has two drawbacks. It is time consuming and the sensitivity of the method is lowered due to the decrease in current with t_d. However, Fig. 6.15 and 6.16 show that extrapolation is essential for such a system, as, otherwise values of α too high by factors as large as 20 may be obtained.

6.6.3.4 Interpretation of the data

The log α values obtained above were interpreted in terms of an average equilibrium quotient, the value of which is in good agreement with that obtained by other methods (Table 6.3). Because of the relative homogeneity of the sample, as well as the relatively small range in which $|L|_t/|M|_t$ was varied (6–25), this average equilibrium quotient approximates to the mean value of the differential equilibrium function, K, under the conditions used. Nevertheless, as discussed in Section 6.2.2.2, recent findings have shown that direct

Table 6.3. Average properties of the complexes formed between Pb(II) and pedogenic fulvic compounds, measured in the range $|L|_t/|M|_t = 7\text{--}30$. All parameters are computed with the average value of molecular weight = 1700

Measured parameter	Voltammetric methods	Pb–ISE	Pb Amalgam electrode
$\bar{\beta}_1 =$ stability constant of dissolved PbL(dm^3 mol^{-1})	$10^{5.3}$	$10^{5.5}$	$10^{5.1}$
$\bar{k}_{5d} =$ dissocation rate constant of PbL(s^{-1})	$> 10^4$		
$\bar{D}_{PbL} =$ diffusion coefficient of the complex (cm^2 s^{-1})	2×10^{-6}		
$\bar{\beta}_1^a =$ stability constant of adsorbed PbL(M^{-1})	$10^{6.7}$		

computation of K is not much more difficult [8, 45] and gives more detailed information.

Table 6.3 also gives the values of other parameters for which voltammetric methods are particularly useful, due to the lack of other techniques: not only the kinetic properties of PbL could be estimated but also its degree of complexation in the adsorbed state. In this respect the mercury–water interface can be used as a well-controlled model interface to provide useful information for better understanding of the reactions occurring at natural interfaces [93]. This ability to study adsorption processes is an additional attribute of the voltammetric method.

The adsorption effect decreases with $|L|_t$ and it might be expected to be weak under natural conditions since the corresponding values of $|L|_t$ are generally lower than those used in this work. However, $|M|_t$ is also smaller, so that ASV techniques must be used. In these techniques the adsorption time, t_a, corresponds to deposition time, i.e. it is much larger than in direct reduction methods. Hence adsorption effects may not be negligible even at low $|L|_t$ [47, 93].

6.6.4   **Conclusions**

The three examples cited above may be considered as three typical cases which may be encountered with aquatic samples. Although the difficulties involved in a correct interpretation of electrode processes depend on the nature of the case under study, all three examples show the great importance of testing the assumptions used for interpreting the complexation data. In practice, these tests are often fairly easy to perform, as many of them can be done by varying the parameters of a given method (scan rate, drop time, deposition time, etc.) without changing the solution. With modern voltammetric instrumentation, to change such parameters or even the method itself is very simple and not time consuming. It must also be emphasized that it is not necessary to use simultaneously all the tests described here. Several of them were performed in Section 6.6.3 mostly to show their capabilities, and because the system was totally unknown. Obviously in the future, the more information that is known about the general properties of natural complexing agents, the less will be the number of tests necessary to characterize a particular system. This is seen in the first two examples, in which the properties of labile complexes forming ligands were well known from previous studies.

Finally, all three examples emphasize the importance of a correct choice of sampling location and time. In the first two, it was possible to justify the validity of the assumption that particulate or adsorbable organic matter was absent in Adriatic sea water. In the third case a sample as homogeneous as possible was available, even before applying any pretreatment. Obviously this is possible mostly when one tries to study a given type of complexing agent. However, even with Objective B, the interpretation of data is possible, thanks to the variation in the relative importance of the various complexing agents as a function of time (season) and location. Good coordination between environmental and electroanalytical considerations is always necessary to obtain meaningful results.

6.6.5   List of symbols

α	degree of complexation of $M = |M|_t/|M|$.
D_M, D_L, D_{ML}	diffusion coefficient of M, L or ML.
K	equilibrium constant for a well-characterized complexation reaction in solution or differential equilibrium function (Section 6.2.2.2).
$\bar{K}$	average equilibrium function for complex formation of M with a mixture of ill-characterized ligands (Section 6.2.3).
$\tilde{K}$	average equilibrium quotient for complex formation of M with a mixture of ill-characterized ligands (Section 6.2.3).
k_d	dissociation rate constant of a complex ML.
k_f	formation rate constant of a complex ML.
$|L|$	free ligand concentration.
$|L|_t$	total ligand concentration.
$|M|$	free metal ion concentration.
$|M|_i$	concentration of the ith species of M or of the ith fraction of M complexes.
$|M|_t$	total concentration of M.
ASV	anodic stripping voltammetry.
DPASV	differential pulse anodic stripping voltammetry.
DPP	differential pulse polarography.
DME	dropping mercury electrode.
HMDE	hanging mercury drop electrode.
NPP	normal pulse polarography.
TMFE	thin mercury film electrode.
ISE	ion selective electrode.
$E_p^c(E_p^{nc})$	peak potential of complexed (uncomplexed) M.
ΔE_p	$E_p^{nc} - E_p^c$
$i_p^c(i_p^{nc})$	peak current of complexed (uncomplexed) M.
t_d	deposition time in ASV techniques/drop time with techniques using DME.
$W_{1/2}$	peak width for $i_p/2$.

6.7 References

1 This volume Section 3.
2 J. Buffle and Cebedeau, p. 165, (May 1979).
3 M. Whitfield and D. Turner, Chemical Modelling in Aqueous Systems. In *Am. Chem. Soc. Symp. Ser.*, **93,** (Ed. E.A. Jenne), Chap. 29. Am. Chem. Soc., (1979).
4 P. Baccini and U. Suter, *Schweiz. Z. Hydrol.*, **41**(2), 291 (1979).
5 W. Stumm and J.J. Morgan, *Aquatic Chemistry*, J. Wiley and Sons, NY (1981).
6 S.A. Huntsman and W. G. Sunda, The Physiological Ecology of Phytoplancton. In *Studies in Ecology*, Vol. 7, (Ed. I. Morris). University California Press, Berkeley (1980).
7 M.M. Benjamin and J.O. Leckie, *Environ. Sci. Technol.*, **15**(9), 1050 (1981).
8 J. Buffle, Circulation of Metals in the Environment. In *Metal ions in Biological Systems*, Vol. 18, (Ed. H. Sigel), chap. 6. M. Dekker, NY (1984).
9 Y.K. Chau, P.T.S. Wong and O. Kramar, *Anal. Chim. Acta*, **146,** 211 (1983).
10 Y.K. Chau, P.T.S. Wong and G.A. Bengert, *Anal. Chem.*, **54,** 246 (1982).
11 M.R. Hofmann, E.C. Yost, S.J. Eisenreich and W.J. Maier, *Environ. Sci. Technol.*, **15**(6), 655 (1981).
12 P. Figura and B. Mc Duffie, *Anal. Chem.*, **52,** 1433 (1980).
13 T.M. Florence and G.E. Batley, *Talanta*, **23,** 179 (1976).
14 T.M. Florence and G.E. Batley, *Talanta*, **24,** 151 (1977).
15 T.A. O'Shea and K.H. Mancy, *Anal. Chem.*, **48**(11), 1603 (1976).
16 O. Zali, *Cycles chimiques dans un lac eutrophe*, Thèse No 2090, University of Geneva (1983).
17 W. Davison, *Limnol. Oceanogr.*, **22,** 746 (1977).
18 W. Davison and S.I. Heaney, *Limnol. Oceanogr.*, **23,** 1194 (1978).
19 R.F. Srna, K.S. Garrett, S.M. Miller and A.B. Thum, *Environ. Sci. Technol.*, **14,** 1482 (1980).
20 Y.K. Chau and K. Lum–Shue–Chan, *Water Res.*, **8,** 383 (1974).
21 M. Plavšić, D. Krznarić and M. Branica, *Marine Chem.*, **11,** 17 (1982).
22 H.W. Nürnberg and P. Valenta, in *Trace Metals in·Sea Water*, (Ed C.S. Wong). Plenum Press, NY (in Press).
23 I. Ružic, *Anal. Chim. Acta*, **140,** 99 (1982).
24 H.W. Nürnberg, P. Valenta, L. Mart, B. Raspor and L. Sipos, *Z. anal. Chem.*, **282,** 357 (1976).
25 J.R. Tuschall and P.L. Brezonik, *Limnol. Oceanogr.*, **25,** 495 (1980).
26 F.L. Greter, J. Buffle and W. Haerdi, *J. Electroanal. Chem.*, **101,** 211 (1979).
27 J. Buffle and F.L. Greter, *J. Electroanal. Chem.*, **101,** 231 (1979).
28 H.W. Nürnberg, *Thalassia Jugoslavica*, **16,** 95 (1980).
29 J. Buffle, *Trends in Anal. Chem.*, **1,** 90 (1981).
30 H.W. Nürnberg, *Pure Appl. Chem.*, **54,** 853 (1982).
31 D.S. Gamble, J.A. Marinsky and C.H. Langford, *Humic trace metal ion equilibria in natural waters*, IUPAC Report, Commission V.6, Analytical Chem. Division, to be published.
32 J. Heyrovsky and J. Kuta, *Principles of Polarography*, Academic Press, London (1966).
33 Z. Galus, *Fundamentals of electrochemical analysis*, Ellis Horwood, Chichester (1976).
34 A.J. Bard and L.R. Faulkner, *Electrochemical Methods*, J. Wiley and Sons, NY (1980).
35 G.H. Nancollas and M.B. Tomson, *Pure Appl. Chem.*, **54,** 2675 (1982).
36 M.S. Shuman, J.C. Bradley, P.J. Fitzgerald and D.L.O. Olson, in *Aquatic and Terrestrial Humic Materials*, (Eds. R.F. Christman and E.T. Gjessing), chap. 17. Ann Arbor Science Pub., Ann Arbor, Mich. (1983).
37 D.E. Wilson and P. Kinney, *Limnol. Oceanogr.*, **22,** 281 (1977).
38 J.A. Marinsky, S. Gupta and P. Schindler, *J. Colloid Interface Sci.*, **89,** 412 (1982).
39 P.W. Schindler, in *Adsorption of Inorganics at Solid–Liquid Interfaces*, (Eds. M.A. Anderson and A.J. Rubin) chap. 1. Ann Arbor Science Pub., Ann Arbor (1981).
40 M.M. Benjamin and J.O. Leckie, *J. Colloid Interface Sci.*, **79,** 209 (1981).
41 J. Buffle, P. Deladoey, F.L. Greter and W. Haerdi, *Anal. Chim. Acta*, **116,** 255 (1980).
42 C.H. Langford, D.S. Gamble, A.W. Underdown and S. Lee, In *Aquatic and Terrestrial Humic Materials*, (Eds. R.F. Christmann and E.T. Gjessing) chap. 10. Ann Arbor Science Pub., Ann Arbor, Mich. (1983).
43 J. Buffle, *Anal. Chim. Acta*, **118,** 29 (1980).
44 D.S. Gamble and M. Schnitzer, In *Trace Metals and Metal-Organic Interactions in Natural Waters*, (Ed. P.C. Singer), chap. 9, Ann Arbor Science Pub. Inc., Ann Arbor (1973).
45 D.S. Gamble, A.W. Underdown and C.H. Langford, *Anal. Chem.*, **52,** 1901 (1980).

46 W. Davison, *J. Electroanal. Chem.*, **87,** 395 (1978).

47 J. Buffle, A. Tessier and W. Haerdi, *Proc. Symposium on Complexation of Trace Metals in Natural Waters*, (May 2–6/1983). Martinus Nijhoff, Dr. W. Junk Pub. (1984).

48 E.M. Perdue and C.R. Lytle, In *Aquatic and Terrestrial Humic Materials*, chap. 14. Ann Arbor Science Pub., Ann Arbor, Mich. (1983).

49 A. Lerman, *Geochemical Processes*, J. Wiley and Sons, NY (1979).

50 S. Ahrland, *Speciation of Trace Metals in Sea Water*, IUPAC Report, Commission V.6, Section 7.

51 D. Turner, M. Whitfield and A.G. Dickson, *Geochim. Cosmochim. Acta*, **45,** 855 (1981).

52 J.R. Tuschall and P.L. Brezonik, In *Aquatic and Terrestrial Humic Materials*, (Eds. R.F. Christman and E.T. Gjessing), chap. 13. Ann Arbor Science Pub., Ann Arbor Mich. (1983).

53 B. Raspor, P. Valenta, H.W. Nurnberg and M. Branica, *Sci. Tot. Environ.*, **9,** 87 (1977).

54 B. Raspor, H.W. Nurnberg, P. Valenta and M. Branica, *J. Electroanal. Chem.*, **115,** 293 (1980).

55 A. Zirino, In *Marine Electrochemistry*, (Eds. M. Whitfield and D. Jagner), chap. 10. J. Wiley and Sons, Chichester, NY (1981).

56 L. Mart, *Z. anal. Chem.*, **296,** 350 (1979).

57 L. Mart, *Z. anal. Chem.*, **299,** 97 (1979).

58 J.P. Riley, In *Chemical Oceanography*, Vol. 3, chap. 19. Academic Press, London, (1975).

59 W. Davison, *Freshwater Biology*, **7,** 393 (1977).

60 P. Baccini, *Schweiz. Z. Hydrol.*, **38,** 121 (1977).

61 R.E. Truitt and J.H. Weber, *Anal. Chem.*, **51,** 2057 (1979).

62 Jr G.T. Wallace, I.S. Fletcher and R.A. Duce, *J. Environ. Sci. Health*, **12,** 493 (1972).

63 R.F.C. Mantoura and J.P. Riley, *Anal. Chim. Acta*, **78,** 193 (1975).

64 M. Ghassemi and R.F. Christman, *Limnol. Oceanogr.*, **13,** 583 (1968).

65 R.F.C. Mantoura and J.P. Riley, *Anal. Chim. Acta*, **76,** 97 (1975).

66 J. Buffle, P. Deladoey and W. Haerdi, *Anal. Chim. Acta*, **101,** 339 (1978).

67 K. Hirose, Y. Dokiya and Y. Sugimura, *Marine Chem.*, **11,** 343 (1982).

68 J. Buffle, P. Deladoey, J. Zumstein and W. Haerdi, *Schweiz. Z. Hydrol.*, **44**(2), 325 (1982).

69 J. Koryta, *Ion Selective Electrodes*, Cambridge University Press, (1975).

70 J. Vesely, D. Weiss and K. Stulik, *Analysis with Ion Selective Electrodes*, Ellis Horwood, NY (1978).

71 J. P. Bernhard, *Application du principe des électrodes à amalgame à la spéciation des métaux lourds dans les eaux naturelles.* Thèse No 2054, Université de Genève (1982).

72 D.M. Mc Knight and F.M.M. Morel, *Limnol. Oceanogr.*, **25,** 62, (1980).

73 G. Sposito and K.M. Hotzclaw, *Soil Sci. Soc. Am. J.*, **43,** 47 (1979).

74 D.D. Perrin and B. Dempsey, *Buffers for pH and Metal Ion Control*, Chapman and Hall, London (1974).

75 J.C. Westall, F.M.M. Morel and D.N. Hume, *Anal. Chem.*, **51,** 1792 (1979).

76 G.J. Moody, N.S. Nassory, J.D.R. Thomas, D. Betteridge, P. Szepesvary and B.J. Wright, *Analyst*, **104,** 348 (1979).

77 J.R. Tuschall and P.L. Brezonik, *Anal. Chem.*, **53,** 1986 (1981).

78 M.S. Shuman, *Anal. Chem.*, **54,** 998 (1982).

79 G.A. Bhat and J.H. Weber, *Anal. Chem.*, **54,** 2116 (1982).

80 J.R. Tuschall and P.L. Brezonik, *Anal. Chem.*, **54,** 2116 (1982).

81 B. Raspor, H.W. Nurnberg, P. Valenta and M. Branica, *J. Electroanal. Chem.*, **115,** 293 (1980).

82 D.D. De Ford and D.N. Hume, *J. Amer. Chem. Soc.*, **73,** 5321 (1951).

83 M. Branica, D.M. Novak and S. Bubic, *Croat. Chem. Acta*, **49,** 539 (1977).

84 A. Zirino and S. Kounaves, *Anal. Chem.*, **49,** 56 (1976).

85 H.P. Van Leeuwen, *J. Electroanal. Chem.*, **99,** 93 (1979).

86 J. Buffle, *J. Electroanal. Chem.*, **125,** 273 (1981).

87 P.L. Brezonik, P. Brauner and W. Stumm, *Water Res.*, **10,** 605 (1976).

88 Z. Kozarac, B. Cosovic and M. Branica, *J. Electroanal. Chem.*, **68,** 75 (1976).

89 H.P. Van Leeuwen, Electroanalysis in Hygiene, environmental, clinical and pharmaceutical chemistry. In *Analytical Chemistry Series Symposia*, Vol. 2, (Ed. W.F. Smyth), Elsevier Pub. Cy, pp. 383–397, Amsterdam (1980).

90 J.H. Weber and K.H. Cheng, *Anal. Chem.*, **51**(7), 796 (1979).

91 B. Cosovic, N. Botina and Z. Kozarac, *J. Electroanal. Chem.*, **113,** 239 (1980).

92 L. Fornaro and S. Trasatti, *Anal. Chem.*, **40,** 1060 (1968).

93 J. Buffle and A. Cominoli, *J. Electroanal. Chem.*, **121,** 273 (1981).

94   Z. Lukaszewski and M.K. Pawlak, *J. Electroanal. Chem.*, **103**, 225 (1979).

95   J. Buffle and G. Nembrini, *J. Electroanal. Chem.*, **76**, 101 (1977).

96   V. Zutic, B. Cosovic and Z. Kozarac, *J. Electroanal. Chem.*, **78**, 113 (1977).

97   H.P. Van Leeuwen, *Anal. Chem.*, **51**, 1322 (1979).

98   S.A. Wilson, T.C. Huth, R.E. Arndt and R.K. Skogerboe, *Anal. Chem.*, **52**, 1515 (1980).

99   T.M. Florence, *Anal. Chim. Acta*, **119**, 217 (1980).

100 N.E. Good, G.D. Winget, W. Witner and T.N. Connolly, *Biochemistry*, **5**, 467 (1966).

101 J. Koutecky and J. Koryta, *Electrochimica Acta*, **3**, 318 (1961).

102 D.R. Turner and M. Whitfield, *J. Electroanal. Chem.*, **103**, 43 (1979).

103 D.R. Turner and M. Whitfield, *J. Electroanal. Chem.*, **103**, 61 (1979).

104 R.A. Horme, *Marine Chemistry*, Wiley, NY. p. 140.

105 B. Raspor, H.W. Nurnberg, P. Valenta and M. Branica, *Limnol. Oceanogr.*, **26**, 54 (1981).

106 B. Raspor, H.W. Nurnberg, P. Valenta and M. Branica, In *Lead in the Marine Environment*, (Eds. M. Branica and Z. Konrad) Pergamon Press (1977), p. 81.

107 B. Raspor and M. Branica, *J. Electroanal. Chem.*, **59**, 99 (1975).

108 J. Buffle, F.L. Greter and W. Haerdi, *Anal. Chem.*, **49**, 216 (1977).

109 E.R. Brown and R.F. Lange, "Electrochemistry methods", In *Techniques of Chemistry*, vol. I *Physical Methods of Chemistry*, Part IIA (Eds A. Weissberger and B.W. Rossiter) Wiley-Interscience, NY (1971) chap. VI.

110 J. Buffle, F.L. Greter, G. Nembrini, J. Paul and W. Haerdi, *Z. anal. Chem.*, **282**, 339 (1976).

111 M. Lovric and M. Branica, *Croat. Chim. Acta*, **53**, 485 (1980).

112 M. Lovric and M. Branica, *Croat. Chim. Acta*, **53**, 477 (1980).

113 H.W. Nurnberg, *Sci. Tot. Environ.*, **37**, 9 (1984).

114 K.W. Bruland, In *Chemical Oceanography*, vol. 8, Academic Press, London (1983), chap. 45.

115 D.P.H. Laxen and I.M. Chandler, *Geochim. Cosmochim. Acta*, **47**, 731 (1983).

116 T.M. Florence and G.E. Batley, *Crit. Rev. Anal. Chem.*, **9**, 219 (1980).

117 T.M. Florence, *Talanta*, **29**, 245 (1982).

118 L. Sipos, P. Valenta, H.W. Nurnberg and M. Branica, In *Lead in the Marine Environment*, (Eds M. Branica and Z. Konrad) Pergamon Press, Oxford (1980), p. 61.

Section 7 Trace Metal Complexation by Inorganic Ligands in Sea Water

By

Commission V.6
IUPAC

S. AHRLAND

Department of Inorganic Chemistry 1
Chemical Center
University of Lund
S-221 00 Lund
Sweden

7.1 Main Constituents of Sea Water

Sea water is essentially a fairly concentrated salt solution. In by far the largest part of the sea, mixing is very efficient and the concentration of the main constituents is, therefore, quite uniform. The ionic medium thus becomes fairly constant, with Na^+, Mg^{2+}, Ca^{2+} and K^+ as predominating cations (concentrations 479, 54.4, 10.5 and 10.4 mmol dm^{-3}, respectively), and Cl^- and SO_4^{2-} as predominating anions (concentrations 559 and 28.9 mmol dm^{-3}, respectively). Other ions contribute very little to the ionic medium, though the concentrations of carbonate, 2.35 mmol dm^{-3} present as HCO_3^- and CO_3^{2-}, and of Br^-, 0.86 mmol dm^{-3}, are certainly large enough to influence the speciation. This may also be the case for borate (total concentration 0.4 mol dm^{-3}) and even for F^- (0.075 mmol dm^{-3}), though hardly for the elements discussed in this Section. The formal ionic strength of sea water of this composition [1] is 0.714 mol dm^{-3}, corresponding to a salinity of 35‰. Owing to excessive evaporation, the salinity may be markedly higher in some cut-off parts of the sea, most notably in the Mediterranean, where it is around 40‰. Much lower salinities are, on the other hand, found in places where large quantities of fresh water are discharged. Such bodies of brackish water are found in estuaries, and, above all, in basins with very restricted communication with the sea, notably the Baltic. The conditions in brackish water will be treated in another part of this survey.

7.2 Value of pH in Sea Water

The value of pH in the ocean varies with depth, temperature and location [2]–[4]. Under normal conditions, however, the value is never far from 8. A value of 8.1 should be representative for the surface water in the major part of the ocean and will be adopted in the following calculations. As $pK_w = 13.8$ at the ionic strength of sea water [5], this means a $p[OH^-] = 5.7$.

Remarkably enough, this value is not far from one of the buffer capacity minima of the carbonate system, the only acid–base system present in sizeable concentration in sea water. To account for the stable value of pH actually found, processes releasing H^+ ions on a vast scale must, therefore, take place [6]. The first hypothesis propounded about the nature of these processes held that extensive ion exchange occurred between the sea water and the silicates transported into the oceans by the rivers. In fresh water, of low salt concentration, the silicates are relatively rich in H^+; once they arrive in the sea, H^+ is exchanged [6, 7] for Na^+ and Mg^{2+}. Closer examination showed, however, that such processes are certainly not able to produce H^+ at the rate needed for the protonation actually taking place. This led to an impasse where no reasonable solution to the problem seemed to be in sight.

Recently, however, a powerful new source of H^+ ions has been found. In the hot springs discovered in large numbers along the submarine ridges where new crust is created from ascending magma, hydrothermal processes take place that release H^+ on a scale that might be large enough [42]. These H^+ originally stem from water penetrating into the rifts where it reacts with the hot magma. Several different processes are thought to contribute to the release of H^+. Certainly hydrolytic reactions involving basalt silicates are very important. The formation of hydrogen sulfide, by the reduction of sulfate, results in a release of H^+ when metal sulfides are subsequently formed.

At the ionic strength of sea water, 0.7 mol dm^{-3}, mainly brought about by a salt presumed to be completely dissociated, viz. sodium chloride, the protonation constant of the carbonate ion $K = [HCO_3^-]/[H^+][CO_3^{2-}]$ is $10^{9.54} \text{ dm}^3 \text{ mol}^{-1}$ at 25°C [1, 8]. In standard sea water, a conditional constant $K_c = 10^{8.95} \text{ dm}^3 \text{ mol}^{-1}$ may be determined, on the assumption that $[CO_3^{2-}]$ and $[HCO_3^-]$ denote the total concentrations of these ions in solution, free as well as bound in soluble complexes. If the difference is assumed to be due to complex formation with Mg^{2+} and Ca^{2+}, the stabilities of these complexes may be estimated. Provided that only complexes with one ligand are formed at these low concentrations, and that the stabilities are the same for Mg^{2+} and Ca^{2+} (which is of course an over-simplification) one arrives at $\beta_1 - 3.9 \, \beta_{1H} = 45 \text{ dm}^3 \text{ mol}^{-1}$ where β_1 and β_{1H} are the stability constants of the complexes MCO_3 and $MHCO_3^+$, respectively. Certainly, $\beta_{1H} \ll \beta_1$, but the minimum value of $\beta_1 = 45 \text{ dm}^3 \text{ mol}^{-1}$ is anyhow somewhat higher than $\beta_1 = 33 \text{ dm}^3 \text{ mol}^{-1}$ suggested previously for $MgCO_3$ in standard sea water [8].

In sea water of pH = 8.1 and $K_c = 10^{8.45}$ the ratio $C_{HL}/C_L = 7.1$. At a total carbonate concentration $= 2.35 \text{ mmol dm}^{-3}$ this means $C_L = 0.29 \text{ mmol dm}^{-3}$ and $C_{HL} = 2.06 \text{ mmol dm}^{-3}$. With the reasonable values $\beta_1 \simeq 50 \text{ dm}^3 \text{ mol}^{-1}$ and $\beta_{1H} \simeq 1 \text{ dm}^3 \text{ mol}^{-1}$, the free ligand concentrations $[CO_3^{2-}] = 0.068 \text{ mmol dm}^{-3}$ and $[HCO_3^-] = 1.92 \text{ mmol dm}^{-3}$ are calculated. The total amount of carbonate bound in complexes would thus be $0.36 \text{ mmol dm}^{-3}$, or $\simeq 15\%$. Other calculations have yielded much the same result [1, 9], $\simeq 12\%$. It should be remembered, however, that these values refer to surface water at 25°C. As K_c varies considerably with temperature [2], the conditions are fairly different in deep water.

The value of $[CO_3^{2-}]$ found is close to that needed to precipitate $CaCO_3(s)$ as calcite or aragonite at the concentration of Ca^{2+} present. Also, the concentration of Mg^{2+} is high enough to make the precipitation of dolomite, $MgCa(CO_3)_2$, thermodynamically possible though this reaction generally does not occur under the conditions prevailing in the oceans [10].

7.3 Oxidation Potential of Sea Water

If certain basins with very poor circulation and mixing are excepted, the oxidation potential of the sea water is obviously high. How high has been a matter of dispute, however. So also has the question of which redox systems are in fact at work [11, 12].

It is usual to ascribe the high oxidation potential to the oxygen/water system, $\frac{1}{2}O_2 + 2H^+ + 2e^- \rightleftharpoons H_2O$, of a standard potential $E^0 = 1229$ mV [13]. The system is not reversible, however, so true equilibria are generally not established with other redox systems present. Formally its oxidative power might be calculated from

$$E = E^0 + \frac{RT}{2F} \ln p(O_2)^{\frac{1}{2}} [H^+]^2 \qquad (7.1)$$

With the atmospheric oxygen pressure $p(O_2) = 0.21$ atm and a value of pH $= 8.1$, eqn (7.1) yields $E = 740$ mV. The corresponding value of pE $= 12.5$ is calculated from

$$pE = E \left(\frac{RT}{F} \ln 10 \right)^{-1} \qquad (7.2)$$

and is another measure of the theoretical oxidative power, often convenient to use in equilibrium calculations [5].

Attempts to measure the actual oxidation potential of sea water by means of the iodate/iodide redox system [12] resulted in a value of pE $= 10.6$, i.e. markedly lower than the value calculated above, though still quite high. A corresponding measurement involving the nitrate/nitrogen system yields pE $= 10.5$, quite close to the iodate/iodide value. Other approaches result in much lower values of pE, however [12].

In areas where strong upwelling occurs, the concentration of oxygen stays practically constant with increasing depth. Generally, however, it decreases to a minimum at around 500 m and then increases again as the decay of organic material causing the oxygen consumption is completed. The minimum concentration is often $< 10\%$ of the surface concentration. However, such a decrease does not mean very much to the oxidative power at equilibrium. The value of pE decreases only to 12.0, even for a drastic reduction of the oxygen to 1% of the initial value. However, the near depletion of the oxidized form of this redox system implies a severe decrease in the redox buffering capacity of the solution. Consequently, a further consumption of oxygen by reducing substances might rather easily bring about a drastic lowering of the oxidation potential. This has indeed happened in the so-called anoxic zones existing in the Baltic and Black Seas, and in the Scandinavian and British Columbian fjords. In these waters, impeded circulation combined with fairly high oxygen consumption results in complete depletion of oxygen. Under such extreme conditions, the oxidation potential may be determined by the sulfide system which is present due to the activity of sulfate reducing bacteria. The reaction $S(s) + 2H^+ + 2e^- \rightleftharpoons H_2S(aq)$ has a very low value [13] of $E^0 = 141$ mV, corresponding to log $K = 4.77$. In anoxic water [14], the sulfide concentration may reach values $\simeq 0.5$ mmol dm^{-3}, while the pH value drops to $\simeq 7.5$. Assuming $K_1 = 10^{6.9}$ dm^3 mol^{-1} for the reaction $SH^- + H^+ \rightleftharpoons H_2S(aq)$, this would mean a very low value of pE $= -3.1$.

7.4 Factors Determining Speciation

The chemical form of a trace metal present in sea water is certainly determined in the main by the variables discussed above, i.e. the concentrations of complexing agents, the pH and the oxidation potential. Temperature and pressure will also influence the equilibria, however, and in a manner which is often not well known. This applies especially to changes due to the high pressures encountered in the deep parts of the sea.

Finally, it should be remembered that sea water contains many complexing agents in minute amounts which vary considerably with time and location. These agents are often of organic origin, stemming from the life or decay processes going on in the ocean or in waters discharged to it. Among those arriving with river water are, for example, humic and fulvic acids. Also some man-made chelates, e.g. NTA, are of interest in this connection. In spite of their low concentrations, some of these agents may be significant because of their high affinity for certain metals. If the latter are present only in trace amounts, they may become more or less completely sequestered, at least locally if the ligand is sufficiently selective. Such effects are naturally very difficult to account for in quantitative terms. In the present review, some probable, or at least possible effects that such ligands may bring about will be discussed. For certain trace metals, a considerable part of their total concentration may even be incorporated in various organisms if these are abundant. This is especially likely to occur with metals vital to life which are taken up even in very minute concentrations by highly efficient mechanisms.

7.5 Trace Metals under Discussion and their Total Concentrations

The following metals will be treated here: cadmium, chromium, cobalt, iron, lead, manganese, mercury, molybdenum, nickel, vanadium and zinc. In addition, the non-metal selenium will also be discussed. The selection has been made primarily from the point of view of their biological significance. In the main, the elements enumerated are those especially important for life and health.

The data found for the total concentrations of these elements in sea water, and the analytical procedures used for their determination, are exhaustively discussed in other parts of this survey. The difficulties in obtaining reliable values for these minute concentrations are formidable [43]–[45]. Nevertheless, a consensus has now been reached.

It has also been found that for certain metals the concentrations vary considerably both with depth and geographical location, even in the open seas. Cadmium provides an especially good example [46]. In the Pacific, the concentration increases about thirty times from surface to bottom (from $\simeq 0.03$ to $\simeq 1$ $mol\,dm^{-3}$); in the Atlantic the surface concentrations are about the same as in the Pacific, but the increase towards the bottom is much less. For other metals, such as copper or lead, the variations are much less [46]. The values listed in Table 7.1 should nevertheless be fairly representative estimates of the average concentrations found in the open sea. Due to human activities, much higher values are regularly found in coastal waters, often some hundred times higher than in the open sea. To a lesser degree, this also applies to many basins which are cut off from oceanic mixing. Thus in the Baltic, most trace metals are present in concentrations 3–5 times higher than in the bulk of the oceans [15].

Table 7.1. Total concentrations ($nmol\,dm^{-3}$) in the open sea of the trace elements discussed in this survey

V	Cr	Mn	Fe	Co	Ni	Cu	Zn
—	1	4	8	0.1	5	2	5

	Mo	Cd	Hg	Pb	Se
	—	0.1	0.02	0.05	1

7.6 Mutual Affinities between Different Classes of Metal Ions and Ligands

From the large number of stability measurements carried out in aqueous solution, general rules about the affinities between metal ions and ligands of different character can be deduced. Metal ions are divided into two classes, (a) 'hard', and (b), 'soft' [16, 17]. Metal ion acceptors of the first class, (a), strongly prefer ligands coordinating via the light donor atoms F, O, and N, while acceptors of the second class, (b), prefer ligands coordinating via the heavier donor atoms of each group. The donors preferred by the hard and soft acceptors are termed hard and soft donors, respectively. The (a)-acceptors have the same affinity sequence towards the various groups of donor atoms, while the sequences of the (b)-acceptors differ between the various groups, as shown in Table 7.2. An important common feature is, however, that the affinity difference between the first and second donor atom is always large, though in the opposite sense for (a)- and (b)-acceptors.

Table 7.2. Characteristic affinity sequences of acceptors of the two classes (a) hard, and (b) soft, for various donor groups

Donor group	Oxidation state	(a) Hard	(b) Soft
7 B	−I	$F \gg Cl > Br < I$	$F \ll Cl < Br < I$
6 B	−II	$O \gg S > Se > Te$	$O \ll S < Se \simeq Te$
5 B	−III	$N \gg P > As > Sb$	$N \ll P > As > Sb$

The (a)-sequences are those expected if the complex formation is mainly due to electrostatic interaction. In concordance with this, the complexes formed by (a)-acceptors are invariably stronger, the higher the charge and the smaller the radius of the acceptor. The most typical (a)-acceptors are, therefore, small multivalent cations.

The (b)-sequences are, on the other hand, evidently not compatible with a mainly electrostatic interaction. In these cases, the bonds formed must be of an essentially covalent character. This conclusion is further corroborated by the observation that complexes formed by (b)-acceptors are generally stronger the larger the radius of the acceptor involved, and that strong complexes are also formed by acceptors of low charge, or even by elements in their zerovalent oxidation state. In most cases, the (b)-character of an element, in fact, increases as the oxidation state decreases.

The metal ion acceptors of class (b) are situated in a triangular area of the periodic system, with its apex at copper. They combine many d-electrons in their outermost shell with a high polarizability, a combination which evidently promotes the formation of bonds of an essentially covalent character. Typical (b)-acceptors are the monovalent oxidation states of copper, silver and gold. These have the outer electron configuration d^{10}, i.e. filled d-shells, and the low ionic charge brings about a high polarizability which increases with the radius. Consequently, markedly covalent bonds can be formed, increasing in strength from copper(I) to gold(I).

Among the trace metals to be discussed here only mercury, present in the divalent state in sea water of ordinary pH and pE, behaves as a typical (b)-acceptor while cadmium(II) and lead(II) are on the borderline between the two classes. The first row transition metals from cobalt to zinc are also present in their divalent oxidation states. At least towards halide ions, these all behave as (a)-acceptors.

The divalent ions mentioned are not perceptibly hydrolysed, even at the fairly high values of pH $\simeq 8$ prevailing in sea water, except for mercury(II) and, to a lesser extent, lead(II) and copper(II) [5, 18]. Vanadium, chromium, manganese and iron are, on the other hand, under these conditions, all present in higher oxidation states and consequently strongly hydrolysed. The result of the hydrolysis differs widely between the various elements, as will be discussed below. The preferential affinity for oxygen over, e.g. chloride is unmistakable, however. These high oxidation states all behave as typical (a)-acceptors.

More or less profound changes in the oxidation potential affect the trace elements discussed very differently. For some of them, notably copper, iron, manganese, chromium and selenium, a lowering of the oxidation state, and a drastic change in the speciation, will take place long before free sulfide appears. Others, e.g. cadmium, zinc, cobalt and nickel, are not reduced.

In the following, the speciation under aerobic conditions will be treated in the first place. For those elements, however, where oxidation state and speciation are strongly influenced by a possible decrease of pE, the behaviour under such conditions will also be discussed.

Discussion of the speciation of trace metals present in sea water based on the principles advanced has in the past been conducted on several occasions

[10]. Recently, another extensive survey has been published, covering most of the trace metals known to be present in sea water [48]. Nevertheless, it seems worthwhile considering the topic again, in the light of the most recent investigations. Not least, the influences of slow acid–base and redox reactions on the actual speciation merit further attention.

This survey will start with the softest of the trace metals considered, viz. mercury, and then proceed to the harder ones.

7.7 Mercury

Mercury(II) is the stable oxidation state under aerobic conditions, and the speciation of mercury is essentially determined by the soft character of this state. Strong complexes are formed with the heavy halides. Moreover, organometallic methylmercury complexes are formed which are not only stable but also inert. Mercury may be methylated in the aquatic environment both by biological and non-biological processes. That such processes do occur in nature was discovered some fifteen years ago [19]. Their importance in the cycling of mercury between air, water and sediments and not least their biological significance was soon recognized [20]. As a consequence they have been studied extensively [21] in recent years.

Both dimethylmercury, $(CH_3)_2Hg$, and monomethylmercury, CH_3Hg^+, are formed. The latter forms a very stable chloride complex [22] CH_3HgCl. At the high chloride concentration of sea water, this species ought to predominate over dimethylmercury, especially as the latter compound is both volatile (bp. 96°C) and very slightly soluble in water. It should, therefore, prefer the atmosphere. In the open sea, the methylation certainly occurs exclusively by biological action. In heavily polluted coastal waters chemical agents may also be involved [21].

Strong complex formation makes the divalent oxidation state very stable in sea water. This applies especially to the methylated species which require much stronger reducing agents to be reduced to metallic mercury than do the simple halide complexes. The latter, but not the former, are reduced by tin(II) chloride solution. This has been utilized in order to differentiate between what has been called reactive and non-reactive mercury [23]. Even where the sea is not polluted, the non-reactive, i.e. the methylated, part is generally found to be at least half, often two-thirds, of the total concentration. The latter is found after reduction of the solution obtained once the methylated species have been destroyed by oxidation with peroxidisulfate. In polluted waters, the non-reactive mercury is even more predominant, which may reflect both higher biological activity and appreciable formation of methylated species via non-biological reactions. On account of the inertness of the methylmercury bond, no equilibrium is established between the reactive and the non-reactive forms.

The concentration of non-reactive mercury seems to vary much more than that of reactive mercury. Independent of place and depth [23, 24] the latter is generally $\simeq 3$ nmol dm^{-3}.

Under these conditions, all statements about the actual speciation of mercury must necessarily be approximate. The values listed in Table 7.3 have been arrived at on the following assumptions. (1) The organometallic mercury is assumed to be 60% of the total, and present only as CH_3HgCl. (2) Any sizeable dissociation of this complex does not take place, nor are any further chloride ligands added [22]. (3) The concentration of $(CH_3)_2Hg$ is presumed to be negligible. For the non-organometallic mercury, present as labile complexes rapidly attaining thermodynamic equilibrium, it should be possible to calculate the distribution from the stability constants characterizing the various equilibria. Generally, the values of these constants are not known for the ionic medium represented by sea water. From determinations performed in related media, however, data accurate enough for the present purpose can be found [5], Table 7.3. The results of the calculations are also shown in Table 7.3. In the first place, these refer to a standard temperature of 25°C and standard atmospheric pressure. They should nevertheless give a fairly good idea of the speciation under other conditions.

Mercury should be present almost exclusively as the complexes CH_3HgCl and $HgCl_4^{2-}$. Beside these, $HgCl_3^-$ and the mixed complex MCl_3Br^{2-} seem to exist in perceptible amounts. Hydrolytic, carbonate and sulfate complexes are virtually absent however.

As a typically soft acceptor, mercury(II) has a very strong affinity for sulfur

Table 7.3. Speciation (α, %) of mercury(II) and cadmium(II) in sea water. The stability constants β_j used for the calculations are also listed[a, b]

Species	Mercury(II) α	Mercury(II) $\log \beta_j$	Cadmium(II) α	Cadmium(II) $\log \beta_j$	Lead(II) α	Lead(II) $\log \beta_j$
M^{2+}	0		3		4	
MCH_3^+	0					
MCH_3Cl	60	5.32[c]				
MOH^+	0	10.1[d]	0	3.6	7	6
MCO_3	0	—	0.2	3	24[e]	5
MCl^+	0	6.74	34	1.36	17	0.9
MCl_2	1	13.22	51	1.79	19	1.2
MCl_3^-	5	14.07	12	1.44	27	1.6
MCl_4^{2-}	26	15.07	0	—	0	—
MBr^+	0	9.05	0.1	1.56	0	1.1
$MClBr$	1	15.9	0.2	2.3	0	—
MCl_2Br^-	2	16.4	0	2.1	0	—
MCl_3Br^{2-}	5	17.2	0	—	0	—
$MClOH$	0	17.4	0	—	0	—
MSO_4	0	1.3	0.3	1	0.4	1

[a] Free ligand concentrations used, p[L]: OH^-, 5.7; CO_3^{2-}, 4.17; Cl^-, 0.25; Br^-, 3.07; SO_4^{2-}, 1.90.
[b] Stability constants of the cumulative equilibria $M + jL \rightleftharpoons ML_j$ (charges omitted), except for MCH_3Cl (see [c]).
[c] Stability constant of the equilibrium $HgCH_3^+ + Cl^- \rightleftharpoons HgCH_3Cl$.
[d] Calculated from the dissociation K_1^* of the equilibrium $Hg^{2+} + H_2O \rightleftharpoons HgOH^+ + H^+$, with $pK_w = 13.8$.
[e] The second carbonate complex has $\alpha = 2$ ($\log \beta_2 = 8$).

donors. Any perceptable concentrations of such donors have not been found in sea water, but they are certainly present in living organisms. When these are abundant, therefore, they may more or less completely sequester the mercury locally. As the organic tissues decay, the sulfur compounds will certainly be oxidized and the mercury will revert to the forms listed in Table 7.3.

Under anaerobic conditions, the formation of very stable sulfide complexes prevents the reduction of mercury(II) [13]. For the expected values of pH and sulfide concentration, a mixture of various sulfide complexes, neutral and anionic, are probably formed. Owing to the extremely low mercury concentration, no precipitation of HgS(s) will occur, however [49].

7.8   Cadmium

Divalent cadmium has a mildly soft character, with a chemistry intermediate between that of the very soft mercury(II) and the fairly hard zinc(II). The formation of metallo-organic complexes of cadmium(II) in nature has never been reported. The complexes present in sea water are seemingly all labile. Though the chloride complexes are only moderately stable [5], they nevertheless dominate, with the second complex $CdCl_2$ as the major species, Table 7.3. The bromide complexes are somewhat stronger, but due to the low bromide concentration even the most abundant of them, the statistically favoured mixed complex CdClBr, still accounts for only $\simeq 0.2\%$ of the total cadmium. Nor are sulfate complexes of any consequence, due to their low stabilities at the ionic strength of sea water.

Cadmium(II) is not very prone to hydrolysis [18] Table 7.3. In sea water, the extensive formation of chloride complexes certainly prevents all hydrolytic reactions.

Cadmium carbonate is a fairly insoluble compound, with a solubility product [25] $K_{so} = 10^{-12.0}$ at $I = 0$ (and $10^{-11.2}$ at $I = 3.0$). For sea water, or for media of a composition similar to sea water, no data seem to exist. In the case of the carbonate [26], however, the value of K_{so} at $I = 0.2$ mol dm^{-3} is almost 10 times larger than at $I = 0$, implying activity coefficients $f \simeq 0.33$ for the divalent ions at $I = 0.2$. For sea water, a further decrease to $f \simeq 0.25$, or 0.2, would be reasonable, which means a further increase in K_s by a factor 2. The total increase in K_{so} between $I = 0$ and sea water would thus be 20 times, or 1.3 log units, resulting in a value of $pK_s = 10.7$. For other 2,2-electrolytes, e.g. magnesium and copper(II) sulfates, where equilibrium constants have actually been determined both at $I = 0$ and for media of an ionic strength comparable to sea water, activity variations of just this magnitude are indeed found [5, 27]. At the levels of $[Cd^{2+}]$ and $[CO_3^{2-}]$ found in sea water, the value of K_{so} cannot be exceeded even in heavily polluted areas. Precipitation of $CdCO_3(s)$ will, therefore, never take place.

The stabilities of the carbonate complexes in solution are not known. However, they should probably be more stable than the magnesium and less stable than the copper(II) complexes. Values of $\log \beta_1 = 3$ and $\log \beta_{1H} = 1.3$

seem to be reasonable estimates. The formation of carbonate complexes would be practically negligible, see Table 7.3.

In the case of cadmium(II), the influence of a polluting ligand of strongly complexing properties, viz. nitrilotriacetate, NTA, has been thoroughly investigated [28]. NTA was added to various salt media, all of the ionic strength of sea water, i.e. $0.7 \ mol \ dm^{-3}$, and all containing $10^{-7} \ mol \ dm^{-3}$ cadmium(II), i.e. a concentration representative of heavily polluted waters. The media chosen were: pure sodium perchlorate; 0.1 and $0.59 \ mol \ dm^{-3}$ sodium chloride; $0.01 \ mol \ dm^{-3}$ calcium chloride; $0.054 \ mol \ dm^{-3}$ magnesium chloride and, finally, both synthetic and natural (Adriatic) sea water. The calcium and magnesium solutions contained these elements in the concentrations found in sea water, and the chloride concentration of the $0.59 \ mol \ dm^{-3}$ sodium chloride solution was also the same as in the sea. As expected, Cl^- competes very markedly with NTA for Cd^{2+}, so for a certain sequestering action (say 50%), a much higher concentration of NTA is needed in a chloride than in a perchlorate solution, see Fig. 7.1. A much larger effect is exerted by Ca^{2+} and Mg^{2+}, however, competing with Cd^{2+} for NTA. When the effects of all micro-elements present in sea water are added, a 50% sequestering of $10^{-7} \ mol \ dm^{-3}$ cadmium(II) solution takes a NTA concentration of $\simeq 10^{-4} \ mol \ dm^{-3}$, irrespective of whether synthetic or natural sea water is used. This means that at least for the Adriatic sea water used here, the trace elements present have no influence on the speciation of cadmium.

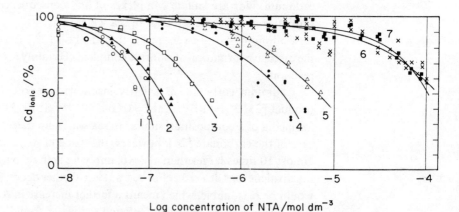

Fig. 7.1. Titration curves of cadmium(II) as a function of the log of the NTA concentration in various media of $I = 0.7 \ mol \ dm^{-3}$ (if necessary adjusted by addition of sodium perchlorate), at values of pH between 7.3 and 8.3 [28]. The media are: $0.7 \ mol \ dm^{-3}$ sodium perchlorate (1), $0.1 \ mol \ dm^{-3}$ sodium chloride (2), $0.59 \ mol \ dm^{-3}$ sodium chloride (3), $0.01 \ mol \ dm^{-3}$ calcium chloride (4), $0.054 \ mol \ dm^{-3}$ magnesium chloride (5), synthetic sea water (6) and natural (Adriatic) sea water (7).

7.9 Lead

Unlike mercury and cadmium, lead might conceivably be oxidized beyond the divalent oxidation state in sea water. The tetravalent state is, *per se*, favoured at high values of pH, on account of its extensive hydrolysis, leading to the oxide PbO_2, and the mixed oxide Pb_3O_4. On the other hand, the divalent state is, as already mentioned, not very prone to hydrolysis. The first dissociation constant of Pb^{2+} is no larger than $K_1^* = 10^{-7.8}$ mol dm^{-3} at the I of sea water [5, 18]. From the standard potentials [5, 13], $E_1^0 = 1455$ mV for $PbO_2(s)$ $+ 4H^+ + 2e^- = Pb^{2+} + 2H_2O$ and $E_2^0 = 295$ mV for $3PbO_2(s) + 2H_2O + 4e^- = Pb_3O_4(s) + 4OH^-$, it can be calculated, however, that a perceptible oxidation to $PbO_2(s)$ or $Pb_3O_4(s)$ cannot possibly take place at the actual pH, not even if the oxidation potential is at the highest value conceivable, i.e. 740 mV.

The carbonate has [5] $pK_{so} = 13.1$ at $I = 0$ which would imply around 11.8 in sea water. Another slightly soluble carbonate of lead(II) is the hydroxide compound $Pb_3(OH)_2(CO_3)_2$ with [5] $pK_{so} = 45.5$ at low ionic strength and $\simeq 42$ in sea water. At the present $p[OH^-] = 5.7$ and $p[CO_3^{2-}] = 4.17$, the $p[Pb^{2+}]$ in equilibrium with these two compounds is about the same, $\simeq 7.5$. As the total concentration of lead is $\simeq 10^{-10}$ mol dm^{-3}, and $[Pb^{2+}]$, on account of the complex formation, is considerably lower, the possibility of precipitation is evidently remote. As to the carbonate complexes in solution, a value of $\beta_1 = 10^7$ at $I = 1$ mol dm^{-3} has been reported [29]. Unfortunately, the solubility measurements yielding this value are not very precise, as is evident from the large spread among the experimental points. Also, the value of $\beta_2 = 10^9$ found in these measurements is about ten times larger than the values $10^{8.2}$ and $10^{7.9}$ found in other investigations [30, 31]. The latter ones moreover apply to media of higher I, viz. 1.7 and 1.8 mol dm^{-3}, which should mean a higher rather than a lower stability. For the following calculations, the values $\beta_1 = 10^5$ and $\beta_2 = 10^8$ have been adopted; they should be regarded as little more than educated guesses.

Lead sulfate has a much larger solubility product than the carbonate [5], viz. $pK_{so} = 6.2$ at $I = 1$ mol dm^{-3}. Even at the fairly high sulfate concentration in sea water, precipitation of $PbSO_4(s)$ is, therefore, excluded. Like other sulfate complexes formed in aqueous solutions, the lead ones are certainly much less stable in a medium of the ionic strength of sea water than they are in pure water [32]. The stabilities in sea water are not known, but the value of β_1 quoted in Table 7.3 seems to be a reasonable estimate.

The stabilities of the chloride complexes are fairly well known for solutions of an ionic strength similar to sea water [5, 33]. The lead(II) complexes are considerably less stable than the cadmium(II) ones, but more stable than those formed by divalent ions of the first row transition metals, Tables 7.3 and 7.4.

The values of β_j selected result in the values of α listed in Table 7.3. The chloride and carbonate complexes predominate. The values actually obtained depend very much upon the values of β_j chosen for the carbonate complexes. It should be stressed once more that these are not very reliable.

Table 7.4. Distribution of various species (α, %) of divalent first row transition metal ions of class (a)

	Log β_1^a			α^b			
	MOH$^+$	MCO$_3$	M^{2+}	MOH$^+$	MCO$_3$	MCl$^+$	MSO$_4$
Co	3.7	4	42	0.4	28	24	5
Ni	3.5	4	42	0.8	28	24	5
Cu	6	5.4	5	9	77^c	3	0.5
Zn	4.4	4	41	2	28	24	5

^{a} $\log \beta_1(SO_4^{2-}) = 1$, $\log \beta_1(Cl^-) = 0$.
^{b} Values of p[L], see Table 7.3.
^{c} The second carbonate complex has $\alpha = 6$ ($\log \beta_2 = 8.5$).

7.10 Cobalt, Nickel, Copper and Zinc

As stated above, the later first row transition metals, from cobalt to zinc, are present in aerobic sea water in their divalent oxidation states. The chemistries of these states are moreover so similar that it seems appropriate to discuss them together. They are all fairly hard acceptors, displaying little affinity for chloride, or bromide, ions [5]. The stability constants β_1 for the complexes MCl$^+$ are all $\simeq 1$. The sulfate complexes MSO$_4$ are of the same modest stability as for cadmium(II) and lead(II), i.e. $\beta_1 \simeq 10$.

The carbonate complexes are, on the other hand, quite stable. Reliable values of β_j have, admittedly, only been determined for the copper(II) system [34], and only at $I = 0$. From the constants found, $\log \beta_1 = 6.73$ and $\log \beta_2 = 9.83$, approximate values pertaining to the ionic strength of sea water can be calculated, however, on the same assumption used above in the recalculation of K_{so} for CdCO$_3$(s) and PbCO$_3$(s), viz. that the activity coefficients for divalent ions in sea water are between 0.2 and 0.25. This means that 1.3 log units should be subtracted from the values of $\log \beta_1$ and $\log \beta_2$ determined at $I = 0$.

The stabilities of complexes formed by the first row transition elements with a certain ligand generally follow the Irving–Williams order [35]. This means that they increase from left to right until a maximum is reached at copper(II), a drop is then observed to zinc(II). Both electrostatic and covalent bonding forces cooperate to establish this order [36]. The carbonate complexes formed by Co^{2+}, Ni^{2+} and Zn^{2+} should, therefore, be less stable than those formed by Cu^{2+}. On the other hand they should, for electrostatic reasons, be more stable than those formed by Cd^{2+}, of the same electron-configuration as Zn^{2+}, but of much larger radius. A value of $\log \beta_1 \simeq 4$ seems reasonable for the three ions, see Table 7.4.

Also for these elements, the main difficulty in determining the stabilities of the carbonate complexes in solution is the ready formation of slightly soluble carbonate phases of varying composition. The precipitation reactions have been carefully investigated for zinc(II) and copper(II) [26, 34]. Zinc(II) forms the simple carbonate $ZnCO_3$, also known as the mineral smithsonite. At the values of pH and $[CO_3^{2-}]$ prevailing in sea water, however, the thermodynamically stable phase is the hydroxide carbonate $Zn_5(OH)_6(CO_3)_2$. Copper(II) forms the hydroxide carbonates malachite, $Cu_2(OH)_2CO_3$, and azurite, $Cu_3(OH)_2(CO_3)_2$, while the simple carbonate $CuCO_3$ does not seem to exist. In equilibrium with sea water, malachite is the stable phase, but the rate of transition is very slow. Also for nickel(II) and cobalt(II), several solid carbonate phases exist, but their solubility products, and rates of transition, are not well known. One might presume, however, that the solubilities are of the same order of magnitude as for the zinc(II) and copper(II) carbonates.

The solubility products determined for the copper and zinc carbonates at $I = 0$ and 0.2 mol dm^{-3} have been recalculated for the ionic strength of sea water by the procedure already described. In Table 7.5, both the values determined at $I = 0.2$ mol dm^{-3} and those arrived at for sea water are listed.

Table 7.5. Slightly soluble carbonate phases of divalent copper and zinc. Values of K_{so} are given both for $I = 0.2$ mol dm^{-3} ($NaClO_4$) and for sea water. The ionic products, IP, have been calculated from $[OH^-]$ and $[CO_3^{2-}]$ given in Table 7.3, and from $[M^{2+}]$ calculated from C_M of Table 7.1 and α of Table 7.4

| | $I = 0.2$ mol dm^{-3} | Sea water | |
	pK_{so}	pK_{so}	pIP
$Cu_2(OH)_2CO_3$	31.4	30.8	35.0
$Cu_3(OH)_2(CO_3)_2$	42.1	41.2	48.8
$ZnCO_3$	9.84	9.5	11.8
$Zn_5(OH)_6(CO_3)_2$	70.1	68.7	80.5

Even at quite low concentrations, $\simeq 10^{-5}$ mol dm^{-3}, hydrolysis of hydrated M^{2+} ions essentially results in polynuclear complexes [18]. The stability constants of the mononuclear complexes which no doubt predominate at the extremely low concentrations present in sea water are, therefore, difficult to determine. Within rather wide limits of error, the values obtained follow, however, the Irving–Williams order, with a very marked maximum at copper(II), see Table 7.4.

The distribution between the various complexes is presented in Table 7.4. For cobalt, nickel, and zinc, the hydrated ions predominate. Due to the high chloride concentration, chloride complexes are also important, in spite of their low β_1 value. They account for about one-fourth of the total. Part of this might be in the form of the second complex MCl_2, but the data available do not warrant any partition between consecutive chloride complexes. The higher affinity of the sulfate ion for the metal ions cannot compensate for their lower concentration, the part present as sulfate complexes amounting only to $\sim 5\%$. For all three metals mentioned, hydrolysis is practically negligible. The carbonate complexes are important, however, though it should be remembered that the values of β_1 are very approximate. The share of the carbonate complexes might, therefore, be much smaller, or larger, than the 28% stated in Table 7.4.

Copper(II), on the other hand, shows a pattern distinctly different from the one common to the three other acceptors.

The carbonate complexes predominate completely. The hydrolytic complex MOH^+ comes next. The hydrated ion is of minor importance, and even more so the chloride and sulfate complexes.

From the value of $[M^{2+}]$ thus found, and the known values of $[OH^-]$ and $[CO_3^{2-}]$, the ionic products referring to the respective solid carbonates have been calculated, see Table 7.5. In no case do they exceed the values of K_{so}. The sea is not saturated with reference to any of these slightly soluble carbonates. Most probably, the amounts of cobalt and nickel present are also far below the limit of saturation.

Copper stands out as the only one of these elements that is reduced at values of pE that can possibly be established in sea water. From the value $E_{12}^0 = 158.6$ mV for the redox system $Cu^{2+} + e^- \rightleftharpoons Cu^+$ and the constants $\beta_2 = 10^{5.0}$ and $K_3 = 10^{-0.35}$ for the equilibria $Cu^+ + 2Cl^- \rightleftharpoons CuCl_2^-$ and $CuCl_2^- + Cl^- \rightleftharpoons CuCl_3^{2-}$, respectively, the amount of monovalent copper is larger than that of divalent for $pE < 6.5$. On the other hand, copper(I) is so highly stabilized by the high $[Cl^-]$ that, assuming a value of $E_{01}^0 = 518.2$ mV for the system $Cu^+ + e^- \rightleftharpoons Cu(s)$, a reduction of a 4 nmol dm^{-3} solution of copper(I) to the metallic state takes place only for $pE < -4.1$. It is then assumed that no further ligands occur in the solution. Under typically anoxic conditions, sulfide is also present, however. At the lowest value of $pE \simeq 7$, where Cu^{2+} still might be of the order 10^{-8} mol dm^{-3}, a concentration of $[S^{2-}] \gtrsim 10^{-28}$ mol dm^{-3} has to be reached in order to bring about precipitation of CuS(s), with $K_{so} = 10^{-36.1}$. The highest value of $[S^{2-}]$ possible at $pE = 7$ can be calculated from the $E_s^0 = 141$ mV of the redox system [13] $S(s) + 2H^+ + 2e^- \rightleftharpoons H_2S(aq)$ and the constant [37] $K_1 \cdot K_2 = 10^{21}$ of the equilibrium $2H^+ + S^{2-} \rightleftharpoons H_2S(aq)$. It turns out to be $[S^{2-}] = 10^{-30.3}$ mol dm^{-3}. Precipitation of CuS(s) will, therefore, not take place before copper(II) is reduced. In order that $Cu_2S(s)$ might precipitate, $K_{so} = 10^{-48.9}$ has to be exceeded [13]. A total concentration of CuI = 4 nmol dm^{-3} means that at the sea water Cl$^-$ concentration, the value of $[Cu^+] = 10^{-13}$ mol dm^{-3} and hence a value of $[S^{2-}] \gtrsim 10^{-23}$ mol dm^{-3} is necessary to bring about precipitation. This value can be reached as soon as $pE \simeq 3.4$. At higher values of $[S^{2-}]$ which are possible at lower values of pE, soluble sulfide complexes may also be formed, analogous to those formed by silver(I) which have been thoroughly investigated [38]. To what extent such complex formation takes place is presently unknown, however.

7.11 Iron and Manganese

At the pH and pE generally prevailing in sea water, these elements predominantly form practically insoluble oxides or hydrous oxides. They will, therefore, be discussed together.

The extensive hydrolysis of iron(III) at the prevailing pH strongly stabilizes this oxidation state relative to iron(II), which is hardly hydrolysed at all. The hydrated Fe^{2+} has an acid dissociation constant [18] $K_1^* \simeq 10^{-10.0}$ mol dm^{-3} at $I = 0.7$, close to those of Co^{2+} and Ni^{2+}, while Fe^{3+} has $K_1^* \simeq 10^{-3.1}$ mol dm^{-3}. The solid phase of iron(III) stable at the pH of sea water is the hydrous oxide $FeO(OH)$. The equilibrium constant of the reaction $FeO(OH)(s) + 3H^+ = Fe^{3+} + 2H_2O$, $K_{so}^* = [Fe^{3+}]/[H^+]^3$, is fairly well determined. A value of $\log K_{so}^* \simeq 3.6$ has been found for the amorphous oxide in equilibrium with 3 mol dm^{-3} $NaClO_4$ solution [18]. With this value, and the standard potential 770 mV for the Fe^{3+}/Fe^{2+} couple, values of $[Fe^{3+}]$ and $[Fe^{2+}]$ in equilibrium with $FeO(OH)(s)$ at the pH and pE of sea water can be calculated, viz. $[Fe^{3+}] = 10^{-20.7}$ mol dm^{-3} and $[Fe^{2+}] = 10^{-20.2}$ mol dm^{-3}. As $[Fe^{2+}]$ is certainly the most important species of iron(II) (cf. Table 7.4), practically no iron(II) is present at equilibrium. Of course, some iron(II) may be present under non-equilibrium conditions, mostly in living organisms.

$[Fe^{3+}]$ is also practically zero. For the soluble hydrolytic complexes $FeOH^{2+}$, $Fe(OH)_2^+$ and $Fe(OH)_4^-$, the constants K_j^* for the dissociation reactions $Fe^{3+} + jH_2O \rightleftharpoons Fe(OH)_j + jH^+$ are at the ionic strength of sea water, at 25°C [18]: $K_1^* = 10^{-3.1}$ $K_2^* = 10^{-7.0}$ and $K_4^* = 10^{-22.5}$ which implies equilibrium concentrations of $[FeOH^{2+}] = 10^{-10.7}$ mol dm^{-3}, $[Fe(OH)_2^+] = 10^{-11.5}$ mol dm^{-3} and $[Fe(OH)_4^-] = 10^{-10.8}$ mol dm^{-3}. These are all much lower than the total iron concentration present, $\simeq 10^{-8}$ mol dm^{-3}, cf. Table 7.1. The polynuclear hydrolytic complexes are of even less importance at this extreme dilution. To what extent an uncharged soluble complex $Fe(OH)_3$ exists is not known. At present, the most reasonable inference seems to be that most of the iron exists as finely dispersed hydrous oxide. At the prevailing low concentration, the dispersion may be at least partly colloidal, despite the high concentration of ions present.

For manganese an even higher (and more strongly oxidized) oxidation state is preferred, namely Mn(IV). The system $MnO_2(s, \beta) + 4H^+ + 2e^- \rightleftharpoons Mn^{2+} + 2H_2O$ has $E^0 = 1230$ mV [13] which in sea water means a concentration $[Mn^{2+}] = 10^{-15.8}$ mol dm^{-3} in equilibrium with the solid β-phase pyrolusite. The hydroxide of manganese(III) is not stable relative to $MnO_2(\beta)$ under the present conditions. On the other hand, the higher oxidation states of manganese are reduced. It may thus be concluded that once equilibrium has been reached, manganese is present in sea water only in the form of dispersed $MnO_2(s)$. Eventually, the iron(III) hydrous oxide as well as the manganese(IV) oxide both reach the bottom sediments. The much discussed iron–manganese nodules represent, in fact, a coprecipitation of these oxides. It may be, however, that the nodules at least mainly originate from the oxides formed as secondary products in the vicinity of the hot springs mentioned above [42]. At lower values of pE both iron and manganese appear in lower oxidation states [39].

The data available for the various solid iron phases are not very precise, but as pE decreases, Fe_3O_4 presumably becomes the stable phase. At still lower pE, iron(II) silicate phases might become important, though little is known about their composition and the condition of their formation. Also the carbonate $FeCO_3$ may possibly be formed. This compound has a fairly large solubility product [5], in sea water pK_{so} should be $\simeq 9.3$. As long as $p[CO_3^{2-}]$ stays at the value found under aerobic conditions, this would imply $[Fe^{2+}] \simeq 10^{-5}\ mol\ dm^{-3}$ at equilibrium. Rather high concentrations of iron(II) in solution would thus be possible. Finally pyrite, FeS_2 comes to the fore at the low values of pE reached in sulfide solution. The conditions are illustrated in Fig. 7.2 which also gives a good idea of the difficulties that still attend such equilibrium calculations. On top of these, the question arises about the kinetics of the different phase transitions.

For manganese [39], MnO_2 seems to be the stable phase over a rather wide range of pE, perhaps down to $pE \simeq 7$. It is not certain whether $MnO(OH)$ and/or Mn_3O_4 appear as stable phases. At lower values of pE, $MnCO_3$ is most probably the stable solid phase. Only at values of $pE \simeq -10$, too low ever to appear, would the sulfide phase MnS be stable.

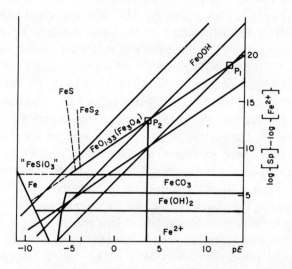

Fig. 7.2. Redox diagram for iron, at pH = 8.1 and 25°C, according to Sillén [39]. The quantity $\log\{Sp\} - \log\{Fe^{2+}\}$, where $\{Sp\}$ and $\{Fe^{2+}\}$ denote the activities of the species Sp and Fe^{2+}, respectively, are plotted versus pE. The various species Sp stated in the figure are all solid phases, so in this particular case $\{Sp\} = 1$. For FeO(OH) three lines, and for Fe_3O_4 two lines are given, corresponding to different data. P_1 and P_2 are two points where both FeO(OH) and Fe_3O_4 may exist in equilibrium. For P_2, at the crossing of the preferred thick lines, pE = 3.7.

7.12 Chromium, Molybdenum and Vanadium

The high values of pE and pH of sea water strongly favour the highest oxidation state of chromium, viz. chromium(VI). Contrary to the prevalent oxidation states of iron(III) and manganese(IV), however, chromium(VI) is present as a soluble species, i.e. the chromate ion, CrO_4^{2-}. Though this ion is a fairly strong base, with a corresponding acid of $pK^* = 5.7$ in the present medium, practically no $HCrO_4^-$ is evidently formed at the prevailing pH = 8.1. At this pH, chromium(III) exists only as the virtually insoluble hydroxide $Cr(OH)_3$. The equilibrium $CrO_4^{2-} + 4H_2O + 3e^- \rightleftharpoons Cr(OH)_3(s) + 5OH^-$ is only slowly established, and the standard potential is, therefore, very difficult to measure. A value of $E^0 = 130\,mV$ has been calculated [13], however, which corresponds to a concentration of chromate $[CrO_4^{2-}] = 10^{-15.6}\,mol\,dm^{-3}$ in equilibrium with $Cr(OH)_3(s)$ at $pE = 12.5$ and pH = 8.1. Evidently, chromium(III) is not thermodynamically stable in sea water. At equilibrium, all chromium will be in the form of CrO_4^{2-}.

In practice, the establishment of equilibrium is so slow that any chromium(III) added to sea water takes a very long time to be oxidized. As a rule, therefore, both chromium(III) and chromium(VI) are found in actual samples [40].

Analogously, the only stable form of molybdenum in sea water is the molybdate(VI) ion MoO_4^{2-}. Also, in this case the oxidation of lower states is slow.

For vanadium, oxidation potentials valid in alkaline solutions are not available. It is, nevertheless, certain that in this case also the highest oxidation state, vanadium(V) is the stable one in sea water. The species present are most probably the mononuclear ions $H_2VO_4^-$ and HVO_4^{2-} (often written as $VO_2(OH)_2^-$ and $VO_2(OH)_3^{2-}$, respectively). With a probable value [5] of $pK \simeq 7.8$, the latter should be the slightly predominating form. No polyvanadates are formed at the very low concentrations present.

The anions formed by these elements are of course also complex-forming ligands. On account of their extremely low concentrations, they are certainly of no consequence in this respect.

7.13 Selenium

Selenium is less easily oxidized to its highest oxidation state than its much more abundant congener sulfur. The high values of pE and pH prevailing in sea water nevertheless favour selenium(VI) relative to selenium(IV) to such an extent that the ratio of their concentrations at equilibrium is $\simeq 10^{12}$, as calculated from the value [13] of $E^0 = 50\,mV$ for the reaction $SeO_4^{2-} + H_2O + 2e^- \rightleftharpoons SeO_3^{2-} + 2OH^-$. In this calculation, a value of $K_1 \simeq 10^7$ is assumed for the protonation constant of the reaction $H^+ + SeO_3^{2-} \rightleftharpoons HSeO_3^-$; the value in pure water [5] is $10^{8.0}$ and in sea water it should be about ten times lower. This means that in sea water selenium(IV) would be present predominantly as SeO_3^{2-}. Hydrogen selenate is quite a strong acid, so selenium(VI) is exclusively present as SeO_4^{2-}. In order that selenium(IV) and selenium(VI) should be present in equal amounts at equilibrium, the value of pE has to be as low as $\simeq 6.5$.

The rate of oxidation of selenium(IV) is so slow, however, that a large proportion of the selenium present in sea water is always found in this state. In surface waters, most of the selenium, up to 80%, is even found in this inherently unstable form; in deeper waters the selenium seems to be about evenly divided between the two oxidation states [41].

7.14 Concluding Remarks

The speciation arrived at above is by necessity an approximation. For several reasons already mentioned exact answers are not possible and, presumably, never will be. First, variations of temperature and pressure influence the equilibria in ways that are presently little known. Secondly, many important reactions take place at a slow rate which means that equilibrium calculations are of restricted use. The rates, moreover, generally depend very much on the temperature. Thirdly, biological activity, both by living organisms, and by substances they excrete, probably plays an important part in the chemical forms of many trace metals. This influence is, moreover, apt to vary strongly from place to place, and from time to time.

The present approach, which combines actual determinations of more or less inert species with equilibrium calculations for labile ones, should nevertheless give a reasonably true picture of the actual conditions. As more data become available, it will of course be possible to refine this picture further.

7.15 References

1   D. Dyrssen and M. Wedborg, In *The Sea* (Ed. E. Goldberg) vol. 5, *Marine Chem.*, J. Wiley and Sons, p. 181 (1974).
2   T. Almgren, D. Dyrssen and M. Strandberg, *Deap-Sea Res.*, **212**, 635 (1975).
3   D. Dyrssen, In *Oceanic Sound Scattering Prediction* (Eds. N.R. Andersen and B.J. Zahuranco), Plenum, New York (1977).
4   R.G. Bates and C.H. Culberson, *The Fate of Fossil Fuel and Carbon Dioxide in the Oceans* (Eds. N.R. Andersen and A. Malahoff) Plenum, New York (1977).
5   L.G. Sillén and A.E. Martell, *Stability Constants of Metal Ion Complexes*, Chemical Society, London, Special Publications No 17 and 25, (1964) (1971).
6   L.G. Sillén, *Oceanography* (Ed. M. Sears), Amer, Ass. Adv. Science, Washington D.C., p. 549 (1961).
7   F.T. Mackenzie, In *Chemical Oceanography* (Eds. J.P. Riley and G. Skirrow), 2nd edn, Academic Press, London (1975).
8   D. Dyrssen and I. Hansson, *Marine Chem.*, **8,** 137 (1973).
9   D. Dyrssen and M. Wedborg, *Chemistry and Biochemistry of Estuaries,* (Eds. E. Clausson and I. Sato), Wiley, New York (1981).

10 S. Ahrland, *The Nature of Seawater* (Ed. E. Goldberg), Dahlem Konferenzen Berlin (1975), and references cited therein.
11 R. Parsons, *The Nature of Seawater*, (Ed. E. Goldberg) Dahlem Konferenzen Berlin (1975), and references cited therein.
12 P.S. Liss, J.R. Hering and E.D. Goldberg, *Nature Phys. Sci.*, **242**, 108 (1973).
13 W.M. Latimer, *Oxidation Potentials*, 2nd edn, Prentice Hall, Englewood Cl iffs, N.J.(1972).
14 T. Almgren, L.G. Danielsson, D. Dyrssen, T. Johansson and G. Nyquist, *Thalassia Jugoslavica*, **11**, 19 (1975).
15 B. Magnusson and S. Westerlund, *Marine Chem.*, **8**, 231 (1980).
16 S. Ahrland, *Structure Bonding*, **5**, 118 (1968).
17 R.G. Pearson, *J. Chem. Educ.*, **45**, 581, 643 (1968).
18 C.F. Baes Jr and R.E. Mesmer, *The Hydrolysis of Cations*, Wiley, New York (1976).
19 G. Westöö, *Acta Chem. Scand.*, **20**, 2131 (1966).
20 S. Jensen and A. Jernelöv, *Nature*, **223**, 753 (1969).
21 J.O. Nriagu, (Ed) *The Biochemistry of Mercury in the Environment*, Elsevier/North-Holland, Amsterdam (1979).
22 M. Jawaid, F. Ingman, D. Hay Liem and T. Wallin, *Acta Chem. Scand.*, *A*, **32**, 7 (1978).
23 C.W. Baker, *Nature*, **270**, 230 (1977).
24 P. Mukherji and D.R. Kester, *Science*, **204**, 64 (1979).
25 H. Gamsjäger, H.U. Stuber and P. Schindler, *Helv. Chim. Acta*, **48**, 723 (1965).
26 P.   Schindler, M. Reinert and H. Gamsjäger, *Helv. Chim. Acta*, **52**, 2327 (1969).
27 O. Johansson, I. Persson and M. Wedborg, *Marine Chem.*, **8**, 191 (1980).
28 B. Raspor, P. Valenta, H.W. Nürnberg and M. Branica, *Sci. Tot. Environ.* **9**, 87 (1978).
29 N.N. Baranova, *Russ. J. Inorg. Chem.*, **14**, 1717 (1969).
30 J. Faucherre and Y. Bonnaire, *C.R. Acad. Sci. Ser.*, *C* **248** 3705 (1959).
31 F. Fromage and S. Fiorina, *C. R. Acad. Sci, Ser. C*, **268**, 1764 (1969).
32 A.M. Bond and G. Hefter, *J. Electroanal. Chem.*, **34**, 227 (1972).
33 A.M. Bond and G. Hefter, *J. Electroanal. Chem.*, **42**, 1 (1973).
34 P. Schindler, M. Reinert and H. Gamsjäger, *Helv. Chim. Acta*, **51**, 1845 (1968).
35 H.M.N.H. Irving and R.J.P. Williams, *J. Chem. Soc.*, 3192 (1953).
36 G. Schwarzenbach, *Chimia*, **27**, 1 (1973).
37 M. Widmer and G. Schwarzenbach, *Helv. Chim. Acta*, **47**, 266 (1964).
38 G. Schwarzenbach and M. Widmer, *Helv. Chim. Acta*, **49**, 111 (1966).
39 L.G. Sillén, *Arkiv Kemi*, **25**, 159 (1966).
40 S. Osaki, I. Osaki, S. Shibata and Y. Takashima, *Bunseki Kagaku*, **25**, 358 (1976) (as cited in literature survey by the Department of Analytical Chemistry, University of Göteborg, Göteborg, Sweden, June 1977).
41 Y. Sugimura, Y. Suzuki and Y. Miyake, *J. Oceangraph. Soc. Jpn*, **32**, 235 (1976).
42 J.M. Edmond and K. Von Damm, *Scientific American*, **228**:4 70 (1983).
43 F. Haber, *Z. angew. Chem.*, **40**, 303 (1927).
44 B. Schaule and C.C. Patterson, *Trace Metals in Sea Water* (Ed. C.S. Wong), Plenum, New York (1983).
45 L. Brügmann, L.G. Danielsson, B. Magnusson and S. Westerlund, *Marine Chem.*, **13**, 327 (1983).
46 H.W. Nürnberg, L. Mart, H. Rützel and L. Sipos, *Chem. Geology*, **40**, 97 (1983).
47 L. Mart, H. Rützel, P. Klahre, L. Sipos, U. Platzek, P. Valenta, and H.W. Nürnberg, *Sci. Tot. Environ.*, **26**, 1 (1982).
48 D.R. Turner, M. Whitfield and A.G. Dickson, *Geochim. Cosmochim. Acta*, **45**, 855 (1981).
49 G. Schwarzenbach and M. Widmer, *Helv. Chim. Acta*, **46**, 2613 (1963).

Section 8 Adsorption of Trace Elements by Suspended Particulate Matter in Aquatic Systems

By

Commission V.6
IUPAC

A. C. M. BOURG

Service Géologique National
Department EAU
Bureau de Recherches Géologiques et Minières
F-45060, Orléans Cédex 2
France

8.1 Introduction

Natural solids are important components of aquatic systems because they influence the water composition. It is now widely accepted that the major ion chemistry of natural waters is related to specific reactions of dissolution (rock alteration) and precipitation (formation of secondary minerals). Trace elements can participate in such reactions, either on their own (weathering of ore deposits or precipitation of minerals) or as part of the geochemical cycles of major ions (dissolution of impurities during rock alteration or coprecipitation with major and minor ion minerals). The significance of adsorption processes in regulating the burden of dissolved trace elements in oxygenated natural systems is, however, increasingly recognized in the scientific community [1–9].

Trace elements are present in aquatic systems in two physical forms: as dissolved species and associated with solid particles. The aqueous phase contains free (hydrated), hydrolysed and complexed species. In solid form, whether in deposited sediments or in suspended particulate matter, trace elements can be linked to various biogeochemical fractions (clays, organic matter, Fe or Mn oxides, carbonates, sulfides or associated to/incorporated in algae, bacteria or other biogenic particles). The mobility and the toxicity of trace elements are strongly dependent upon their physical form (dissolved versus particulate). Understanding the processes which can transfer trace elements across the solid–solution interface (especially adsorption/desorption) is thus of fundamental significance for the fields of environmental biogeochemistry and hydrogeochemical prospecting for hidden mineral deposits. Particle surfaces are important scavengers and sinks for heavy metal ions in natural water systems because of their ability to compete with soluble complexing agents for metal ions [7]. Moreover, where the uptake processes are reversible, particles will act as buffers for the dissolved trace metal content of natural waters.

Fluxes of trace elements across the solid–water interface occur in response to changing environmental conditions as the system progresses towards equilibrium. Solid particles buffer the water composition. An increase in dissolved trace elements will be partly compensated by adsorption or precipitation. Conversely, a decrease in dissolved load (such as during mixing of waters of different compositions) will provoke some remobilization.

The chemical speciation of trace elements in the dissolved phase of natural aquatic systems can be investigated by iterative computer program calculations assuming that all acid/base and complexation/dissocation reactions are governed by thermodynamics [10]. A knowledge of the total concentrations of the dissolved components of the system and of the pertinent equilibrium constants, allows the calculation of the distribution of the dissolved species of all components. The degree of saturation of the solution with respect to any given mineral can then be obtained.

However, calculations using these models imply that the dissolved phase of natural waters is in the thermodynamic sense a closed system. But this is definitely not so and fluxes across the water–solid particulate matter (by adsorption/desorption and precipitation/dissolution), the air–water and the water–biota interfaces should also be considered. Precipitation and dissolution of minerals can be accounted for, assuming constant reacton rates, by a mathematical model of geochemical mass transfer [12–16]. The present review is an attempt to describe adsorption/desorption processes in such a way that they can be added to the classical calculations of speciation in the dissolved phase.

Adding a new component, i.e. an adsorbing surface, to the system as demonstrated in Table 8.1, can have a considerable effect on the speciation in the aqueous phase. This is exemplified by the speciation of dissolved Cu(II) in the presence of the adsorbable organic ligand 2,2′-bipyridyl with (Fig. 8.1b) and without (Fig. 8.1a) solid amorphous silica. At low pH values when Cu is not adsorbed, the speciations are almost identical. But this is no longer the case when the metal and the organic ligand react with the silica surface (high pH) [17].

Various models have been proposed to explain the adsorption of ions by solid surfaces [18], but in this chapter only one of them will be described, the surface complex formation model. Examples of applications of the adsorption of metal cations, of inorganic and organic ligands and of metal complexes will be presented. Finally, it will be shown how, by using this model, one can extend the calculated description of the speciation of the aqueous phase to the solid–solution interface.

Table 8.1. Dissolved speciation of a trace metal M in the presence of a complexing agent L and of an adsorbing surface S

Complexation of M (dissolved)	$M + L \rightleftarrows ML$ with $K_{ML} = [ML]/[M][L]$ (1)
Adsorption of M	$M + S \rightleftarrows MS$ with $K_{MS} = [MS]/[M][S]$ (2)

Dissolved speciation of M (as fraction of free M over total dissolved M)
 Using eqn (1):

$$\alpha_0 = \frac{[M]}{[M]+[ML]} = \frac{1}{1 + K_{ML}[L]} \tag{3}$$

α_0 is independent of free metal ($[M]$), of total dissolved metal ($[M]+[ML]$) and thus also of the adsorption of M on S.

I. *No adsorption of L on S*
 $[L]$ is independent of S and thus α_0 is completely independent of S.

II. *Adsorption of L on S*
 Adsorption of L $L + S \rightleftarrows LS$ with $K_{LS} = [LS]/[L][S]$ (4)

 Eqn (3) becomes:

$$\alpha_0 = \frac{1}{1 + \dfrac{K_{ML}[LS]}{K_{LS}[S]}}$$

α_0 is still independent of free and total dissolved metal, but not of adsorbed ligand.

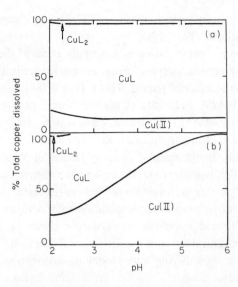

Fig. 8.1. Speciation of dissolved Cu(II) in the presence of the organic ligand 2,2'-bipyridyl (L). ($Cu_{total} = L_{total} \times 1.1 = 1.57 \times 10^{-4} \, mol \, dm^{-3}$) (a) in the absence, and (b) in the presence of solid amorphous silica ($0.050 \, kg \, dm^{-3}$) (adapted from Bourg *et al.* [17]).

8.2 Theory of Adsorption at the Water/Solid Particulate Matter Interface

Two schools of thought dominate the theories of adsorption of ions at the water/solid particulate matter interface.

(a) The chemical approach: dissolved species react with chemical entities on the surface, e.g. by ion-exchange or by coordination reactions. Physical phenomena, e.g. electrostatic forces, are considered only as a correction factor [19–25].

(b) The physical approach: the adsorption is interpreted in terms of a combination of (1) electrostatic interactions between ions and the surface charge, and (2) of ion/solvent interactions. A correcting factor including all other interactions at the interface, e.g. Van der Waals forces, is added [26–28].

Vuceta [29] compared model calculations of the adsorption of Cu(II) and Pb(II) on α-quartz in the absence and in the presence of carbonate, citrate and EDTA, according to the two approaches described above and, within experimental error, found that the data could be explained by both concepts. Vuceta and Morgan [30] used the physical approach to introduce adsorption reactions into mathematical modelling of chemical speciation.

In this review the chemical approach is preferred because, at least in the case of solid hydrous oxides, cations and anions seem to form inner layer surface complexes [18, 24]. The adsorption of metal complexes can also be interpreted, analogously to solution coordination chemistry, in terms of the formation of ternary surface complexes [17, 31, 32]. These justifications will be detailed later on.

The following presentation is a simplification of the various theories of adsorption by surface complex formation; the reader interested in more detail is referred to the original papers. This review is based mostly on the work of Stumm (EAWAG), Schindler (Univ. of Bern), Leckie (Stanford) and co-workers on the adsorption of cations and anions on hydrous oxide surfaces [18, 22–25, 33–44] and on the extension of these concepts by Schindler, Leckie, Davis (USGS), Benjamin and Murray (Univ. of Washington), and Bourg (BRGM) and co-workers to (a) the adsorption of metal complexes [17, 31, 32, 45–51] and (b) to natural suspended particulate matter [4–6, 52–58]. Although this oversimplified presentation may perhaps shock many surface chemists, it should hopefully give solution chemists, geochemists, limnologists, oceanographers and environmental chemists a ready means of understanding, quantifying and modelling adsorption. The application of laboratory adsorption studies on simple solid hydrous oxides to environmental and geochemical problems will also be discussed.

8.3 Surface Complex Formation Model

8.3.1 Surface sites

The surface of natural particulate matter almost always contains ionizable functional groups. They can be –OH (for solid hydrous oxides, clay minerals, dead and living organic matter), –COOH and –NH$_2$ (organic matter) or even –OPO$_3$H$_2$ [25]. In an aqueous medium these surface groups are susceptible to proton transfer (which depends upon the pH of their surroundings). For example, all hydrous oxides behave amphoterically:

$$\equiv S\text{-OH}_2^+ \underset{}{\overset{K_1}{\rightleftharpoons}} \equiv S\text{-OH} \underset{}{\overset{K_2}{\rightleftharpoons}} \equiv S\text{-O}^- \tag{8.1}$$

Therefore, surface charges are positive at low pH and negative at high pH. (Most of them are negatively charged at pH values of natural waters.)

8.3.2 Adsorption of metal cations

The adsorption of cations can be explained in terms of competition with protons for surface sites, according to the following equations

$$\equiv SOH + M^{z+} \underset{}{\overset{*K_1^s}{\rightleftharpoons}} \equiv SOM^{(z-1)+} + H^+ \tag{8.2}$$

with

$$*K_1^s = \frac{[\equiv SOM^{(z-1)+}][H^+]}{[\equiv SOM][M^{z+}]} \tag{8.3}$$

260   SECTION 8

$$\begin{matrix} =\text{SOH} \\ | \\ =\text{SOH} \end{matrix} + M^{z+} \underset{\longleftarrow}{\overset{*\beta_2^s}{\rightleftharpoons}} (=\text{SO})_2 M^{(z-2)+} + 2H^+$$ (8.4)

with

$$*\beta_2^s = \frac{[(=\text{SO})_2 M^{(z-2)+}][H^+]^2}{[=S_2(\text{OH})_2][M^{z+}]}$$ (8.5)

It has also been suggested by Davis and Leckie [34] that eqn (8.4) should be replaced by:

$$\equiv \text{SOH} + M^{z+} \underset{\longleftarrow}{\overset{*K_{\text{OH}}^s}{\rightleftharpoons}} \equiv \text{SOM(OH)}^{(z-2)+} + 2H^+$$ (8.6)

It is, however, usually not possible to differentiate between reactions (8.4) and (8.6) because (a) they liberate the same number of protons per metal ion adsorbed, and (b) the concentration of $\begin{matrix} =\text{SOH} \\ | \\ =\text{SOH} \end{matrix}$ is, in the absence of steric or coordinating bond length restrictions, equal to the concentration of single $\equiv$ SOH surface groups.

Adsorption constants can only be compared when they are based on the same model [59], therefore only a few values determined under the same conditions by Schindler and co-workers for amorphous silica are presented here (Table 8.2).

Table 8.2. Formation constants of surface complexes with amorphous silica (25°C)

Metal	Ionic strength	Log $*K_1^s$	Log $*\beta_2^s$†	Reference
Fe^{3+}	3	-1.77	-4.42	[23]
Pb^{2+}	1	-5.09	-10.68	[23]
Cu^{2+}	1	-5.52	-11.19	[23]
Cu^{2+}	0.1	-4.89	-10.18	[17]
Cd^{2+}	1	-6.09	-14.20	[23]
Mg^{2+}	1	-7.70	-17.15	[38]
Ca^{2+}	1	-8.1	-16.7	[39]

$$^\dagger \beta_2^s = \frac{[(\equiv\text{SO})_2 M^{(z-2)+}][H^+]^2}{[\equiv S(\text{OH})]^2 [M^{z+}]}$$

Typical examples of the adsorption of metals are shown in Fig. 8.2 for various oxide surfaces. The adsorption curves present a sharp increase for the pH range 1–1.5. This part of the curve is called the pH-adsorption edge by Benjamin and Leckie [41]. Its position is characteristic of each metal, but it depends on adsorbent and also, even if less, on adsorbate concentrations [40, 43] and ionic strength [54]. Typically, a tenfold increase in surface site concentration shifts the adsorption edge by 1 pH unit towards the acidic range (Fig. 8.3).

The consequences of the strong pH dependence on adsorption and of the

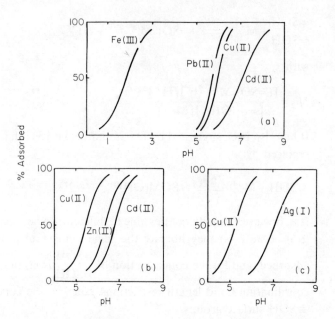

Fig. 8.2. Adsorption of trace metals on (a) amorphous SiO_2 (from Schindler *et al.* [23]), (b) γ-Al_2O_3, (c) amorphous $Fe_2O_3 \cdot H_2O$ (from Davis and Leckie [47]).

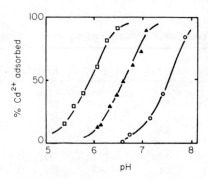

Fig. 8.3. Adsorption of cadmium on amorphous $Fe(OH)_3$ for various concentrations of solids: $\square$, 1.3×10^{-2}; $\blacktriangle$, 1.3×10^{-3}; $\bigcirc$, 1.3×10^{-4} mol $Fe \cdot dm^{-3}$; from Benjamin and Leckie [40].

influence of environmental parameters on the position of the adsorption edge are that:

(a) observations such as 'cadmium adsorbs onto iron oxide at pH 7' are meaningless if the adsorbate concentration and the chemical characteristics of the system considered are not given [44],

(b) the pH variations (diurnal, seasonal and spatial) typical of natural systems are thus potentially significant with respect to the adsorption (particle scavenging)/desorption (mobilization) of any trace metal (see Section 8.5).

8.3.3 Adsorption of inorganic and organic ions

Similarly, the adsorption of inorganic and organic anions can be explained by the competition with OH^- for surface sites [24, 36, 37]

$$\equiv SOH + A^{z-} \underset{}{\overset{K_1^s}{\rightleftharpoons}} \equiv SA^{(z-1)-} + OH^- \tag{8.7}$$

$$\begin{matrix} =SOH \\ | \\ =SOH \end{matrix} + A^{z-} \overset{\beta_2^s}{\rightleftharpoons} (=S)_2 A^{(z-2)-} + 2OH^- \tag{8.8}$$

For protonated ligands, the ligand exchange can be followed by deprotonation:

$$\equiv SOH + HA^{(z-1)-} \rightleftharpoons \equiv SHA^{(z-2)-} + OH^- \rightleftharpoons \equiv SA^{(z-1)-} + H_2O \tag{8.9}$$

In the same way as for the adsorption of metal cations, the formation of bidentate complexes (Fig. 8.4) can be prevented for geometric reasons. The distance between the OH^- groups of geothite and the length of the O–O bond of a phosphate ion are almost equal [eqn (8.10)], but the binding of one aromatic dicarboxylic acid to two neighbouring surface OH groups of γ-Al_2O_3 is unlikely [eqn (8.11)] [24].

$$\begin{matrix} =FeOH \\ | \\ =FeOH \end{matrix} + H_3PO_4 \rightleftharpoons (=Fe)_2 PO_4^- + H^+ + 2H_2O \tag{8.10}$$

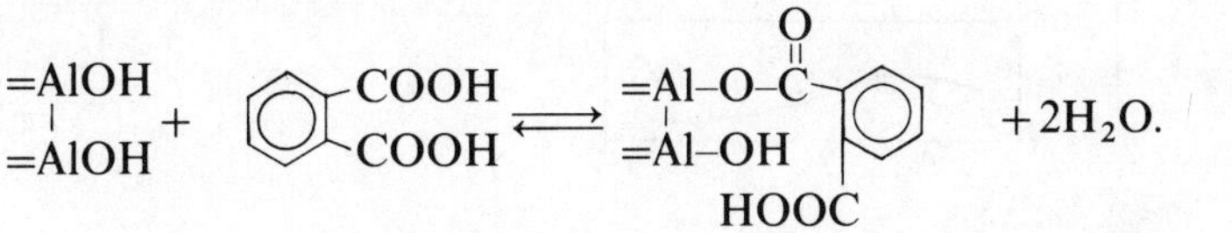

$$\tag{8.11}$$

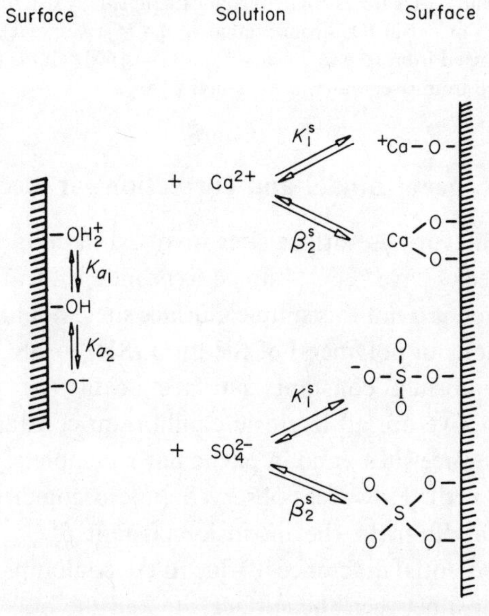

Fig. 8.4. Interaction of cations and anions with suspended particulate matter at surface sites (adapted from Stumm *et al.* [24]).

Examples of the adsorption of inorganic and organic anions are given in Fig. 8.5 and the corresponding surface stability constants are listed in Table 8.3.

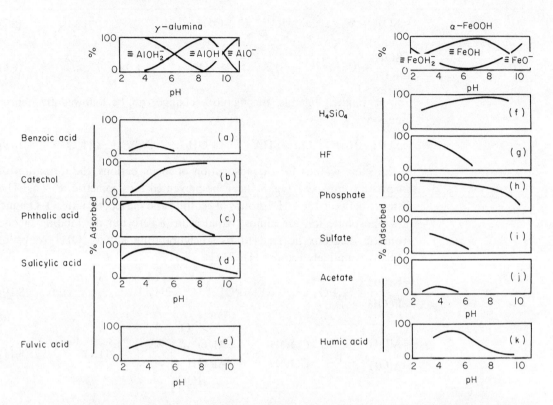

Fig. 8.5. Adsorption of (a)–(e) organic acids on γ-Al$_2$O$_3$ and (f)–(k) anions on α-FeOOH. Curves (a)–(d) and (f)–(j) are calculated from experimentally determined stability constants, at 22°C in 0.1 mol dm^{-3} NaClO$_4$ (from Stumm *et al.* [24]). Curve (e), experimental curve for fulvic acid extracted from lake sediments (from Davis[60]). Curve (k), experimental curve for humics collected from river water (from Tipping [61]).

8.3.4   Chemical concept model and correction for electrostatic effects

Until now the various surface sites involved in coordination reactions with dissolved species have been assumed to be independent of other neighbouring sites. However, it is not this simple. Surface sites should be treated as *polymers* (or associations of polymers) of the form (SH)$_y$ or (SOH)$_y$ (see Fig. 8.6).

The equilibrium constants (surface acidity or surface complexation) mentioned above are all intrinsic equilibrium constants (i.e. no effect from neighbour surface sites – and in particular a completely chargeless surrounding – is assumed). However, observed (microscopic) equilibrium constants contain, in addition to the intrinsic constant $\beta^s_{n(intr.)}$, a component which reflects the potential difference Ψ (due to the coulombic energy of the charged surface groups) between the surface site and the bulk solution:

$$\beta^s_n = \beta^s_{n(intr.)} \exp(F\Psi/RT) \tag{8.12}$$

Table 8.3. Coordination of α-FeOOH (Goethite) and γ-Al$_2$O$_3$ with anions or their conjugate acids (from Stumm *et al.* [24])

	Intrinsic constant (22°C)* log K
1. *Goethite*	
=FeOH$_2^+$ $\rightleftarrows$ =FeOH + H$^+$	-6.4
=FeOH $\rightleftarrows$ =FeO$^-$ + H$^+$	-9.25
=FeOH + F$^-$ $\rightleftarrows$ =FeF + OH$^-$	-4.8
=FeOH + SO$_4^{2-}$ $\rightleftarrows$ =FeSO$_4^-$ + OH$^-$	-5.8
2=FeOH + SO$_4^{2-}$ $\rightleftarrows$ =Fe$_2$SO$_4$ + 2OH$^-$	-13.5
=FeOH + HAc $\rightleftarrows$ =FeAc + H$_2$O	2.9
=FeOH + H$_4$SiO$_4$ $\rightleftarrows$ =FeSiO$_4$H$_3$ + H$_2$O	4.1
=FeOH + H$_4$SiO$_4$ $\rightleftarrows$ =FeSiO$_4$H$_2^-$ + H$_3$O$^+$	-3.3
=FeOH + H$_3$PO$_4$ $\rightleftarrows$ =FePO$_4$H$_2$ + H$_2$O	9.5
=FeOH + H$_3$PO$_4$ $\rightleftarrows$ =FePO$_4$H$^-$ + H$_3$O$^+$	5.1
=FeOH + H$_3$PO$_4$ $\rightleftarrows$ =FePO$_4^{2-}$ + 2H$^+$ + H$_2$O	-1.5
2=FeOH + H$_3$PO$_4$ $\rightleftarrows$ =Fe$_2$PO$_4$H + 2H$_2$O	8.5**
2=FeOH + H$_3$PO$_4$ $\rightleftarrows$ =Fe$_2$PO$_4^-$ + H$^+$ + 2H$_2$O	4.5
2. *γ-Al$_2$O$_3$*	
=AlOH$_2^+$ $\rightleftarrows$ =AlOH + H$^+$	-7.4
=AlOH $\rightleftarrows$ =AlO$^-$ + H$^+$	-10.0
Benzoic acid	
=AlOH + HA $\rightleftarrows$ =AlA + H$_2$O	3.7 ± 0.3
Catechol	
=AlOH + H$_2$A $\rightleftarrows$ =AlAH + H$_3$O$^+$	3.7 ± 0.3
=AlOH + H$_2$A $\rightleftarrows$ =AlA$^-$ + H$_3$O$^+$	< -5
Phthalic acid	
=AlOH + H$_2$A $\rightleftarrows$ AlAH + H$_2$O	7.3 ± 0.3
=AlOH + H$_2$A $\rightleftarrows$ =AlA$^-$ + H$_3$O$^+$	2.4 ± 0.4
Salicylic acid	
=AlOH + H$_2$A $\rightleftarrows$ =AlAH + H$_2$O	6.0 ± 0.4
=AlOH + H$_2$A $\rightleftarrows$ =AlA$^-$ + H$_3$O$^+$	-0.6 ± 0.6

*Equilibrium constants are defined as:

$$K_{a1}^s = \frac{[\text{=FeOH}][\text{H}^+]}{[\text{=FeOH}_2^+]}$$

All concentrations are given in mol dm^{-3}
**Bidentate equilibria are defined as:

$$K = \frac{[\text{=Fe}_2\text{PO}_4\text{H}]}{[\text{=FeOH}][\text{H}_3\text{PO}_4]}$$

The value of Ψ cannot be measured directly. For acid-base surface reactions, the intrinsic constants can be evaluated by extrapolating, to conditions of zero surface charge, the values of the microscopic constants measured for different surface charges (Fig. 8.7). For surface complex formation reactions, it is often assumed that the electrostatic term is equal to unity [22, 23] or is constant at a given ionic strength (and is included in the value of the intrinsic constant which is, therefore, an apparent constant). A more rigorous treatment is possible [33–35], but being cumbersome it can be avoided, especially for environmentally related purposes. It has nevertheless been used by some researchers,

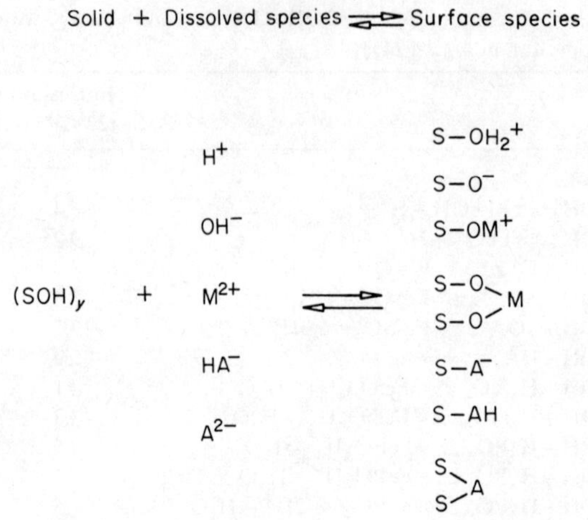

Solid $+$ Dissolved species $\rightleftharpoons$ Surface species

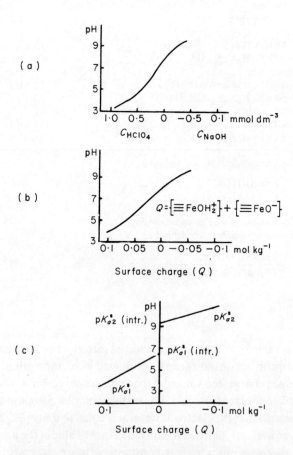

Fig. 8.6. Polymer surface complexation model (adapted from Stumm *et al.* [24]).

Fig. 8.7. (a) Acid–base titration of the amphoteric α-FeOOH. (b) Surface charge as a function of pH. (c) Microscopic acidity constants (extrapolation to zero charge gives intrinsic $pK_{a_1}^s$, $pK_{a_2}^s$ values) (all diagrams are from Stumm *et al.* [24]).

e.g., on model sea water by Balistrieri and Murray [62] and a much simpler version [63] on model lake water by Baccini et al. [64].

8.3.5 **Generalization of the determination of surface stability constants**

Measuring the surface stability constants for all types of surfaces would be fastidious. Fortunately, two generalizations have already been noticed. Correlations exist for:

(a) trace metals, between surface constants and solution hydrolysis constants (Fig. 8.8) and,

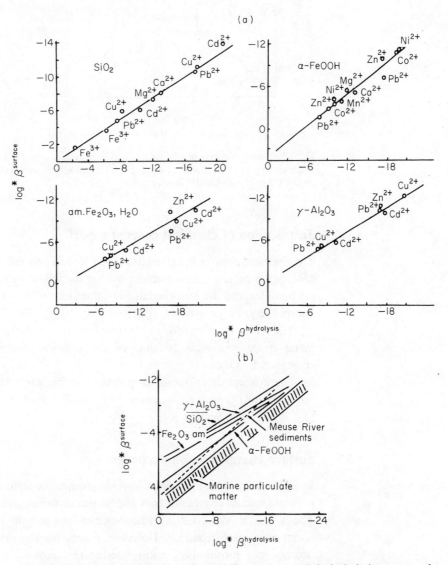

Fig. 8.8. Correlations between the surface stability constants and the hydrolysis constants of metals. (a) SiO_2 from Schindler et al. [23]; marine particulate matter and α-FeOOH from Balistrieri et al. [55]; Fe_2O_3 amorphous and γ-Al_2O_3 from Davis and Leckie [34, 47]; (b) Meuse River sediments from Mouvet and Bourg [5].

(b) ligands, between surface complexes and solution complexes of the metal or metalloid involved in a given surface (e.g. Al, Si, Fe or Mn) (Fig. 8.9).

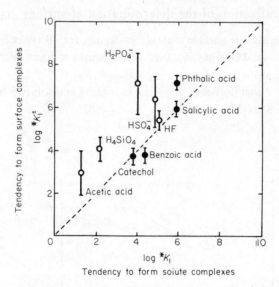

Fig. 8.9. Correlation between the tendency to form surface complexes and that to form solute complexes (from Stumm *et al.* [24]). (O) with FeOOH and FeOH^{2+}, respectively; (●) with Al$_2$O$_3$ and AlOH^{2+}, respectively.

8.3.6   Justification of chemical concept model

The chemical approach, rather than the physical concept, is used here for the description of adsorption phenomena for the following reasons.

(a) Some cations are adsorbed at pH values for which surfaces are positively charged (e.g. Pb on γ-Al$_2$O$_3$) [22].

(b) The tendency of ligands to form surface complexes with a given hydrous metal oxide is similar to that of the solution complexation with the corresponding metal (Fig. 8.9).

(c) The adsorption of metal complexes can be interpreted, as in solution coordination chemistry, in terms of the formation of ternary surface complexes [17, 31, 32] (see Section 8.4).

8.3.7   Surface charge and ion exchange

In addition to the ionization of surface groups, the solid particle can have a pH-independent charge due to isomorphous substitution, e.g. in clay minerals. The present model takes this phenomenon into account only if electrostatic corrections are introduced. However, except for the case of clay minerals (mainly montmorillonite), natural sediments seem to behave like hydrous oxides or organic matter [8, 52, 55, 65–67] (see Section 8.5). It can thus be assumed that in natural composite solids, this permanent surface charge is masked by other components of the solid and/or that it acts as a modification of microscopic surface constants by demobilizing the charge imbalance towards surface sites.

It should be noted that the concept of heterogeneous complexation is not, in principle, different from the ion exchange theory commonly used for alkali and alkaline earth metals. Here, metal adsorption constants are measured with a reference cation, the proton. Therefore, the exchange between cations A and B on a surface site S is considered as the sum of the two reactions:

$$SH + B^+ \rightleftharpoons SB + H^+ \tag{8.13}$$

$$SA + H^+ \rightleftharpoons SH + A^+ \tag{8.14}$$

$$SA + B^+ \rightleftharpoons SB + A^+ \tag{8.15}$$

A fundamental difference, however, is that the surface complexation model presented here assumes that all surface reactions are of the inner sphere type [6]. This is not, however, completely true: for example Stumm *et al.* explained the adsorption of robust Co(II) complexes of ammonia by purely electrostatic interaction [18].

8.3.8 Heterogeneity of surface sites

Studies at various adsorption densities and competitive adsorption experiments are consistent with the existence of several distinct types of site on the surfaces of γ-FeOOH, γ-Al$_2$O$_3$, Fe$_2$O$_3 \cdot$H$_2$O (am) [40, 41, 43, 57]. At very low adsorption densities, all sites are available in excess and the surface behaves as though it were composed of identical sites. However, when only a few percent of the surface sites are occupied, the average binding constant decreases (Fig. 8.10). This site heterogeneity is discussed in the section on modelling (Section 8.6).

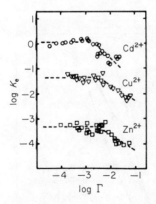

Fig. 8.10. Apparent equilibrium constants K_e for adsorption by goethite in major ion sea water as a function of adsorption density (Γ = moles of metal adsorbed/moles of surface sites) (from Balistrieri and Murray [57]).

8.4 Adsorption of Metal Complexes

8.4.1 Existence of ternary surface complexes

It was thought until recently that the presence of an excess of complexing agent in solution prevented the adsorption of trace metals. The introduction of chelating agents such as natural organic matter in an aquatic system could, therefore, mobilize adsorbed trace metals. A mathematical model study describing the chemical speciation of trace metals as a competition between adsorption and solution complexation was even presented recently by Vuceta and Morgan [30].

This concept is correct for some systems [29, 46], but they are special cases (see Cu-EDTA in Fig. 8.11). The Cu-EDTA complex is not adsorbed on silica. However, the presence of equimolar concentrations of Cu(II) and of complexing organic molecules does not necessarily prevent adsorption of the metal on silica (Fig. 8.11). Oxalate is a weaker chelant than EDTA and at pH values > 6 it cannot compete successfully with the silica surface sites. The Cu-oxalate curve in Fig. 8.11 corresponds to the adsorption of 'free' Cu(II). But, in the presence of 2,2'-bipyridyl (bipy) and ethylenediamine (en) the behaviour of Cu(II) cannot be explained without involving the adsorption of the metal complexes [17, 31, 32]. Studies on the stability of ternary complexes in solution [68, 69] have shown that Cu(bipy)$^{2+}$ exhibits a higher affinity for oxygen donor ligands (such as oxalate) than Cu^{2+}. Similarly, at the silica–water interface Cu-bipy complexes form stronger bonds with surface silanol groups than Cu^{2+} (Bourg *et al.* [17]). These results suggest that the rules that govern the stability constants of complexes in solution are applicable to the water–solid interface.

The ternary surface complexes mentioned above are of the form

Surface–Metal–Ligand   (I)

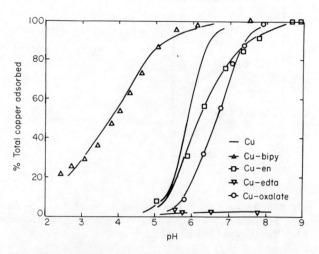

Fig. 8.11. Adsorption of Cu on amorphous silica in the presence of equimolar quantities of complexing agent (from Bourg and Schindler [32]).

but, the existence of a second kind has also been demonstrated by Davis and Leckie [47]:

Surface–Ligand–Metal   (II)

The adsorption pattern of Ag(I) on amorphous iron oxide in the presence of thiosulfate (Fig. 8.12a) is explained by the formation of ternary surface complexes of the form (II) (Fig. 8.12b).

Other ternary systems were studied, including inorganic complexing agents, but interpretations of a qualitative nature only were given (Table 8.4). Additional work, especially on ubiquitous inorganic ligands (OH^-, Cl^-, CO_3^{2-}, HCO_3^-, SO_4^{2-}), is urgently required.

Table 8.4. Some ternary systems for which only a qualitative model is presented

Metal	Ligand	Surface	Reference
Cu(II)	NTA, glycine and aspartic acid	γ-Al_2O_3	[71]
Cu(II)	Amino acids	γ-Al_2O_3 and activated carbon	[72]
Cu(II)	NTA and glycine	Alumino-silicates (of various Si/Al ratios)	[73]
Cd, Cu, Zn and Pb	Variety of complexing agent (including humic acids)	kaolinite, illite and montmorillonite	[74]–[82]
Cd	Cl, SO_4 and S_2O_3	α-SiO_2, γ-$FeOOH$, γ-Al_2O_3 and $Fe_2O_3 \cdot H_2O$ (am)	[49]

(a)

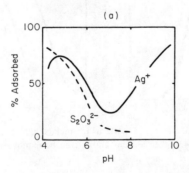

(b)

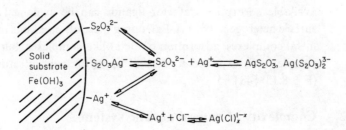

Fig. 8.12. Ternary surface complex of the form Surface–Ligand–Metal (simultaneous adsorption of Ag and $S_2O_3^{2-}$ on amorphous $Fe(OH)_3$) (from Davis and Leckie [47]).

The modification of the adsorption of a trace metal when it is complexed with a ligand or of a ligand when it is complexed with a trace metal can be caused by several phenomena. Bourg and Schindler [6, 32, 70] demonstrated the existence of the following:

(a) statistical

(b) electrostatic

(c) steric

(d) electronic (π electrons)

(e) weak interactions (Van der Waals and H-bonds).

Only the statistical component can be predicted accurately and analogy with solution chemistry allows, in the best cases, estimates of the magnitude of the other effects. But the experimental determination of the stability constant is necessary in most systems. This was done only for a few organic ligands, Cu(II), Fe(III) and silica, and indirectly for dissolved lake organic matter, Cu(II) and lake suspensions. The constants are evaluated by obtaining the best fit of the experimental data for several values of the ratio [total metal] to [total ligand] [6, 32].

A conceptual model for metal–ligand–surface interactions was presented by Benjamin and Leckie [48]. The model considers that complexed species behave 'metal-like' or 'ligand-like' depending on whether adsorption of the complex increases or decreases with increasing pH (Fig. 8.13). This qualitative model incorporated into classical binary complex adsorption models could yield quantitative predictions.

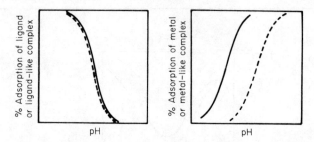

Fig. 8.13. Typical adsorption patterns for metals, ligands and complexes. --, low adsorbate concentration;——, higher concentration (from Benjamin and Leckie [48]).

There is sufficient evidence to suggest that the surfaces of most oxides are composed of several types of site (see Section 8.3.8). By contrast, most of the available adsorption data for ligands can be explained without invoking surface heterogeneity [35]. Extending these observations to the adsorption of metal complexes, adsorption of metal-like complexes, but not of ligand-like complexes, is expected to depend upon the concentration of the complex (Fig. 8.13) [48].

8.4.2   Complexity of natural aquatic systems

The complexity of natural organic matter, trace metals and solid surface associations in natural waters, including interstitial waters, has been demonstrated unequivocally by an experiment investigating the simultaneous adsorption of fulvic acid (FA), Cd, Zn and Cu on γ-Al$_2$O$_3$ (Bourg [54]) (Fig. 8.14) which showed that the adsorption of FA is not much affected by the presence of the three metals and is similar to the adsorption of FA alone, shown in Fig. 8.5e. However, this is not the case for the adsorption of Cd, Zn and Cu. For example, at low pH, Cu and to a lesser extent Cd are adsorbed *via* FA ternary surface complexes of the form (II). It is not possible to determine whether or not FA is adsorbed as a ternary surface complex of the form (I) at high pH because it is present in excess compared to the trace metals.

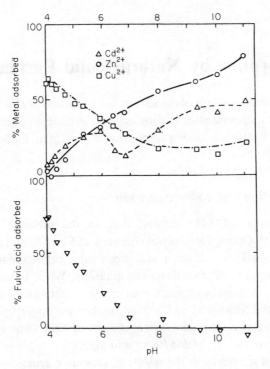

Fig. 8.14. Simultaneous adsorption of Cd, Cu, Zn and fulvic acid (FA) on γ-Al$_2$O$_3$. Metal total $= 5 \times 10^{-6}$ mol dm^{-3}; FA $= 80$ mg dm^{-3} (from Bourg [54]).

The coating of natural suspended particulate matter by organic matter such as humic substances and anthropogenic surfactants and the subsequent adsorption of trace metals correspond to the formation of ternary surface complexes of the form (II). In this case, the coordination behaviour of the coating with the trace metals is modified by its attachments by covalent bonds, etc. to the solid surface.

In natural aquatic systems, ternary surface complexes of forms (I) and (II) certainly occur simultaneously (Fig. 8.15). It is, therefore, somewhat difficult to quantify precisely the adsorption of naturally occurring ternary surface complexes. However, it is still possible, by studying systems such as the one presented in Fig. 8.14, to improve our understanding of general trends.

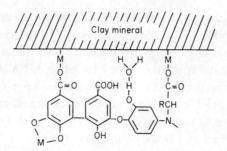

Fig. 8.15. Surface–Metal–Ligand association in natural systems (adapted from Stevenson and Ardakani [83]).

8.5 Adsorption by Natural Solid Particulate Matter

Several investigations of the adsorption behaviour of natural fluvial, estuarine or marine suspended and deposited solid particles and soils have been carried out, but only a few among them present conceptual modelling of the systems studied.

8.5.1 Adsorption and co-precipitation

Benjamin *et al.* [44] showed that, in the presence of 10^{-3} mol dm^{-3} $Fe_2O_3 \cdot H_2O$(am), the removal efficiency of 5×10^{-7} mol dm^{-3} Cd is essentially identical whether the iron is precipitated before the trace metal is added (adsorption) or afterwards (coprecipitation). When Laxen [84] performed a similar experiment with much smaller concentrations (5–50 nanomolar) of Cd, Pb, Cu and Ni, he observed that the uptake was greater, by a factor of 2, during coprecipitation than adsorption, suggesting that some adsorption sites are blocked as the precipitate forms and ages.

The discrepancy in the results of these two experiments probably arises from the different practical conditions used. Benjamin *et al.* separated the solids by centrifugation after 1 h experiments, whereas Laxen filtered the solution through 0.05 μm filters and, 'when necessary' results were corrected for a small amount of colloidal iron passing through the filters. This phenomenon is most likely to occur for a fresh coprecipitation procedure than for adsorption of an aged precipitate. The presence of humic substances in Laxen's experiment is perhaps more significant. Trace metals could be more competitive for adsorption of hydrous iron oxides when exposed to the surface at the same time as humics. If the humics are already present, they may block some higher energy adsorption sites, see also Baccini *et al.* [64] in Section 8.5.5.

8.5.2 Total number of exchangeable surface sites (Ns)

For solid hydrous oxides, Ns can be determined by classical acid–base titration (Fig. 8.7) but, for natural solids, dissolution occurs more readily when the pH of the titration medium is brought away from the range of natural waters [108]. It can, however, be measured by the tritium exchange method of Yates and Healy [112] or by a classical cation exchange capacity measurement via displacement of major cations by NH_4^+. These two methods gave values in good agreement for deep ocean interfacial sediments, viz. 2.7 ± 0.2 and 3.0 ± 0.2 equivalents kg^{-1} of solids, respectively [58].

8.5.3 The marine environment

The significance of sinking particles in the removal of trace elements from the deep ocean is often put forward in the literature. Recent studies of removal rates estimated from sediment trap data and from diffusion–advection models indicate scavenging in which dissolved trace elements are in adsorptive equilibrium with the surfaces of sinking particles [55, 66, 85]. The deep-sea distribution of Th isotopes between dissolved and particulate forms is not consistent with chemical scavenging models which assume irreversible uptake of Th on particle surfaces. The results can, however, be explained if continuous exchange of Th isotopes between sea water and particle surfaces is assumed. The estimated rate of exchange is fast compared to the residence time of suspended matter in the deep ocean (0.36–10 years depending on the method of evaluation) suggesting that particle surfaces are in equilibrium with respect to the reversible exchange of metals with sea water [86, 87].

The adsorption of 13 trace metals from sea water on deep oceanic interfacial sediments from the Guatemala Basin (MANOP site H) has been studied. For some elements, equilibrium was reached only after about 20 days. As for hydrous Fe and Al oxides, the binding energy was dependent on total metal concentration at high adsorption densities. The pH-adsorption edge of most metals was similar to that for simple oxides. However, for the weakest binding constants, viz. Cd, Ba and Co, it was not so sharp [58] (Fig. 8.16). It seems to correspond to binding of the metal with a negatively charged pH dependent site on a protonated polymer [88, 89] rather than surface coordination.

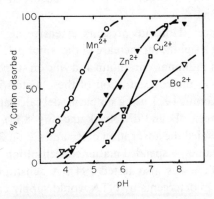

Fig. 8.16. Adsorption on deep oceanic interfacial sediments (from Balistrieri and Murray [58]).

8.5.4 Coastal and estuarine environments

For coastal areas where particle residence times are shorter than in the deep oceans, a more precise definition of adsorption kinetics is required. Sorption rate constants on particles from surface sediments and sediment traps collected at various oceanic locations, e.g. Narragansett Bay, San Clemente Basin, deep-sea red clay, were measured in batch experiments for several radioactive trace elements. For Zn, Se, Sr, Cd, Sn, Sb, Cs, Ba, Hg, Th and Pa (group A) equilibrium was reached after a few days, whereas for Be, Mn, Co and Fe (group B) uptake still increased after 108 days. This time-dependence was interpreted by an adsorption/desorption equilibrium (group A) followed by slow migration in the lattice structure (group B). The dependence of the (particulate)/(dissolved) ratio with the particle concentration was explained by the coagulation of radioactively tagged colloidal particles onto larger (collectable by filtration) particles [90, 91].

Lion *et al.* [52] evaluated the adsorption behaviour of Cu, Cd and Pb on salt marsh estuarine sediments, and came to the conclusion that it is controlled by Fe–Mn hydrous oxides and organic surface coatings.

8.5.5 Lakes and rivers

Baccini *et al.* [64] investigated the adsorption characteristics of pyrolized (at 550°C) particles from a pre-alpine Swiss lake. They observed under lake conditions that the adsorption of Cu and Zn is reduced significantly in the presence of natural organic matter. They postulated that this is due to the saturation of the binding sites of the mineral surface by organic matter, thus preventing other species from being adsorbed. This result is quite unexpected and it contradicts the commonly accepted idea that organic surface coatings enhance adsorption capacity and strength. The discrepancy observed could be due to two things. (i) Only a very small fraction of the total natural organic matter is adsorbed on the mineral particles while the rest is capable of keeping a sizeable amount of Zn and Cu in solution by complexation. (ii) The particulate organic matter destroyed by pyrolysis to obtain mineral particles is a more efficient adsorber than newly adsorbed organic matter. This definitely needs more work.

Lake Ijsselmeer (Holland) presents extensive algal blooms during the summer causing a significant increase in the water pH. This in turn provokes an adsorption of cadmium, chromium and zinc on suspended matter together with a decrease in the concentration of these metals in the surface waters. Salomons and Kerdijk [92], using simple model calculations, observed a good correlation between pH and dissolved zinc (Fig. 8.17).

Salomons studied the adsorption of Zn and Cd on Rhine River sediments for various pH values, suspended matter concentrations and chlorinities [93]. The effect of NTA was also tested [94]. A substitution, even partial, of polyphosphates in detergents by NTA would supply aquatic systems with a very strong complexing agent which is a good competitor with surface sites for binding trace metals. Even though metal-NTA complexes can be adsorbed

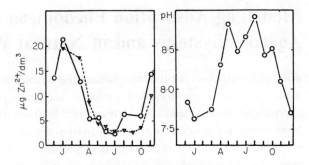

Fig. 8.17. Seasonal cycle for zinc and pH in the Lake Ijsselmeer. O, measured; △, calculated (from Salomons and Kerdjik [92]).

[71] the uptake is only partial. Not only does NTA make trace metals much less susceptible to adsorption on natural sediments, it also promotes their mobilization [94].

8.5.6 Aquifers

Because of more intense weathering, groundwaters are usually more mineralized than surface waters. Their larger ionic strength results in lower activities for trace elements, thus increasing their solubility. Moreover, larger concentrations of dissolved alkaline earth elements also decrease the quantity of adsorbed cationic trace elements. On the other hand this may be offset by, larger (solid surface)/(liquid volume) ratios in aquifers which should enhance adsorption.

Christensen [95–99] has carried out an extensive study of the adsorption of Cd in the μl range on two Danish soils (loamy sand and sandy loam). Even though the interpretation is not conceptual, the various observations are worth mentioning here. More than 95% of the sorption takes place within 10 minutes, equilibrium is reached in 1 h, and exposures up to 67 weeks did not reveal any long term changes in Cd sorption. The loamy sand exhibited full sorption reversibility in 10^{-3} mol dm^{-3} CaCl$_2$ at pH = 6.0, while under the same conditions a partial irreversibility of 1 μg Cd g^{-1} sandy loam was observed. This partial retention cannot be explained by any strong binding since no irreversibility was observed in 10^{-2} mol dm^{-3} CaCl$_2$ for the same pH. Ageing of the soil at a temperature of 1°C (after exposure to Cd) for 35 and 67 weeks did not produce any changes with respect to desorption of Cd in 10^{-1} mol dm^{-3} CaCl$_2$ at pH = 6.0. Relative solute velocity equations, based on Freundlich and linear isotherms were applied successfully to the prediction of Cd mobility under intermittent flow conditions in laboratory soil columns.

8.6 Modelling Adsorption Phenomena in Complex Aqueous Systems and in Natural Waters

With the exception of montmorillonite, the adsorption of trace metals on natural aquatic suspensions and sediments and on clay minerals is, to a first approximation, very similar to that on solid hydrous oxides [65] (Fig. 8.18). Most natural suspended particulate matter usually contains some montmorillonite, but the large permanent surface charge of this alumino-silicate mineral is not noticeable in the mixtures of clays and other components that make up most natural suspended particulate matter (Figs. 8.18 and 8.19).

The extrapolation to natural sediments of adsorption phenomena studied on hydrous oxides is a hypothetical exercise but, as shown for the Meuse River, it can prove useful for the understanding of processes in natural aquatic systems.

Several sophisticated computer programs are available for the determination of the equilibrium speciation of chemical components of aqueous systems [10]. Simple models are reported in the literature [100, 101], but calculations of dissolved speciation need to be incorporated into them if they are to be used for complex aqueous systems.

Such comprehensive computer programs already exist. The REDEQL group (REDEQL 2 [102], MINEQL [11], MICROQL [63, 103]) can model the adsorption of major and trace metals by solid oxides and hydroxides, such as silica, alumina, $Fe_2O_3 \cdot H_2O$ and goethite. The program GEOCHEM [104], developed from REDEQL 2, is adapted to soil solutions. It contains a subroutine which accounts for cation exchange on constant charge surfaces.

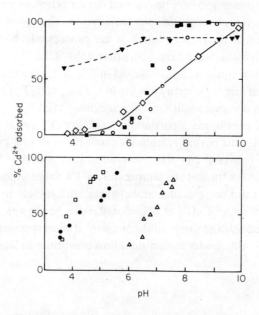

Fig. 8.18. Adsorption of Cd (5×10^{-6} mol dm^{-3}) on various surfaces (1 g dm^{-3}) (from Bourg [54]).

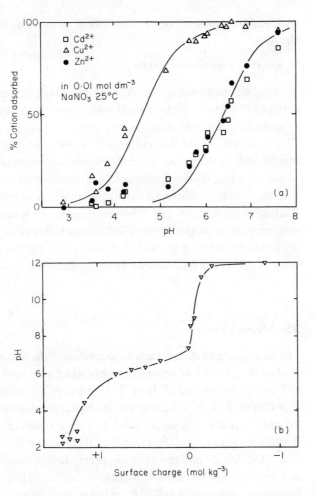

Fig. 8.19. Surface charge and trace metal adsorptive behaviour of Meuse River bottom sediments (from Mouvet and Bourg [5]).

The more recent, ADSORP can include ternary surface complexes [105]. Not as general as the REDEQL family, but written in BASIC for micro computers, and using a much simpler method of convergence, ADSORP is within the reach of water scientists less familiar with the intricacies of mathematics. ADSORP takes into consideration ternary surface complexes of the form (I) (Surface–Metal–Ligand), assuming that the ligand perturbs the formation of bonds between the surface and the metal, but that it is still the metal which regulates the chemistry of the adsorption. Ternary surface complexes of the form (I) are thus conceptualized as ternary complexes in solution. In natural systems, the negative charge of suspended particulate matter surfaces is thought to be caused by their surface coating of natural organic matter [106, 107]. In this case, the attraction between the surface and the ligand is so strong and independent of pH, e.g., hydrophobic forces of organic macro-molecules, that form (II) ternary surface complexes can, for practical purposes, be simply considered as pseudo-binary complexes [5] of the form:

(SURFACE AND COATING)-METAL

Existing mathematical models of adsorption do not take the multiple site surface concept (see Section 8.3.8) into account and thus they are not capable of modelling adsorption accurately over wide ranges of adsorption density. The inclusion of site non-uniformity is not a problem mathematically but much is still to be learned about the chemical concept of surface sites [41]. The situation is not, however, completely intractable because, as demonstrated in Section 8.7, trace elements should normally represent only a very small fraction of surface sites and if their binding energy is much greater than that of other species, competitive side effects should be negligible. The approach used in modelling the trace metal distribution in the Meuse River is a composite of conceptual model calculations and of experimental data fitting.

The Meuse River

The adsorption of trace metals by sediments of the Meuse River (Holland) was studied (Fig. 8.19) and the constants obtained were used to model the chemical speciation of the same river [5]. Adsorption constants were obtained experimentally for Cu, Cd and Zn and by extrapolation (Fig. 8.8b) for Ca, Mg, Ni and Pb (Table 8.5). After calculating the speciation of the dissolved metals it was found that, with the exception of calcite for Ca and possibly willemite (Zn_2SiO_4) for Zn in a few samples, precipitation/dissolution processes could not account for the dissolved concentrations [5]. The calculations of the computer program ADSORP indicate that adsorption is probably the controlling process in these Meuse River water samples.

Difficulties encountered in the study of natural solids, such as the determination of surface site concentration and of adsorption constants, will be discussed in detail elsewhere [108].

Table 8.5. Pseudo-adsorption* constants on Meuse River sediments (in $0.01\ mol\ dm^{-3}$ $NaNO_3$)

Element	$Log_{10}\ *K^S$
Mg^{2+}	-5.2
Ca^{2+}	-6.5
Cd^{2+}	-3.7
Cu^{2+}	-1.8
Ni^{2+}	-3.8
Pb^{2+}	-1.7
Zn^{2+}	-3.6

* See Bourg and Mouvet [108]
for a discussion of the term
'pseudo'.

8.7 Speciation on Hydrous Oxide Surfaces in Natural Waters

The knowledge of the pertinent surface and solution constants and the utilization of an iterative computer program permit the determination of the distribution of species adsorbed on a given surface in the same way as the calculation of speciation in solution. Such calculations are, however, still rare, mostly because of the scarcity of surface constants of simple surfaces and, even more, of dissolved organic components typical of natural waters and of natural sediments. Generalizations such as those in Figs. 8.8 and 8.9 should help. Nevertheless, two examples for goethite of such calculations appeared recently in the literature [62, 109].

The speciation of the goethite surface in a natural lake water is presented in Table 8.6 [109]. The calculations were limited to inorganic components because of the lack of values of surface constants for natural dissolved organic matter. Nevertheless, by considering that natural lake organic matter and phthalic and salicylic acids show similar adsorption patterns on γ-Al_2O_3 and by using adsorption constants for organic acids on alumina, Sigg and Stumm [109] estimated that $\sim 10\%$ of the surface sites may be occupied by adsorbed organic matter. This value is probably conservative because in natural waters dissolved organic molecules are present in polymeric form [60]. Apart from these considerations, the results detailed in Table 8.6 demonstrate the importance of phosphate and silicate surface species, even though these species are minor components of the dissolved phase. Consequently, the significance of such theoretical model calculations can be illustrated by

Table 8.6. Speciation of the goethite surface in a natural lake water* (pH = 7.5 and =FeOH$_{total}$ = 1×10^{-6} mol dm^{-3}**)

Dissolved components	Total concentration in the aquatic system (dissolved and adsorbed) (mol kg^{-1})	Surface species	Log $K_{intr.}$	% of surface sites
—	—	=FeOH	—	25
H^+	3.2×10^{-8}	=FeOH$_2^+$	−6.4	16
OH^-	4.9×10^{-7}	=FeO$^-$	−9.25	0.05
SO_4^{2-}	1.0×10^{-4}	=FeSO$_4^-$	−5.8	0.001
$H_2PO_4^-$	1.0×10^{-6}	=FeHPO$_4^-$	7.2	35
H_4SiO_4	5.0×10^{-5}	=FeH$_3$SiO$_4$	4.1	15
		=FeH$_2$SiO$_4^-$	−3.3	2.4
HCO_3^-	5.0×10^{-3}	=FeCO$_3$	2.5†	4.9
Mg^{2+}	2.0×10^{-4}	=FeOMg$^+$	−6.2	0.8
Ca^{2+}	1.0×10^{-3}	=FeOCa$^+$	−8†	0.05
Pb^{2+}	1.0×10^{-8}	=FeOPb$^+$	−3†	0.05

* From Sigg and Stumm [109]; ** =FeOH$_{total}$ = 1×10^{-3} g dm^{-3} with =FeOH$_{total}$ = 1×10^{-3} mol g^{-1}; † estimated.

indicating the potential importance of surfaces such as goethite in the control of dissolved phosphate and silicate concentrations of natural lake systems.

A similar investigation was carried out on the same goethite surface in sea water by Balistrieri and Murray [62]. The calculations were limited to the major inorganic components. The prevalent surface species were Mg and SO_4 and to a lesser extent, Ca and Cl (Table 8.7). Na and K did not contribute much to the occupation of surface sites. The adsorption of carbonate species was not considered in this study, but by analogy with the data in Table 8.6 and with sea water speciation in the dissolved phase, it can nevertheless be safely predicted that a substantial fraction of surface sites should, in fact, be occupied by carbonates. Balistrieri and Murray [62] did not investigate the adsorption of trace metals but, as in lake water, they would not be important surface species with respect to the total number of surface sites.

In laboratory adsorption experiments, investigators usually try to work under conditions where the adsorbable species of interest are present in concentrations low enough with respect to the total number of surface sites that they do not disturb the overall surface speciation. This might not always be true in natural systems, but Table 8.6 illustrates that it is true for the adsorption of Pb on goethite in a model lake water.

It follows that even if the adsorbed fraction of total trace metals in an aquatic system is high, trace metals should not contribute much to the particle surface site speciation (see discussion in Section 8.6).

Table 8.7. Speciation of the goethite surface in sea water (major ions only)* (pH = 8 and $=FeOH_{total} = 1.65 \times 10^{-3}$ mol dm^{-3}**)

Dissolved components	Total concentration in sea water (mol kg^{-1})	Surface species	Log $K_{intr.}$	% of surface sites
—	—	$=FeOH$	—	37
H^+	1×10^{-8}	$=FeOH_2^+$	-5.57	0.2
OH^-	$6.5 \times 10^{-6\dagger}$	$=FeO^-$	-9.52	1.8
Na^+	0.468 47	$=FeONa$	-8.40	1.4
K^+	0.010 20	$=FeOK$	-8.40	<0.01
Mg^{2+}	0.053 07	$=FeOMg^+$	-5.45	17
		$=FeOMgOH$	-14.25	9.3
Ca^{2+}	0.010 28	$=FeOCa^+$	-5.0	8.1
		$=FeOCaOH$	-14.50	0.9
SO_4^{2-}	0.028 23	$=FeSO_4^-$	-9.10	22
		$=FeHSO_4$	-14.40	<.01
Cl^-	0.545 90	$=FeCl$	-7.00	3.6

* From Balistrieri and Murray [62]; ** $=FeOH_{total} = 4.25 \times 10^{-6}$ mol m$^{-2} = 387.5$ m^2 dm^{-3}; $^\dagger pK_w = 13.19$ for sea water at 25°C (Dyrssen and Hansson [110]).

8.8   Conclusions

The adsorption of trace metals, of ligands and of their complexes can be interpreted by their coordination to surface sites. The concepts of solution chemistry are applicable to the water-suspended particulate matter interface, with the restriction that surfaces really behave like polymers and not like independent ligands.

Modelling a natural system can be done by using either the stability constants of model surfaces, such as silica, alumina or goethite or constants obtained from real sediments of the aquatic system investigated. The latter may be less satisfying for the physical chemist, but they are more realistic. For the reasons explained below, these methods should give compatible results.

Going back to the conclusions of Table 8.6 it can be objected that goethite is not often present as a significant fraction of natural suspended particulate matter. The question arises, therefore, of the significance of the control of dissolved phosphate and silicate concentrations of natural waters by this mineral (see Section 8.7). However, looking at Figs. 8.8 and 8.9, it can be seen that the relative adsorbing strength of all the adsorbable species on a given surface will remain of the same order. The extrapolation of results from the model surfaces of hydrous oxides is thus not as extravagent as one might think. More fastidious work on the determination of surface stability constants, especially on systems of environmental significance is, however, badly needed.

The difficulties encountered in the interpretation of studies with natural solids are demonstrated by the contradictory results sometimes observed. These contradictions may be apparent rather than real and come from either poor characterization of the system studied or different chemical conditions.

It is, however, clear that dissolved and particulate organic matter, as well as solid Fe and Mn hydrous oxides, are the most significant components, together with pH for the adsorption patterns of trace elements.

In a recent review concerning oxidized estuarine sediments, but quite applicable to all oxidized natural solid particles, Luoma and Davis [111] list the following requirements for modelling trace metal partitioning between dissolved and solid phases:

(a) the determination of binding intensities and capacities for important sediment components;
(b) the determination of relative abundance of these components;
(c) the assessment of the effect of particle coatings and of multi-component aggregation on the available binding capacity of each substrate;
(d) the consideration of the effect of major competitors, e.g. Ca and Mg;
(e) the evaluation of kinetics of metal redistribution among sediment components.

Since adsorption is a potentially important geochemical control of dissolved trace metals in natural waters, fulfillment of the requirements described above will certainly prove to be worth the effort in the development of comprehensive models.

31 A.C.M. Bourg and P.W. Schindler, *Chimia*, **32,** 166 (1978).

32 A.C.M. Bourg and P.W. Schindler, *J. Colloid Interface Sci.*, in preparation.

33 J.A. Davis, R.O. James and J.O. Leckie, *J. Colloid Interface Sci.*, **63,** 480 (1978).

34 J.A. Davis and J.O. Leckie, *J. Colloid Interface Sci.*, **67,** 90 (1978).

35 J.A. Davis and J.O. Leckie, *J. Colloid Interface Sci.*, **74,** 32 (1980).

36 R. Kummert, *Doctoral Thesis*, Swiss Federal Inst. Technol., ETH Zürich (1979).

37 L. Sigg, *Doctoral Thesis*, Swiss Federal Inst. Technol., ETH Zürich (1979).

38 A. Gisler, *Doctoral Thesis*, Univ. of Bern (1980).

39 L. Sigg, *Lizentiatsarbeit*, Univ. of Bern (1974).

40 M.M. Benjamin and J.O. Leckie, *J. Colloid Interface Sci.*, **79,** 209 (1981).

41 M.M. Benjamin and J.O. Leckie, in *Contaminants and Sediments*, (Ed. R.A. Baker) Ann Arbor Science, Ann Arbor, Michigan, 305 (1980).

42 M.M. Benjamin and N.S. Bloom, in *Adsorption from Aqueous Solutions*, (Ed. P.H. Tewari) Plenum Press, N.Y., 41 (1981).

43 J.O. Leckie, M.M. Benjamin, K. Hayes, G. Kaufman and S. Altman, Report EPRI RP-910, Electric Power Res. Inst., Palo Alto, Calif. (1980).

44 M.M. Benjamin, K.F. Hayes and J.O. Leckie, *J. Water Pollution Control Fed.*, **54,** 1472 (1982).

45 J.A. Davis and J.O. Leckie, in Chemical Modelling in Aqueous Systems (Ed. E.A. Jenne) *Am. Chem. Soc. Symposium Ser.*, **93,** ACS, Washington, D.C., 299 (1979).

46 A.C.M. Bourg and P.W. Schindler, *Inorg. Nucl. Chem. Lett.*, **15,** 225 (1979).

47 J.A. Davis and J.O. Leckie, *Environ. Sci. Technol.*, **12,** 1309 (1978).

48 M.M. Benjamin and J.O. Leckie, *Environ. Sci. Technol.*, **15,** 1050 (1981).

49 M.M. Benjamin and J.O. Leckie, *Environ. Sci. Technol.*, **16,** 162 (1982).

50 A.C.M. Bourg, *Proc. Internat. Conf, Heavy Metals in the Environment (London)*, 365 (1979).

51 A.C.M. Bourg, *J. Français Hydrol.*, **10,** 159 (1979).

52 L.W. Lion, R.S. Altmann and J.O. Leckie, *Environ. Sci. Technol.*, **16,** 660 (1982).

53 A.C.M. Bourg, H. Etcheber and J.M. Jouanneau, *Biol. Ecol. Méditerranéenne*, VI(3/4), 161, (1979).

54 A.C.M. Bourg, in *Trace Metals in Sea Water*, (Eds. C.S. Wong, E. Boyle, K. Bruland, J.D. Burton and E.D. Goldberg) Plenum Press, N.Y., 195 (1983).

55 L. Balistrieri, P.G. Brewer and J.W. Murray, *Deep Sea Res.*, **28A,** 101 (1981).

56 J.A. Davis, *Geochim. Cosmochim. Acta*, **48,** 679 (1984).

57 L.S. Balistrieri and J.W. Murray, *Geochim. Cosmochim. Acta*, **47,** 1091 (1983).

58 L.S. Balistrieri and J.W. Murray, *Geochim. Cosmochim. Acta*, **48,** 921 (1984).

59 J.C. Westall and H. Hohl, *Adv. Colloid Interface Sci.*, **12,** 265 (1980).

60 J.A. Davis, in *Contaminants and Sediments* (Ed. R.A. Baker) Ann Arbor Science, Ann Arbor, Michigan, 279 (1980).

61 E. Tipping, *Geochim. Cosmochim. Acta*, **45,** 191 (1981).

62 L. Balistrieri and J.W. Murray, *Am. J. Sci.*, **281,** 788 (1981).

63 J.C. Westall, in Particulates in Water, (Eds. M.C. Kavanaugh and J.O. Leckie), *Am. Chem. Soc. Adv. Chem. Ser.*, **189,** ACS, Washington D.C., 33 (1980).

64 P. Baccini, E. Grieder, R. Stierli and S. Goldberg, *Schweiz. Z. Hydrol.*, **44,** 99 (1982).

65 A.C.M. Bourg, *C.R. Acad. Sci. (Paris)*, **294,** 1091 (1982).

66 K. Hunter, *Deep Sea Res.*, **30(6A),** 669 (1983).

67 B.T. Hart, *Hydrobiologia*, **91,** 299 (1982).

68 H. Sigel, *Angew. Chem.*, **11,** 391 (1975).

69 G.A. L'Heureux and A.E. Martell, *J. Inorg. Nucl. Chem.*, **28,** 481 (1966).

70 A.C.M. Bourg and P.W. Schindler, *Proc. Symposium on Natural Lakes (Chambéry)*, French Limnological Assoc., 158 (1978).

71 H.A. Elliott and C.P. Huang, *J. Colloid Interface Sci.*, **70,** 29 (1979).

72 H.A. Elliott and C.P. Huang, *Environ. Sci. Technol.*, **14,** 87 (1980).

73 H.A. Elliott and C.P. Huang, *Water Res.*, **15,** 849 (1981).

74 H. Farrah and W.F. Pickering, *Aust. J. Chem.*, **29,** 1167 (1976).

75 H. Farrah and W.F. Pickering, *Aust. J. Chem.*, **29,** 1177 (1976).

76 H. Farrah and W.F. Pickering, *Aust. J. Chem.*, **29,** 1649 (1976).

77 H. Farrah and W.F. Pickering, *Aust. J. Chem.*, **30,** 1417 (1977).

78 H. Farrah and W.F. Pickering, *Water, Air, Soil Pollut.*, **8,** 189 (1977).

79 H. Farrah and W.F. Pickering, *Chem. Geol.*, **25,** 317 (1979).

80 D. Hatton and W.F. Pickering, *Water, Air, Soil Pollut.*, **14,** 13 (1980).

81 A. Beveridge and W.F. Pickering, *Water, Air, Soil Pollut.*, **14,** 171 (1980).

285   REFERENCES

82   J. Slavek and W.F. Pickering, *Water, Air, Soil Pollut.*, **16**, 209 (1981).
83   F.J. Stevenson and M.S. Ardakani, in *Micronutrients in Agriculture*, Amer. Soc. Agron, 79 (1972).
84   D.P.H. Laxen, *Proc. Internat. Conf. Heavy Metals in Environment (Heidelberg)*, 1082 (1983).
85   P.G. Brewer and W.M. Hao, in Chemical Modelling in Aqueous Systems, (Ed. E.A. Jenne) *Am. Chem. Soc. Symposium Ser.* **93**, ACS, Washington, D.C., 261 (1979).
86   Y. Nozaki, Y. Horibe and H. Tsubota, *Earth Planet. Sci. Lett.*, **54**, 203 (1981).
87   M.P. Bacon and R.F. Anderson, *J. Geophys. Res.*, **87**, 2045 (1982).
88   J.R. Disnar, *Doctoral Thesis*, Univ. of Orléans, France (1982).
89   R.F.M.J. Cleven and H.P. Van Leewen, in *Complexation of Trace Metals in Natural Waters*, (Eds. C.J.M. Kramer and J.C. Duinker) Martinus Nijhoff and Dr. W. Junk Publ., The Hague, 371 (1984).
90   U.P. Nyffeler, Y.H. Li and P.H. Santschi, *Geochim. Cosmochim. Acta*, **48**, 1513 (1984).
91   P.H. Santschi, U.P. Nyffeler, Y.H. Li and P. O'Hara, *Proc. 3rd. Internat. Symposium Interactions between Water and Sediments (Geneva)*, 18 (1984).
92   W. Salomons and H.N. Kerdijk, *Proc. Internat. Conf. Heavy Metals in Environment (Heidelberg)*, 880 (1983).
93   W. Salomons, *Sci. Tot. Environ.*, **16**, 217 (1980).
94   W. Salomons and J.A. Van Pagee, *Proc. Internat. Conf. Heavy Metals in Environment (Amsterdam)*, 694 (1981).
95   T.H. Christensen, *Proc. Internat. Conf. Heavy Metals in the Environment (Amsterdam)*, 214, (1979).
96   T.H. Christensen, *Water, Air, Soil Pollut.*, **21**, 105 (1984).
97   T.H. Christensen, *Water, Air, Soil Pollut.*, **21**, 115 (1984).
98   T.H. Christensen, *Water, Air, Soil Pollut.*, submitted.
99   T.H. Christensen, *Water, Air, Soil Pollut.*, submitted.
100 R.D. Guy, C.L. Chakrabarti and L.L. Schramm, *Can. J. Chem.*, **53**, 661 (1975).
101 S.M. Oakley, P.O. Nelson and K.J. Williamson, *Environ. Sci. Technol.*, **15**, 474 (1981).
102 R.E. McDuff and F.M.M. Morel, Tech. Report EQ-73-02, W.M. Keck Lab., Calif. Inst. Technol., Pasadena, Calif. 75 p (1974).
103 J.C. Westall, Tech. Report, Swiss Federal Inst. Technol., EAWAG, Dübendorf, (1979).
104 S.V. Mattigod and G. Sposito, in Chemical Modelling in Aqueous Systems, (Ed. E.A. Jenne) *Am. Chem. Soc. Symposium Ser.* **93**, ACS, Washington, D.C., 837 (1979).
105 A.C.M. Bourg, *Environ. Technol. Lett.*, **3**, 305 (1982).
106 R.A. Neihof and G.I. Loeb, *J. Marine Res.*, **32**, 5 (1974).
107 K.A. Hunter and P.S. Liss, *Limnol. Oceanogr.*, **27**, 322 (1982).
108 A.C.M. Bourg and C. Mouvet, *Water Res.*, in preparation.
109 L. Sigg and W. Stumm, *Colloids and Surfaces*, **2**, 101 (1980).
110 D. Dyrssen and I. Hansson, *Marine Chem.*, **1**, 137 (1972–73).
111 S.N. Luoma and J.A. Davis, *Marine Chem.*, **12**, 159 (1983).
112 D.E. Yates and T.W. Healy, *J. Colloid Interface Sci.*, **55**, 9 (1976).

Section 9 Physicochemical Speciation of Trace Elements in Oxygenated Estuarine Waters

Commission V.6

IUPAC

A. C. M. BOURG

Service Géologique National
Department EAU
Bureau de Recherches Géologiques et Minières
F-45060 Orléans Cédex 2
France

9.1 Introduction

9.1.1 Geochemical and biological importance of trace element speciation in estuaries

Estuaries are one of the several pathways for fluxes of trace elements between continents and oceans. Materials leached from the surface and subsurface of the continents are carried via streams and groundwaters towards the mouths of rivers. Since rivers supply about ten times more dissolved and particulate matter to the oceans than do glaciers and about 100 times more than the atmosphere [1], the fate of trace elements in estuaries is very important for the understanding of global geochemical fluxes. The extent and manner in which passage through the estuarine environment affects the speciation of trace metals must be determined. Dissolved species may be adsorbed, co-precipitated or co-flocculated in estuaries while particulate forms may be mobilized upon encountering increasing salinities. Similarly, organic complexes of trace elements may be displaced by inorganic ligands, such as Cl^-, CO_3^{2-}, HCO_3^- and SO_4^{2-}, present in greater concentrations in sea water than in river water.

It is now well recognized that it is not necessarily the total amount of a given trace element which is most important for the understanding of geochemical and biological cycles. The key to the flux of an element is its physico-chemical form. This is especially important for trace elements because they are greatly influenced by major components of aquatic systems such as hydroxides, carbonates, chlorides, organic matter or solid surfaces.

If a given trace element can be taken up by an organism only, for example, as the free ionic form M^{z+}, the availability of M^{z+} will not only be regulated by the uptake of M^{z+} but also by the thermodynamics and the kinetics of the dissociation of other forms into M^{z+}:

$$M_{bound} \leftrightharpoons M^{z+}{}_{free} \xrightarrow{uptake} M_{in\ biological\ species} \qquad (9.1)$$

Thus, it could happen that even if the equilibrium concentration of M^{z+} were very low, the total amount of M could become available to the organism by equilibria displacement.

This paper describes overall trace element speciation and is an attempt to complement existing reviews of the speciation [2–4] of dissolved species with present knowledge of solid–solution interfacial phenomena with increasing salinities [5–8] in estuarine waters.

9.1.2 General characteristics of estuaries

Estuaries are highly dynamic and very complex aquatic systems which represent the transition between river water and sea water. They are characterized by strong gradients in chemical parameters caused by mixing of waters of low and high salinities. Some elements are more abundant in sea water than in fresh water, e.g. Na, K, Mg, Ca, Cl, SO_4, HCO_3 and CO_3 (Table 9.1), while others are more abundant in river water, e.g. nutrients such as P, N and Si, and Al and many of the transition metals. The composition of sea water is generally constant throughout the oceans, at least for major ions and pH. The composition of river waters, on the other hand, is variable. For example, the estuarine pH gradient may be negative (typically from 8.4 to 8.1) for high alkalinity river water, or positive (typically from 7.3 to 8.1) for low alkalinity river water (Fig. 9.5). In general, the organic matter burden decreases during the transition from river to sea water (Table 9.2). The suspended matter concentration usually decreases also, but it may exhibit some relatively high turbidities [10] (Fig. 9.1) at intermediate salinities.

The total trace element load of dissolved components and particulate matter of each 'bucket' of water going down the estuary is subjected to a continually changing chemical environment. In addition, the dissolved load moving with the aqueous phase travels faster downstream than the particulate matter. The presence of these two dynamic factors raises the question of the existence of equilibrium or non-equilibrium reactions between trace elements

Table 9.1. Average concentrations (in $mol\,dm^{-3}$) of dissolved inorganic constituents in estuarine end members (river water with concentrations typical of high and low alkalinities, and sea water)*

	River water		
Constituent	Low alkalinity	High alkalinity	Sea water
Na	0.000 30	0.000 30	0.479 32
K	0.000 065	0.000 065	0.010 45
Mg	0.000 030	0.000 20	0.054 39
Ca	0.000 087	0.000 57	0.010 53
Cl	0.000 20	0.000 20	0.558 62
SO_4	0.000 15	0.000 15	0.028 89
$HCO_{3\,total}$	0.000 1	0.001 36	0.001 86
$CO_{3\,total}$	9.4×10^{-8}	0.000 018	0.000 275
pH	7.30	8.43	8.12
ionic strength	0.001	0.004	0.7
salinity (‰)	0.041	0.144	35

* From Dyrssen and Wedborg [3].

Table 9.2. Typical concentration ranges of organic carbon in estuarine end members*

Concentration of organic carbon (in $mg\,dm^{-3}$)	River water	Coastal sea water	Open ocean water
Dissolved	10–20	1–5	0.5–1.5
Particulate	5–10	0.1–1.0	0.003–1.0

* From Head [9].

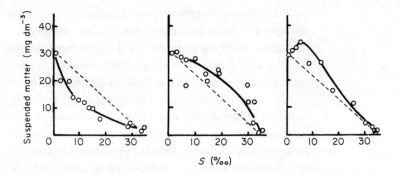

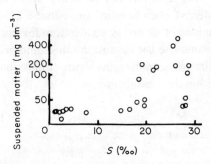

Fig. 9.1. Suspended matter concentrations versus salinity in surface waters of the Rhine Estuary (from Duinker and Nolting [11, 12]).

and solid surfaces or dissolved complex formers at any given salinity. At this point it should be noted that since, for any given geographical location, the mixing of river and sea water depends on the characteristics of river flow and on the tide, a given 'bucket' of water will be most conveniently referred to by its salinity and not by its location.

9.2 Important Processes Affecting the Speciation of Trace Elements

9.2.1 Introduction

Estuarine processes are composed of geochemical, hydrodynamic and biological components. In this review, only the first aspect is discussed but, mention will be made of the last two in relation to their indirect effects on the geochemistry of the system.

The geochemical component is important with respect to the solution coordination chemistry of trace elements as well as the processes leading to solid ⇆ liquid phase changes such as adsorption/desorption, precipitation/dissolution or flocculation-coagulation/stabilization of colloids. Flocculation and coagulation and the reverse phenomenon cannot be considered as phase changes in the strict sense of the term. However, because of technical limitations in their separation, the partition between dissolved and particulate forms may be operationally defined and the dissolved fraction is usually considered to contain fine colloids in addition to the species in true solution.

The sedimentation and resuspension of solid particles, i.e. the hydro-dynamic components, will be briefly mentioned when the differences in residence times of the particulate and the aqueous phases affect the trace element fluxes through estuarine systems. The biological component *per se* is not discussed. However, indirect effects such as E_h or pH changes, due to photosynthesis or respiration, and dissolved ligand concentration changes due to microbial consumption or excretion can be identified.

All reactions can be written as formal chemical reactions. The reader is referred elsewhere in this volume [13] for a detailed justification of this statement as far as adsorption/desorption processes are concerned. In the strict sense, the complexation should probably be a polymer type reaction for particulate matter, involving surface ligands and organic macromolecules as in humic substances

$$M \quad + \quad L \quad \rightleftharpoons ML \quad\quad\quad (9.2)$$

trace element (including surface
ligands)

Two types of effect can modify the speciation of M.
(a) Direct effect: an increase in the concentration of L causes more M to be complexed.
(b) Indirect effect: an increase in the concentration of a competitor with M for complexation causes less M to be complexed.

It is thus evident that if we are to understand the speciation of trace elements, a knowledge of both ligands and competitors for ligands is necessary. In practice, many direct and indirect effects will compete.

9.2.2 Changes in water composition from river to sea

The strong gradients in geochemical parameters in an estuary are caused by the mixing of river and sea water. If no phenomenon other than mixing is involved, the concentration gradients of Na, Cl, K, Ca, SO_4, and total alkalinity are linear with salinity. But other parameters such as activity coefficients, pH, Mg, suspended matter, organic matter and trace elements vary in a non-linear mode. The influence of salinity is described first and the interpretation of such observations is discussed in Section 9.3.

First of all, an index of conservative mixing is needed. Chlorinity and salinity are usually used. Chlorinity is measurable directly, for example, by titration with silver nitrate, and there is no evidence that Cl behaves non-conservatively during estuarine mixing. Salinity is also easily determined, by conductivity measurement, but the question of the variations of individual elements contributing to the salinity, especially in the river end member, has been raised. There is practically no such variation for salinities > 1‰ (Fig. 9.2)

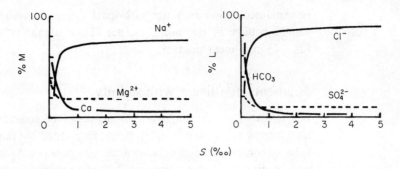

Fig. 9.2. Fraction of the total concentration of the major components of estuarine waters (formed by mixing average river water and average seawater) (from Millero [14]).

and salinity is, therefore, a useable parameter for these waters. However, the significance of processes occurring in regions of very low salinity [8, 15] makes it necessary to use chlorinity as the reference index in the river end of estuaries, contrary to the suggestion by Burton [16].

9.2.2.1 Two chemical zones of estuaries

Two zones can be identified [14] based on evolution of the composition of the major dissolved components of estuarine waters from the river to the ocean (Fig. 9.2).

(a) A mixing zone (salinity < 1‰) with water of varying composition and thus of varying chemical reactivity. The special importance of this zone has already been pointed out [8, 15]. It should probably even be considered as the real river–ocean interface, because the variations in medium composition which occur in a somewhat spatially limited area have a greater chemical and biological impact, e.g. ionic strength and osmotic pressure, than in the rest of the estuary. For example, variations in ionic strength from 0.005 to 0.05 and from 0.05 to 0.5 have a similar influence on the physical chemistry of water. It is, therefore, quite unfortunate that many scientists define the river end member of their estuaries as occurring at a salinity of 0.5‰, where, moreover, chlorinity is a more sensitive mixing index.

(b) A dilution zone (salinity > 1‰) where for major cations and anions one can consider, for all practical purposes, the 1‰ salinity water to be diluted sea water.

9.2.2.2 Gradients linear with salinity

In the absence of contrary evidence, Cl is assumed to show conservative behaviour. The investigations of Hosokawa *et al.* [17] on the Chikugogawa Estuary and of Carpenter (cited in Liss [18]) on the Potomac River suggest that most of the major components of sea water, viz. Na, K, Ca and SO_4, behave conservatively during mixing. Mg exhibits physical mixing behaviour in some estuaries [17], but not in others [19, 20]. If this behaviour of Mg is to be explained by ion exchange with species on particulate matter surfaces, Ca should behave in the same manner. Total alkalinity varies with salinity [21–24] in a linear manner.

9.2.2.3 Gradients non-linear with salinity

Before discussing the major and minor dissolved constituents which are not inert during estuarine mixing, it should be pointed out that even if an element behaves conservatively, its chemical activity does not. As can be seen in Fig. 9.3 and as discussed later (Section 9.3.5), the activity coefficients do not vary linearly with increasing ionic strength (I). The value of I for sea water is taken as somewhere between 0.7 and 0.5 depending on the ion pairing involved.

The pH of the river water entering an estuary may be higher or lower than that of sea water (Table 9.1). If the hydrogen ions behaved conservatively, one would obtain a variation of pH versus salinity such as that plotted in Fig. 9.4. But this is not so because the pH is controlled by the alkalinity. The peculiar pH behaviour is caused by the increase of the first and second apparent dissociation constants of carbonic acid with salinity [22]. The pH distribution as a function of salinity depends on the pH of the river water (Fig. 9.5).

9.2.2.4 Natural organic matter

There is, in general, 2–10 times more 'dissolved', i.e. $\leqslant 0.45\ \mu m$, organic matter than particulate organic matter in rivers, estuaries and coastal waters (Table 9.2). The bulk of the natural organic matter consists of biopolymers, polypeptides and polysaccharides, and geopolymers such as humic substances [26]. The biopolymers are more easily degraded and as a result they make up

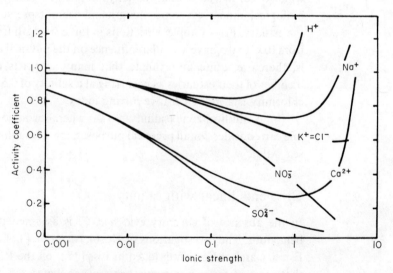

Fig. 9.3. Activity coefficients versus ionic strength for some common ions (using the Debye–Hückel equation for ionic strength <0.1 and the mean salt method above) (from Garrels and Christ [25]).

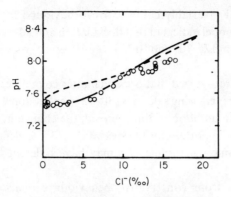

Fig. 9.4. pH distribution in the Western Scheldt Estuary. O, observed values; ———, calculated values using fresh and sea water data (from Mook and Koene [22]; ———, calculated values assuming conservative mixing of H$^+$ (the activity coefficient of H$^+$ is calculated according to the extended Debye–Hückel expression).

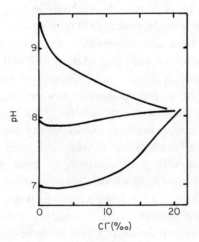

Fig. 9.5. Theoretical pH profiles for estuaries of rivers with various pH values (from Mook and Koene [22]).

<10% of the bulk of the organic matter of estuaries. For the same reason, oceanic organic matter, which can be found in estuaries, is mostly formed from refractory humic substances.

Dissolved organic carbon (D.O.C.) usually behaves conservatively in estuaries [21, 27–29].

In six sampling cruises over a two year span in the Delaware Estuary (USA), dissolved organic carbon was observed to behave conservatively, as opposed to the variability and non conservativity of dissolved nutrients, indicating that it is thus less biogeochemically reactive [21]. This was not true as a whole since the humic acid fraction showed a distinctly non-conservative behaviour with minimum concentration in the middle of the estuary.

The concentration of dissolved organic carbon investigated over $2\frac{1}{2}$ years in the turbid slowly flushing ($\sim$ 200 days) Severn Estuary and Bristol Channel

(UK) showed invariably conservative behaviour [28]. A dissolved organic carbon flux model indicated that flocculation and adsorption processes would remove, at most, 10 and 0.2% of dissolved organic matter in rivers, respectively.

It is well recognized that organic matter coagulates and flocculates upon encountering increasing salinities [30–32] (see Section 9.3 for a more elaborate discussion). In addition to this removal, the change in nature of the organic matter can be important for the speciation of trace elements because different sorts of organic macromolecules may have different complexation characteristics.

Deviations from conservative behaviour can also be observed if the dissolved organic carbon is mainly allochthonous in origin or in the presence of large variations in water chemistry, such as a pH gradient from 7 to 9 observed by Laane in the Ems-Dollart Estuary [33].

The very rapid decrease in D.O.C. observed in the Rhine River estuary [34] and in the Gironde Estuary [35, 36] might be due to local pollution [34] or to heavy flocculation or microbial activity in the very low salinity zone. Morris *et al.* [15] observed a peak in D.O.C. at a salinity of 0.2‰ in the Tamar Estuary, followed by a non-conservative decrease which they assumed to result from the microbiological degradation of dissolved organic matter to CO_2 and NO_3^-. Such reactions could account for up to 90% of the sharp decrease in dissolved oxygen between fresh water and the 1.00‰ isohaline. The coincidence of the oxygen sag and the D.O.C. peak with a chlorophyll fluorescence minimum in the 0.10–1.00‰ mixing zone suggest the involvement of a common mechanism. It was postulated to be a sequence of phenomena starting with mass mortality of fresh water halogenophobic phytoplankton not capable of withstanding the sharp osmotic increase and water chemistry change at the fresh water–sea water interface (see Section 9.2.2.1). This would release easily degradable dissolved organic matter supporting an active population of O_2-utilizing bacteria in a very localized area on the salinity scale.

Dyrssen and Wedborg compared the stability of inorganic forms of dissolved trace metals to that of organic species [3]. From calculations with a mathematical speciation model under equilibrium conditions, they concluded that for the usually low concentrations of dissolved organic matter, the presence of macromolecules or polymers is needed to produce a significant complexation of trace metals. In a similar study, Mantoura *et al.* [2] calculated that dissolved humic acid complexes of trace metals are important only in the case of Cu and, for salinities up to 25‰, of Hg. Their investigation of the extent of copper–humic interactions as a function of the nature of humic substances demonstrated the importance of particulate humics. Even trace metals with humic complexation constants lower than those of Cu and Hg could complex (adsorb) with particulate organic matter. The organic matter is present in interstitial waters at levels one or two orders of magnitude greater than in free flowing waters. Therefore, even though D.O.C. must compete with sulfides for trace metal complexation, it could be an important component in trace element speciation.

9.2.2.5 Natural suspended particulate matter and sediments

Suspended matter does not seem to behave conservatively during estuarine mixing (Fig. 9.1). This is, however, an artifact. In mixing experiments, whose principles are described in Section 9.3.3, Duinker and Nolting showed that the suspended matter concentration is diluted by the addition of filtered sea water [12]. Flocculation and coagulation of colloidal organic matter may have occurred in that experiment, but the amount of non-colloidal particulate material formed, if any, was not significant compared to the total suspended matter. In a real estuary, however, some of the suspended matter may flocculate and coagulate and some of the sediments can be resuspended because of estuarine currents or tidal bores, all possibly helped by changes in physico-chemical parameters such as salinity and pH. Suspended matter, i.e. surface ligands, represents an important component of trace element speciation (Section 9.2.3). Consequently, drastic changes in the concentration of surfaces, such as in a turbidity maximum, will significantly disturb the concentrations of all adsorbable constituents. Since trace components are, in general, more powerful adsorbents than major ions and ligands, their speciation will be greatly affected, demonstrating the potential significance of suspended matter dynamics (deposition/resuspension).

As for organic matter, the nature of suspended matter and bottom sediments varies with increasing salinity. The variation with salinity of relative amounts of clay minerals in the Gulf of Mexico [37] and in the Pamlico River estuary, and laboratory experiments on clay stabilities [38] indicate that colloidal forms of illite are more stable than those of kaolinite, which are, in turn, more stable than those of montmorillonite.

In the Gironde Estuary, France, it was observed that particles contained 5–15% particulate organic matter and 5–6% Fe and Mn oxy-hydroxides in the very low salinity zone and lower than 5% particulate organic matter and 6–7.5% Fe and Mn oxy-hydroxides in the middle of the estuary [39]. Particles from the zone of turbidity maximum (middle of the estuary on a log scale of salinity) presented specific surfaces twice as large as particles from either end of the estuary [8].

Variations may also be seasonal [36, 40]. Generally, in winter, particles tend to be derived from continental weathering, whereas in summer they originate primarily from estuarine biological processes.

In a review of estuarine coagulation, Burton [16] summarized the mechanisms involved in the process: (i) neutralization of the negative charge of colloids by specific adsorption, (ii) compression of the electrical double layer due to the increase in ionic strength, (iii) formation of bridging bonds between particles by organic molecules.

For all of the reasons discussed earlier in this section, coagulation and flocculation are expected to be important processes with respect to the physical (particulate/dissolved) speciation of trace elements in estuaries. Hahn and Stumm [37] have presented a model of coagulation and sedimentation of colloidal matter in natural aquatic systems and they applied it with success to the northern Gulf of Mexico. Such models should be used to improve mathematical equilibrium speciation calculations.

9.2.3 Reactions at the water/sediment interface

Reactions at the water/solid particulate matter interface are discussed elsewhere in this volume. Here, only the aspects of adsorption processes applicable to the specific characteristics of estuaries are covered.

The adsorption/desorption behaviour of trace elements and small ligands can be explained by the formation of surface complexes. The phenomenon may be understood in terms of competition for surface sites between metal cations and protons or between ligands and hydroxyls. Reactions such as:

$$SOH + M^{z+} \xrightleftharpoons{*\beta^{surf}} SOM^{(z-1)+} + H^+ \tag{9.3}$$

can thus be written. The question of interest here is whether adsorbed trace elements can be mobilized by increasing concentrations of either a competitor for adsorption (9.4) or a complexing agent in solution (9.5)

$$SOCa^{(z-1)+} + Ca^{2+} \xrightleftharpoons{} SOCa^+ + M^{z+} \tag{9.4}$$

$$SOM^{(z-1)+} + HL \xrightleftharpoons{} SOH + ML^{(z-1)+} \tag{9.5}$$

Duinker [5] reviewed contradictory reports from the literature suggesting, depending on the aquatic systems, that trace metals are either scavenged or mobilized upon encountering sea water. This summary, together with more recent and additional older data, is presented in Table 9.3. With the exception of Fe, which is always removed from the dissolved phase, field observations show no general trend for any given trace element (Figs. 9.6 and 9.7).

Laboratory experiments have been criticized, but they can provide carefully controlled conditions under which one can study single phenomena [7, 8, 73–80]. For example, in laboratory mixing experiments, Kharkar et al. [73] desorbed Co, Se and Ag from selected clays and solid oxides, and Van den Weijden et al. [76] showed mobilization of Cr, Mn, Co, Ni, Cu, Zn and Cd, but not Pb and Fe, from River Rhine sediments using artificial sea water. Recently, two detailed investigations of the adsorption of Zn, Cd and Cu on model surfaces and natural suspended matter, (SiO_2, γ-Al_2O_3, kaolinite, montmorillonite, Rhone River, Garonne River, Rhine River and Gironde estuary), were reported [7, 8, 77, 78]. There follows a brief review of these observations and conclusions.

9.2.3.1 Effect of ionic strength (I)

As in solution chemistry, ionic strength can change the extent of surface reactions and thus the value of the corresponding constants. However, one can see that in reactions (9.3) or (9.4) the effect should not be large for $I < 0.1$. This is observed on amorphous silica for the adsorption of Cu but not for Cd (Fig. 9.8). The difference can readily be explained by the much lower surface constants of the second metal. The adsorption of Cu is not affected by a greater number of Na^+ and NO_3^- support electrolyte ions. But Cd is not as competitive with Na^+ with respect to adsorption. In the case of Cd, for a given pH value, an increasing amount of Na^+ will cause desorption.

Table 9.3. Summary of field observations of geochemical reactivity of trace elements in estuaries (updating the review of Duinker [5])

Estuary	Element	Process	Authors
Alaskan fjord	Zn	Removal	Burrell (1973) [41]
Satilla, Ogeechee,	Fe	Removal	Windom et al., (1971) [42]
Altamata	Mn	Removal to some extent	
Back River	Mn, Fe, Ni, Cu, Zn, Pb, Cd	Removal of waste water plant discharge	Helz et al., (1975) [43]
Corpus Christi Bay	Cd, Mn	Removal, remobilizaion from bottom in winter	Holmes et al., (1974) [44]
Var	Zn	Release at higher salinities	Fukai et al., (1975) [45]
Beaulieu	Fe	Removal	Holliday and Liss (1976) [46]
	Mn	Conservative, possibly in situ production at low salinities	
	Zn	Conservative	
Fraser	Zn	Removal and release at higher salinities	Grieve and Fletcher (1977) [47]
Clyde	Zn	Removal	Mackay and Leatherland (1976) [48]
St. Lawrence	Fe	Removal	Subramanian and d'Anglejan (1976) [49]
	Mn	Conservative	
Strait of Georgia off Fraser River	Cu, Zn	Release bottom sediment lower estuary in spring and early summer	Thomas and Grill (1977) [50]
Merrimack	Fe	Removal	Boyle et al. (1974) [51]
4 Scottish rivers	Mn, Fe, Al	Removal	Sholkovitz (1979) [52]
Columbia	^{54}Mn, ^{65}Zn	Release	Evans and Cutshall (1973) [53]
Rhine	Fe, Cd, Cu	Removal	Duinker and Nolting (1976, 1977) [11, 54]
	Pb	Conservative	
	Zn	Conservative or removal	
	Mn	Cycling; in situ production at low salinities from bottom sediment	Duinker et al. (1979) [55] Wollast et al. (1979) [56]
	Se	Conservative or slight removal	Sloot, Van der, et al. (1978) [57]
	Sb	Conservative	
	Mo	Conservative	
Newport	Mn	Cycling; in situ production at low salinities from bottom sediment	Evans et al. (1977) [58]
Scheldt	Mn	Cycling; in situ production at low salinities from bottom sediment	Duinker et al. (1979) [55] Wollast et al. (1979) [56]
	Sb	Removal	Sloot, Van der, et al. (1978) [57]
	Mo	Conservative	
	Se	Conservative	
Godavari			
La Have	Hg	Removal	Cranston (1976) [59]
Mullica	Fe	Removal	Conley et al. (1971) [60]
Solent	Se	Conservative, for both Se total and Se(IV)	Measures and Burton (1978) [61]
Coastal waters, off Takai-mura	Se	Release at high salinities	Sugimara et al. (1976) [62]

Table 9.3. (*continued*)

Estuary	Element	Process	Authors
Amazon	Ni, Cu	Conservative, with possible small release at low salinities	Boyle *et al.* (1982) [63]
	Cd	Release	
Hudson	Cu	Conservative, with possible small release at intermediate salinities (most likely due to sewage in New York Harbor)	Klinkhammer (1983) [64]
	Ni	Release	
Savannah	Cu	Release and removal depending upon salinity	Windom *et al.* (1983) [65]
Ogeechee	Cu	Release	
Scheldt	Zn, Cu, Cd, Ni, Mn	Release	Kerdijk and Salomons (1981) [66]
	Fe	Removal	
Gironde	Cu, Pb, Zn, Ni	Release	Jouanneau (1982) [67]
Seine	Cd, Cu	Release, associated with turbidity maximum	Boust (1981) [68]
	Cr, Fe, Mn, Pb, Zn	Same as above but only on some occasions	
Conway	Mo	Release	Jones (1974) [69]
Southampton water	Mo	Conservative	Head and Burton (1970) [70]
Gironde	Rare earths	Removal	Martin *et al.* (1976) [71]
Chesapeake Bay	Zn	Release from bottom sediments	Bradford (1972) [72]

9.2.3.2 Effect of inorganic complexing agent

The adsorption edge of Cu on α-SiO_2 is displaced towards higher pH in the presence of increasing chloride concentration [81]. The formation of chloro-complexes competes with adsorption (Fig. 9.9). The same effect is observed for the systems Cd/γ-Al_2O_3 (Fig. 9.10) and Zn, Cd/River Rhine sediments (Figs. 9.11 and 9.22).

9.2.3.3 Effect of competing adsorbing species

The addition of $CaCl_2$ to the medium provides the additional effect of the competition of Ca^{2+} for surface sites, displacing the adsorption curve to the right (Fig. 9.10). Of the three metals investigated in this study, the effect was greatest for Cd because it is the metal with the smallest surface complex formation constants and the largest chloro-complex formation constants.

This behaviour is not limited to Al and Si oxides. Estuarine and river sediments behave in the same fashion (Fig. 9.12). The Cl^- complexation and the Ca^{2+} competition are felt more strongly by Zn than by Cu because for Cu, surface constants are greater and the chloro-complexes constants are smaller. It is interesting to note that the Garonne River and the Gironde sediments

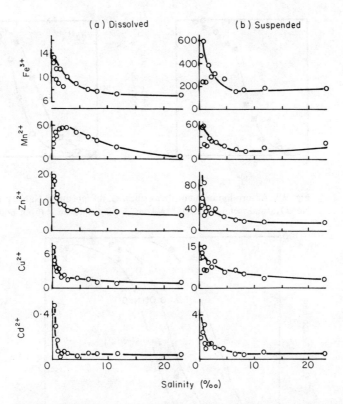

Fig. 9.6. Dissolved (a) and particulate leachable* (b) metals in the Rhine Estuary (all concentrations in $\mu g\,dm^{-3}$; *leached for 18 h in 0.1 $mol\,dm^{-3}$ HCl) (from Duinker and Nolting [12]).

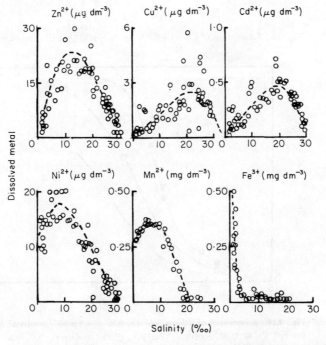

Fig. 9.7. Dissolved metals in the Scheldt Estuary (from Kerdijk and Salomons [66]).

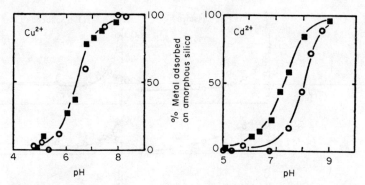

Fig. 9.8. Adsorption on amorphous silica. O, $0.1\ mol\ dm^{-3}$ $NaNO_3$; ■, $0.001\ mol\ dm^{-3}$ $NaNO_3$; total metal $=5\times10^{-6}\ mol\ dm^{-3}$; 1.0 g solid dm^{-3} (from Bourg [7]).

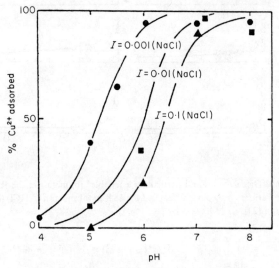

Fig. 9.9. Adsorption of Cu on α-SiO_2 (quartz) ($Cu_{total}=10^{-6}\ mol\ dm^{-3}$) (from Vuceta [81]).

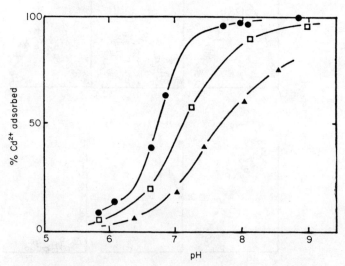

Fig. 9.10. Adsorption of Cd on γ-Al_2O_3 ($Cd_{total}=5\times10^{-6}\ mol\ dm^{-3}$; 1 g solid dm^{-3}) (from Bourg [7]). ●, $0.001\ mol\ dm^{-3}$ $NaNO_3$; □, $0.1\ mol\ dm^{-3}$ $NaCl$; ▲, $0.1\ mol\ dm^{-3}$ $NaCl+$ $0.01\ mol\ dm^{-3}$ $CaCl_2$.

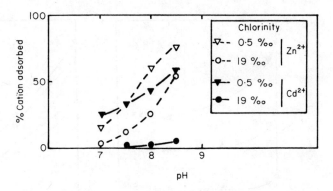

Fig. 9.11. Adsorption on Rhine River sediments (adapted from Salomons [77]).

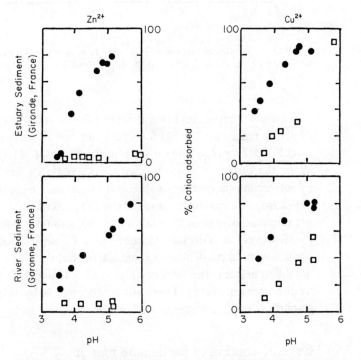

Fig. 9.12. Adsorption of Zn and Cu on natural sediments. Total metal $= 5 \times 10^{-6}$ mol dm^{-3}; 1 g sediment dm^{-3}; ●, 0.001 mol dm^{-3} NaNO$_3$; □, 0.1 mol dm^{-3} NaCl + 0.01 mol dm^{-3} CaCl$_2$ (from Bourg [7]).

exhibit similar surface behaviour. The surrounding waters of greatly different salinities do not seem to affect the surface behaviour.

9.2.3.4 Effect of dissolved organic matter

Salicylic acid is an organic ligand which exhibits some adsorption properties similar to natural dissolved organic matter. Figure 9.13 shows (a) that increasing the ionic strength provokes important adsorptive changes, and (b)

303 PROCESSES AFFECTING SPECIATION

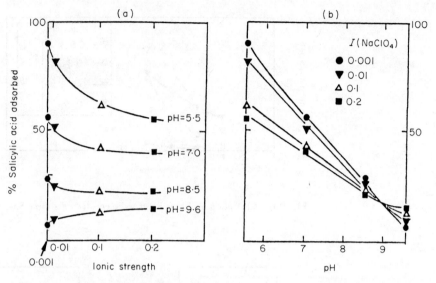

Fig. 9.13. Adsorption of salicylic acid on γ-Al_2O_3 at various ionic strengths (salicylic acid total $= 2 \times 10^{-4}$ mol dm^{-3}; 4.9 g γ-Al_2O_3 dm^{-3}) (from Kummert [82]).

that the pH is important in regulating surface/solution distribution. The value of pH_{ZPC} represents the pH for which the overall surface charge is equal to zero. For $pH < pH_{ZPC}$, the surface is positive and for $pH > pH_{ZPC}$ it is negative. When the surface is positively charged, i.e. $pH < 8.8$, for γ-Al_2O_3, increasing salinity promotes desorption. Natural surfaces are negatively charged for the pH range of natural waters and thus, in the presence of increasing salinity, the organic matter, especially small molecules, should become more adsorbed.

However, the Gironde, Garonne and Rhine sediments investigated do contain natural particulate organic matter, but their adsorption behaviour is not different from other surfaces (Figs. 9.11, 9.12 and 9.22) except in some cases by displacement of the pH curve, or in other words, by different, larger, surface stability constants or number of surface sites.

9.2.3.5 Organic coating of particulate matter

Neihof and Loeb [83, 84] studied the electrical charge of model and real estuarine and coastal particulate matter. In artificial sea water, containing only Na, K, Mg, Ca, Cl, SO_4 and HCO_3, and in photo-oxidized sea water, for the same salinity, the electrophoretic mobilities of a given model particle were very similar. Depending on the nature of the particle, electrophoretic mobility values ranged from $+1.4 \times 10^{-8}$ to -1.8×10^{-8} m^2 s^{-1} V^{-1}. However, in the presence of untreated natural sea water, the electrophoretic mobility of the model particles became negative if they were positive, and less negative if they were already negative (Fig. 9.14). This behaviour was explained by the adsorption of soluble organic matter as it is well known that an adsorbed coating dominates the electrophoretic behaviour of particles with different overlying surfaces [84]. In agreement with the preceding results, Hunter and

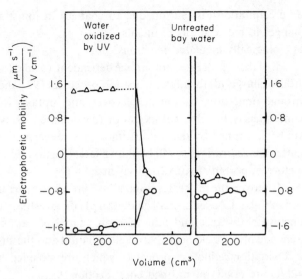

Fig. 9.14. Electrophoretic mobilities of particles in Chesapeake Bay sea water. △, anion-exchange resin; ○, quartz (from Neihof and Loeb [84]).

Liss [85] found that the electrophoretic mobility of suspended particles became less negative with increasing salinity in four estuaries.

Pravidic [86] and Martin *et al.* [87] observed that suspended particles, originally negatively charged in river water, reversed the sign of their charge for salinities ~20‰. This observation could explain the coagulation of particles. However, they were not able to reproduce their charge reversal even with water samples of similar origin [88]. Also, it is now well recognized that coagulation phenomena are most important at salinities ≪20‰ [89].

The more or less identical value of electrophoretic mobility for the various particles immersed in untreated sea water by Neihof and Loeb [84] seems to indicate that it is not the nature of the surface but that of the coating which is important. And the adsorption of natural organic matter may, therefore, make all natural surfaces behave in the same manner. More work is badly needed on the adsorption properties of bulk particulate organic matter and organic coatings. Indication of this importance has been shown by Balistrieri *et al.* [90] and by Bourg [91].

9.2.4 **Processes affecting trace element speciation: emphasis on water/solid particulate matter interface**

In estuaries many phenomena may affect the distribution of trace elements between dissolved and particulate phases:
(a) flocculation and coagulation of colloidal particles,
(b) desorption in the presence of increasing concentrations of competitors, e.g. Ca, Mg, Na, K, and of complexing agents such as Cl, CO_3,
(c) adsorption/desorption of organic matter (carrying some metals along in the phase change),

(d) precipitation/dissolution due to changes in ion associations, including changes in redox status of metals,

(e) biologically mediated reactions.

All of these processes are surface dependent: (b) and (c) directly, (a) through surface charge, (d) through nucleation processes [92] and (e) at least partially through ingestion by detrital feeders and uptake by biological tissues. Consequently, the physical speciation (particulate/dissolved) of trace elements will be controlled by the solid surface sites concentration, i.e. the suspended matter concentration. It will be shown later (Section 9.3.5) that this parameter is more important than variations in salinity.

Moreover, the processes listed above are all very pH-dependent. As can be seen in Fig. 9.15, even a small change in pH (in an estuary the pH may vary as a function of distance and time by more than 2 pH units [33]) will affect the exchange of trace elements between the liquid and the suspended solid phase.

Oxidation/reduction processes which are included here in processes (d) and (e) are briefly mentioned later (Section 9.3.5).

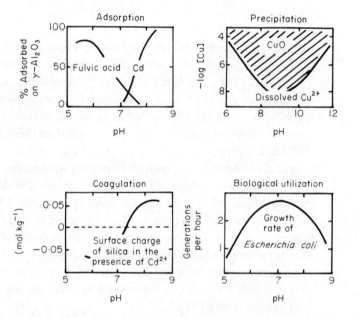

Fig. 9.15. Effect of pH on solid/solution interface phenomena (from Bourg [91]).

9.3   Critical Evaluation of Existing Approaches

This section discusses two methods of interpretation of field observations, one laboratory experiment, the product approach, and mathematical model calculations of speciation.

9.3.1   Conservative or non-conservative behaviour of dissolved constituents

A classical method for determining the existence in estuaries of processes other than physical dilution is the study of the relationship between the concentration of the dissolved trace element of interest and a conservative mixing index, such as salinity or chlorinity (Fig. 9.16, see Section 9.2.2). If the interpretation of such mixing curves is straightforward in the case of physical dilution, the other types of experimental curves can be due to artefacts other than actual estuarine geochemical processes. The main problems are the selection of incorrect end members, the presence of other water inputs such as tributaries within the estuarine system and the non-linear variation with salinity or chlorinity of parameters such as pH and suspended matter concentration. Typical curves of this type are presented in Figs 9.6 and 9.7 for the Rhine and Scheldt estuaries. Dissolved Mn is produced and this is coherent with a corresponding decrease in suspended particulate matter (Fig. 9.7b). But the removals (below the theoretical physical dilution line, see Fig. 9.16), of dissolved Fe, Zn, Cu and Cd in Fig. 9.7b do not correspond to additions in the suspended load of these metals. The reader is referred to the review of Liss [18] for a more detailed critical evaluation of the method.

From a review of the field observations of the estuarine behaviour of dissolved and suspended trace elements, it seems that for a given element, the three types of processes (conservative, removal, release) can exist (Table 9.3). Fe is however an exception, because it is always removed from the dissolved phase. For example, in the compilation of Duinker [5] in nine papers dealing with Zn there were three cases of removal, three of mobilization, one of

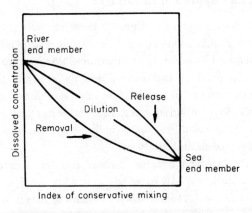

Fig. 9.16. Variation of dissolved trace element concentration as a function of an index of conservative mixing.

conservative behaviour, one of removal and release depending on the salinity and one of removal or conservativity.

9.3.2 Behaviour of particulate constituents

Another method of investigation of the chemical reactivity of trace elements is the study of the concentration in suspended or bottom sediments as a function of salinity. Two kinds of process have been identified in this manner:
(a) mobilization by desorption,
(b) physical mixing of trace element-rich fluviatile sediments with sediments of lower trace element content. The latter may be sediments of marine origin or 'reworked' estuarine sediments. It should, however, be pointed out that these 'reworked' estuarine sediments must have lost part of their trace element load.

The decrease in trace metal concentrations in bottom sediments as one moves downstream in the Rhine estuary has been interpreted by de Groot [93, 94] in terms of desorption in presence of the major ions of sea water. Bradford [72] observed a dramatic increase in zinc concentration in Chesapeake Bay just above the sediment–water interface at a time when the chlorinity increased from the usual 0.2‰ to a value of 3.5‰ and he concluded that Zn was displaced from the sediment by Ca and Mg. Other field evidence for the occurrence of desorption is summarized in Table 9.3.

In some estuaries, such as the Rhine and the Elbe, the physical mixing of fluviatile and low metal concentration marine sediments may explain this decrease [95]. This interpretation was supported by studies of the dynamics of bottom sediments in the Elbe estuary.

Jouanneau [67] suggested a physical mixing with old estuarine particles followed by a slower desorption to explain the decrease in trace metal concentrations in the suspensions of the Gironde estuary.

The analysis of both particulate and dissolved fractions helps in eliminating some of the problems inherent in the two preceding methods [7, 11, 12, 52, 54–66].

9.3.3 The product approach (Sholkovitz [31])

It is often impossible to distinguish between the various possible origins of estuarine particulate matter, e.g. river-borne, resuspended estuarine sediment, sea-derived and brought by the incoming tide, or recently flocculated in the estuary. The product approach allows the study of the formation and the extent of occurrence of flocculation and coagulation of colloidal particles.

Methodology. Filtered river and sea water are mixed to give various salinities. After equilibration, usually < 1 h, the newly formed suspended matter is removed and analysed.

Mn, P, dissolved organic carbon and humates are flocculated under estuarine conditions. All or most occurs for salinities between 0 and 20‰. Sholkovitz concludes that dissolved organic matter is important in controlling the behaviour and thus the speciation of trace elements by co-coagulation and co-flocculation.

The work of Eckert and Sholkovitz [32] suggests that flocculation is due to electrostatic and chemical interactions of the major sea salts with river-borne colloidal humic substances.

It is important to note that the conclusions from the three preceding methods do not necessarily agree.

9.3.4   ## Mathematical modelling of chemical speciation

Scientists confronted by the limited performance of the present state-of-the-art direct analytical techniques and puzzled by the sometimes conflicting results of field studies can turn towards another course of action: mathematical modelling. In order to be able to perform calculations, one must have a knowledge as detailed and as precise as possible of the various phenomena which might occur. The recognition of the occurrence of a given process must come from observations of field experiments. The next step is to obtain a description by mathematical laws. This is usually achieved by performing laboratory simulation experiments under carefully controlled conditions. Until now, the mathematical modelling of processes in natural aquatic systems, including estuaries, has been largely limited to the description of the chemical speciation of dissolved and adsorbed species under equilibrium conditions.

It must, therefore, be remembered that if a chemical reaction is prevented from taking place for kinetic reasons, the calculated overall speciation picture may not represent the real system. Also, uncertainties in the mathematical formulation of existing processes, e.g. uncertainty in the determination of activity coefficients at high ionic strength, or if the stability constant of an important complex is not well known, will be reflected by poor reliability of the calculated results. In spite of these two restrictions, mathematical models provide general trends one can expect to occur. However, a model cannot be better that what is put into it. It does not invent processes. For example, a mathematical determination of the chemical speciation of the dissolved phase will not be able to provide insights into the total fluxes of trace elements through estuarine systems because (a) fluxes also occur in particulate forms, and (b) a continuous exchange between the aqueous phase and the surface of suspended particulate matter takes place.

Investigations of speciation in the aqueous phase are not useless because they can provide useful information on toxicity. In addition, they were the first step in speciation calculations.

Many studies of the speciation of the end members of estuaries, viz. sea water and river water, have been published in the literature, but here four models which follow along the salinity gradient are presented. They will be introduced by a short presentation of the calculation of activity coefficients and followed by an example of the influence of hydrodynamics and redox chemistry.

9.3.4.1 Effect of ionic strength (or salinity) on activity coefficients

As mentioned earlier, not only does the activity coefficient of a given species vary with ionic strength (I), but this dependence is not linear (Fig. 9.3):

$$a = \gamma \cdot c, \tag{9.6}$$

where a, γ and c stand for activity, activity coefficient and concentration, respectively. The activity coefficient depends on changes:
- in salinity (non-specific effects)
- in composition (specific interactions which include ion pairing and chelation).

The theories developed to quantify these effects have been fully discussed by Millero [96], but only the brief summary by Burton [16] is presented here.
(a) Bjerrum's ion pair treatment: the activity coefficient of a single ion is subdivided in two components, one related only to the ionic strength and one affected by specific interactions. For example:

$$\gamma_{Ca^{2+}(total)} = \gamma_{Ca^{2+}(free)} \cdot y \tag{9.7}$$

where y stands for the fraction of free-to-total calcium. The importance of the correction term 'y' increases with ion pairing and, therefore, with salinity.
(b) The Brønsted–Guggenheim hypothesis of specific ion interactions: deviations from activities expected on the basis of ionic strength are explained by non-coulombic effects.

The reader is referred elsewhere [16, 96–98] for a more elaborate discussion.

In practice, simpler methods are often used. The calculation of γ-coefficients for non-specific effects is usually performed according to one of the methods listed in Table 9.4. Method 3 is used for major ions and methods 1 and 2 for trace ions. Approximations are then made to account for ion-pairing. A complete method, presented and used by Mantoura et al. [2], is discussed below.

The Davies equation (Table 9.4) is used with some modifications and

Table 9.4. Calculation of activity coefficients

Method of calculation	Equation	Limitations
(1) Extended Debye–Hückel	$\log \gamma = -A\, z^2 \dfrac{\sqrt{I}}{1 + Ba\sqrt{I}}$	for $I \leq 0.1$
(2) Davies [99]	$\log \gamma = -A\, z^2 \dfrac{\sqrt{I}}{1 + \sqrt{I}} - 0.3\, I$	for $I < 0.5$ (no adjustable parameter to account for the size of the ion)
(3) Mean salt method [25]	based on the assumption that $\gamma_{K^+} = \gamma_{Cl^-} = \gamma_{KCl}$	

I (ionic strength) $= \frac{1}{2} \Sigma c_i\, z_i^2$; $z =$ charge of ion; $c =$ concentration; $A = 1.82 \times 10^6 (\varepsilon T)^{-3/2}$ where $\varepsilon =$ dielectric constant; $A = 0.509$ for water at 25°C ; $B = 50.3\,(\varepsilon T)^{-1/2}$, $B = 0.329$ for water at 25°C; $a =$ "effective diameter" of the ion (in Ångstroms) (for values of a, see Kielland [100]).

assumptions. A semi-theoretical equation [109] gives the activity coefficients of ion pairs (IP).

$$\log \gamma_{IP} = -BI, \tag{9.8}$$

with $B = 0.1, 0.3$ and 0.5 for 1–1, 1–2 and 2–2 ion pairs, respectively. Activity coefficients of humics and humic complexes [2] and of surface ligands and surface complexes [8] are assumed to be equal

$$\gamma_{humic}/\gamma_{metal-humic} = \gamma_{surface\ site}/\gamma_{metal\ adsorbed} = 1. \tag{9.9}$$

Accounting for ion pairs does not greatly affect speciation calculations [2]. It must also be realized that in the computation of the fraction of a given species, the activity coefficients of the metals cancel out.

Turner et al. [4] avoided using activity coefficients by experimentally fitting the stability constants of nearly 500 complexes of 58 trace elements as a function of ionic strength (I):

$$\log \beta^* = \log \beta^\circ + 0.511\ [\Delta z^2 I^{\frac{1}{2}}/(1 + BI^{\frac{1}{2}})] + CI + DI^2 \tag{9.10}$$

where $\Delta z^2 = \Sigma z^2$ (products) $- \Sigma z^2$(reactants); z is the charge and B, C and D are adjustable parameters.

9.3.4.2 Behaviour of manganese in the Rhine and Scheldt estuaries (Wollast et al. [56])

The equilibrium concentration of dissolved manganese species was calculated using the following equations:

$$\left| \begin{array}{l} Mn^{2+} + HCO_3^- \rightleftarrows MnCO_3(s) + H^+ \\ \text{with } pK = -0.366 \end{array} \right. \tag{9.11}$$

$$\left| \begin{array}{l} MnO_2(s) + 4H^+ + 2e \rightleftarrows Mn^{2+} + 2H_2O \\ \text{with } E^0 = 1.229\ V \end{array} \right. \tag{9.12}$$

$$\left| \begin{array}{l} MnOOH(s) + 3H^+ + e \rightleftarrows Mn^{2+} + 2H_2O \\ \text{with } E^0 = 1.495\ V \end{array} \right. \tag{9.13}$$

$$\left| \begin{array}{l} Mn^{2+} + Cl^- \rightleftarrows MnCl^+ \\ \text{with } pK = -0.607 \end{array} \right. \tag{9.14}$$

$$\left| \begin{array}{l} Mn^{2+} + SO_4^{2-} \rightleftarrows MnSO_4^0 \\ \text{with } pK = -2.30 \end{array} \right. \tag{9.15}$$

obtained from laboratory experiments by Morgan [101] for eqns (9.11), (9.12) and (9.13), and from Crerar and Barnes [102] for eqns (9.14) and (9.15). E_h was obtained from the equation of Breck [103] for oxygenated waters:

$$E_h = 1.012 - 0.059\ pH + 0.030\ \log pO_2. \tag{9.16}$$

The activity coefficients were evaluated for each value of the ionic strength (I) according to the extended Debye–Hückel equation, and I was given by the relation of Lyman and Fleming [104]

$$I = 0.001\ 47 + 0.035\ 92(Cl^-) + 0.000\ 068(Cl^2) \tag{9.17}$$

311   EVALUATION OF EXISTING APPROACHES

where (Cl^-) is the chlorinity (in g Cl^- dm^{-3}).

The dissolved Mn calculated from experimental values of chlorinity, pH, dissolved oxygen and bicarbonate concentration, the latter being estimated from the end members, assuming conservative mixing, agreed well with observed values (Fig. 9.17). The authors concluded that high turbidity, i.e. active surfaces, and microbiological activity probably enhanced the rate of some of the usually rather slow manganese redox reactions.

This model is limited to Mn and its dependence upon pH, salinity, HCO_3 and dissolved oxygen, but it has the advantage of comparing well with experimental observations.

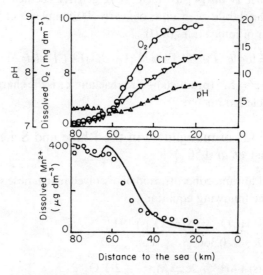

Fig. 9.17. Comparison between experimental data of dissolved Mn concentration in the Scheldt Estuary (circles, lower diagram) and values computed with the thermodynamic equilibrium model using the experimental profiles of chlorinity, oxygen concentration and pH smoothed as shown in the upper diagram (from Wollast *et al.* [56]).

9.3.4.3 Chemical speciation of dissolved major and minor elements (Dyrssen and Wedborg [3])

A system containing only the metal M and the ligand L can be described by the following mass balance equations:

$$M_t = [M] + [ML] + \cdots + [ML_n] = \sum_0^n [ML_i] = \sum_0^n \beta_i [M][L]^i = \sum_0^n \beta_i m l^i \tag{9.18}$$

$$L_t = [L] + [ML] + \cdots + n[ML_n] = [L] + \sum_1^n i\beta_i [M][L]^i = l + \sum_1^n i\beta_i m l^i. \tag{9.19}$$

In real systems, hydrolysis reactions of M and protonation reactions of L should be included, but they are omitted here for the sake of clarity. If all the stability constants are known, this system of two equations and two unknowns m and l can be solved and the speciation is then easily determined: $[ML_n] = \beta_n m l^n$.

For more complex systems of x metals and y ligands such calculations must be performed by iterative computer calculations.

An important simplification can, however, be used. Since constituents of natural aquatic systems occur in two ranges of total concentrations, differing by several orders of magnitude, a trace element can complex only a negligible fraction of a major constituent. For example, the concentrations of the Zn chloro-complexes can be neglected in the Cl mass balance for sea water, but not in the Zn mass balance. It is, therefore, sufficient to use computer calculations to determine the speciation of the major constituents. The concentrations of free species, ligands for example, are then used to calculate the speciation of trace constituents, i.e. trace metals, according to:

$$[ML_n]/M_t = \beta_n l^n / \sum_0^n \beta_i l^i \qquad (9.20)$$

where l denotes all important ligands.

The problem of changes in activity coefficients as a function of salinity can be treated by the conditional constant approach [105]. The decrease in the values of conditional stability constants is most pronounced for very low salinities. The increase in the concentrations of the free forms of the major ligand constituents is, therefore, very sharp (Fig. 9.18) for very low salinities ($< 2‰$). Since CO_3^{2-} and OH^- are some of the most important ligands for trace elements such as Pb, Ni and Cu, the different evolution of estuarine speciation of high and low alkalinity river waters can be predicted from Fig. 9.18.

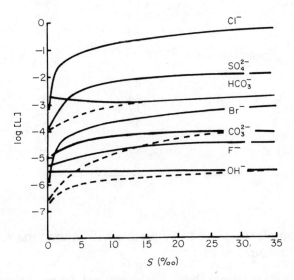

Fig. 9.18. Concentrations of inorganic ligands in estuarine waters plotted as log [L] versus salinity. When separate, the low alkalinity river water curves are dashed (from Dyrssen and Wedborg [3]).

313   EVALUATION OF EXISTING APPROACHES

To obtain the speciation of a given trace element, it is enough to compare the values of $\beta_n l^n$ for the different species (Table 9.5). Dyrssen and Wedborg [3] presented calculations for many other trace metals and major inorganic ligands. They studied the effect of dissolved organic matter by investigating the complexing of trace elements by organic ligands such as salicylic acid. They concluded that, for the usually low concentrations of natural organic matter, the presence of macromolecules or polymers is needed to produce significant complexation.

Table 9.5. Values of $\log \beta_n l^n$ for hydroxo-, carbonato-, chloro- and sulfato-complexes of copper and lead (from Dyrssen and Wedborg [3])

	Cu^{2+}	Pb^{2+}
Sea water		
M^{2+}	0	0
MOH^+	0.3	0.3
$M(OH)_2$	0.2–1.7	−0.8
$MHCO_3^+$	−0.6	−0.6
MCO_3	1.2	0.4
$M(CO_3)_2^{2-}$	0.3	−1.4
MCl^+	−0.4	0.9
MCl_2		1.0
MSO_4	−1	−1
High alkalinity river water		
M^{2+}	0	0
MOH^+	0.8	0.8
$M(OH)_2$	1.0–2.6	0.1
$MHCO_3^+$	0.3	0.3
MCO_3	1.7	0.9
$M(CO_3)_2^{2-}$	−0.2	−1.9
MCl^+	−3.3	−2.1
Low alkalinity river water		
M^{2+}	0	0
MOH^+	−0.4	−0.4
$M(OH)_2$	−1–0.3	−2.2
$MHCO_3^+$	−0.8	−0.8
MCO_3	−0.3	−1.2
$M(CO_3)_2^{2-}$	−4.3	−6.0
MCl^+	−3.3	−2.1

9.3.4.4 Complexation of trace metals with humic materials (HA) (Mantoura *et al.* [2])

The complexing ability of humic and fulvic substances increases with increasing pH [106]. Therefore, the recent determination by gel filtration chromatography of metal–HA stability constants at pH 8.0 (a value more realistic than that of previous measurements) justifies the inclusion of these complexes into an estuarine mathematical speciation model [2]. For the metal investigated, the most important changes in speciation occur for salinities < 5‰.

9.3.4.5   Physico-chemical speciation in estuaries (Bourg [6–8])

The preceding model can be improved by the inclusion of adsorption/desorption reactions, including the adsorption of ternary surface complexes, viz. surface–metal–HA [107]. See elsewhere in this volume for a detailed discussion of such complexes.

Figure 9.19, for a constant suspended matter concentration of $30\,mg\,dm^{-3}$ and pH of 8.0, is not very different from Fig. 2.4 of Mantoura et al. [2]. This is also true for the other metals studied. The advantage of the physico-chemical speciation model ADSORP [107] is that it can easily be utilized for real environments where one can compare calculated and observed values of the total adsorbed and total dissolved fractions.

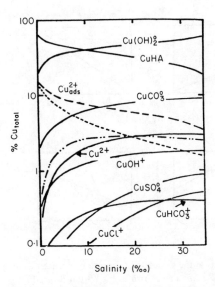

Fig. 9.19. Equilibrium physico-chemical specification of Cu in a theoretical estuary (constant pH and suspended matter concentration, HA: humics) (from Bourg [6]).

Preliminary results on the Gironde Estuary (France) show reasonable agreement between the model and the observed values [8]. The model can be improved since the calculations assumed constant surface properties of the suspended matter throughout the estuary, but, as mentioned earlier [39], this is not so.

The physico-chemical speciation model confirms the importance of surfaces (Section 9.2.4). Figure 9.20 shows that at a given pH value the calculated adsorption of, for example, Cu and Zn follows the variations of the suspended matter concentration and not those of salinity.

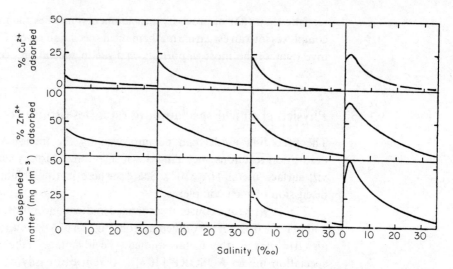

Fig. 9.20. Adsorption of Zn and Cu in a theoretical estuary for several patterns of variation of the concentration of suspended matter (constant pH of 8.0 throughout the estuary) (from Bourg [6, 7]).

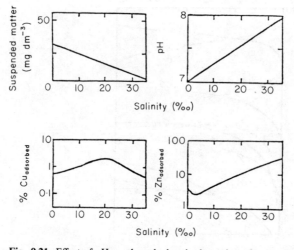

Fig. 9.21. Effect of pH on the calculated adsorption of trace metals (pH is on a log. basis and therefore Cu and Zn are also shown as log.) (from Bourg [7]).

The water pH is even more important (Fig. 9.21). The increase in adsorption is due to the increase in pH. The decrease in Cu adsorbed at high salinity and, more important, at high pH, is caused by competition for Cu atoms between adsorption and formation of dissolved hydroxo- and carbonate-complexes.

These calculations, which must certainly be carried out on many other estuaries, have the potential to explain the diversity in the geochemical behaviour of trace elements in various estuaries under different river regimes and seasons. The complexity and variability of chemical conditions could thus be accounted for, assuming rapid equilibrium. The model does take into account the most important parameters for the calculation of the adsorbed/dissolved distribution: pH, suspended matter and chlorinity (Fig. 9.22).

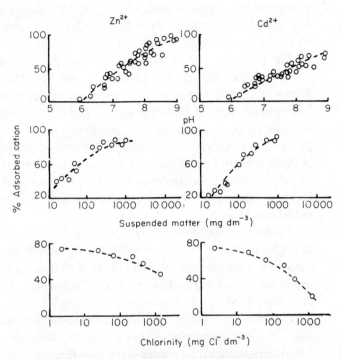

Fig. 9.22. Influence of pH, turbidity and chlorinity on the adsorption of Zn and Cd on Rhine River sediments (from Salomons and Mook [78]).

9.3.4.6 Hydrodynamics, biochemistry and geochemical speciation (Donard and Bourg [108])

In the Gironde Estuary, France, which is a macrotidal estuary, characterized by a large turbidity maximum, 80% of the upstream trace metal input is in solid particulate form. On the other hand, the metal output towards the Atlantic Ocean takes place mostly in dissolved form (80%) [67]. A similar release associated with the turbidity maximum was observed in the Seine River [68]. A qualitative model was proposed, indicating the complementary nature of hydrodynamic processes related to neap-tide–spring-tide cycles and of biological and chemical processes taking place in the estuary [108]. Two main events can be distinguished: (1) sedimentation and biogeochemical 'maturation' of the turbidity maximum, viz. bacterial decay producing slightly anaerobic conditions favouring the dissolution of iron coatings and of their associated trace metals, and (2) resuspension of deposited particles, i.e. precipitation of previously dissolved iron and readsorption and/or coprecipitation of trace metals on/with freshly formed solids/coatings.

9.4   Conclusions and Recommendations

The chemical speciation of dissolved trace elements in estuaries seems to be dominated by organic matter, CO_3^{2-} and Cl^-, depending on the element. Natural organic matter, such as humic substances, complexes only with Cu and, for salinities $<20‰$, Hg.

The physico-chemical speciation of trace elements should be controlled more by surface ligands and pH than by salinity. The model presented in Section 3.5.5 of this chapter offers the potential for explaining much of the contradictory data of field observations and of laboratory experiments. The importance of the solid–solution interface and the significance of water parameters, especially of pH and organic matter is recognized. Consequently, both the adsorbed and dissolved fractions and system parameters, mainly pH and suspended matter concentrations, but also dissolved organic matter and alkalinity, should be monitored in all estuarine field surveys and laboratory model experiments.

The importance of very low salinity regions of estuaries for chemical (and biological) reactions has been indicated by Morris *et al.* [15]. This is in agreement with the various observations of changes in ligand concentration, in adsorption and in dissolved speciation. This fact has, however, not yet been fully recognized by estuarine scientists (Section 9.2.2.1).

Estuaries are transient systems. They should thus be modelled using the laws of thermodynamics of open systems. Kinetics of adsorption and desorption on natural particles must, therefore, be quantified.

Further studies needed to model trace element physico-chemical speciation and fluxes in estuaries are the determination by laboratory experiments of:

(a) extent and kinetics of desorption,
(b) kinetics of adsorption,
(c) simultaneous surface reactions of trace elements and organic matter, in environments of increasing salinity; and the inclusion of coagulation phenomena and hydrodynamics of water and sediments. Finally, the amplitude and kinetics of biological transformations, including their effect on water redox chemistry, should be quantified and incorporated into the models.

Acknowledgements

This work was supported in part by the Swiss National Foundation for Scientific Research and the French National Geological Survey. Many thanks are expressed to Paul Schindler for his interest in this review and to Anna Kay Bourg for typing the manuscript.

9.5   References

1 J.M. Martin and M. Meybeck, *Marine Chem.*, **7**, 173 (1979).
2 R.F.C. Mantoura, A. Dickson and J.P. Riley, *Estuar. Coastal Marine Sci.*, **6**, 387 (1978).
3 D. Dyrssen and M. Wedborg, in *Chemistry and Biogeochemistry of Estuaries*, (Eds. E. Olausson and I. Cato) Wiley and Sons, New York, 71 (1980).
4 D.R. Turner, M. Whitfield and A.G. Dickson, *Geochim. Cosmochim. Acta*, **45**, 855 (1981).
5 J.C. Duinker, in *Chemistry and Biogeochemistry of Estuaries*, (Eds. E. Olausson and I. Cato) Wiley and Sons, New York, 121 (1980).
6 A.C.M. Bourg, *Proc. Internat. Conf. Heavy Metals in Environment (Amsterdam)*, 355 (1981).
7 A.C.M. Bourg, in *Trace Metals in Sea Water*, (Eds. C.S. Wong, E. Boyle, K.W. Bruland, J.D. Burton and E.D. Goldberg) Plenum Press, New York and London, 195 (1983).
8 A.C.M. Bourg, *Document B.R.G.M. No. 62*, National Geological Survey, Orléans, France, 171 p (1983).
9 P.C. Head, in *Estuarine Chemistry*, (Eds. J.D. Burton and P.S. Liss) Academic Press, London, 53 (1976).
10 H. Postma, in *Chemistry and Biogeochemistry of Estuaries* (Eds. E. Olausson and I. Cato) Wiley and Sons, New York, 153 (1980).
11 J.C. Duinker and R.F. Nolting, *Neth. J. Sea Res.*, **10**, 71 (1976).
12 J.C. Duinker and R.F. Nolting, *Neth. J. Sea Res.*, **12**, 205 (1978).
13 A.C.M. Bourg, this volume. Section 8, p. 257–284.
14 F.J. Millero, *Geochim. Cosmochim. Acta*, **45**, 2085 (1981).
15 A.W. Morris, R.F.C. Mantoura, A.J. Bale and R.J.M. Howland, *Nature*, **274**, 678 (1978).
16 J.D. Burton, in *Estuarine Chemistry*, (Eds. J.D. Burton and P.S. Liss) Academic Press, London, 1 (1976).
17 I. Hosokawa, F. Oshima and N. Kondo, *J. Oceanogr. Soc. Jpn*, **26**, 1 (1970).
18 P.S. Liss, in *Estuarine Chemistry*, (Eds. J.D. Burton and P.S. Liss) Academic Press, London, 93 (1976).
19 T.B. Warner, *J. Geophys. Res.*, **77**, 2728 (1972).
20 K.L. Russell, *Geochim, Cosmochim. Acta*, **34**, 893 (1970).
21 J.H. Sharp, C.H. Culberson and T.M. Church, *Limnol. Oceanog.*, **27**, 1015 (1982).
22 W.G. Mook and B.K.S. Koene, *Estuar. Coastal Marine Sci.*, **3**, 325 (1975).
23 J. Lebel and E. Pelletier, *Atmosphere Ocean*, **18**, 154 (1980).
24 L.G. Danielson and D. Dyrssen, *Report on the Chemistry of Sea Water XVI.*, Dept. Anal. Chem. Univ. of Göteborg (1975).
25 R.M. Garrels and C.L. Christ, *Solutions, Minerals and Equilibria*, Harper and Row, New York (1965).
26 J.H. Reuter, *Chesapeake Sci.*, **18**, 120 (1977).
27 R.W.P.M. Laane, *Neth. J. Sea Res.*, **14**, 192 (1980).
28 R.F.C. Mantoura and E.M.S. Woodward, *Geochim. Cosmochim. Acta*, **47**, 1293 (1983).
29 R.M. Moore, J.D. Burton, P.J. LeB. Williams and M.L. Young, *Geochim. Cosmochim. Acta*, **43**, 919 (1979).
30 M.E. Hair and C.R. Bassett, *Estuar. Coastal Marine Sci.*, **1**, 107 (1973).
31 E.R. Sholkovitz, *Geochim. Cosmochim. Acta*, **40**, 831 (1976).
32 J.M. Eckert and E.R. Sholkovitz, *Geochim. Cosmochim. Acta*, **40**, 847 (1976).
33 R.W.P.M. Laane, *Neth. J. Sea Res.*, **15**, 331 (1982).
34 D. Eisma, G.C. Cadee and R.W.P.M. Laane, *Mitt. Geol. Paläont. Inst. Univ. Hamburg*, **52**, 483 (1982).
35 G. Cauwet, F. Elbaz, C. Jeandel, J.M. Jouanneau, Y. Lapaquellerie, J.M. Martin and A. Thomas, *Bull. Inst. Géol. Aquitaine, Bordeaux*, **27**, 5 (1980).
36 H. Etcheber, *Doctoral Thesis*, Univ. of Bordeaux, France (1983).
37 H.H. Hahn and W. Stumm, *Am. J. Sci.*, **268**, 354 (1970).
38 J.K. Edzwald, J.B. Upchurch and C.R. O'Melia, *Environ. Sci. Technol.*, **8**, 58 (1974).
39 H. Etcheber, A.C.M. Bourg and O. Donard, *Proc. Interant. Conf. Heavy Metals in Environment (Heidelberg)*, 1200 (1983).
40 A.C. Sigleo and G.R. Helz, *Geochim. Cosmochim. Acta*, **45**, 2501 (1981).
41 D.C. Burrell, in *Radioactive Contamination of the Marine Environment*, IAEA, Vienna, 89 (1973).
42 H.L. Windom, K.C. Beck and R. Smith, *Southeast. Geol.*, **12**, 169 (1971).
43 G.R. Helz, R.J. Huggett and J.M. Hill, *Water Res.*, **9**, 631 (1975).

44 C.W. Holmes, E.A. Slade and McLerran, *Environ. Sci. Technol.*, **8**, 255 (1974).

45 R. Fukai, C.N. Murray and L. Huynh-Ngoc, *Estuar. Coastal Marine Sci.*, **3**, 165 (1975).

46 L.M. Holliday and P.S. Liss, *Estuar. Coastal Marine Sci.*, **4**, 349 (1976).

47 D. Grieve and K. Fletcher, *Estuar. Coastal Marine Sci.*, **5**, 415 (1977).

48 D.W. Mackay and T.M. Leatherland, in *Estuarine Chemistry*, (Eds. J.D. Burton and P.S. Liss) Academic Press, London, 185 (1976).

49 V. Subramanian and B. d'Anglejan, *J. Hydrol.*, **29**, 341 (1976).

50 D.J. Thomas and E.V. Grill, *Estuar. Coastal Marine Sci.*, **5**, 421 (1977).

51 E.A. Boyle, R. Collier, A.T. Dengler, J.M. Edmond, A.C. Ng and R.F. Stallard, *Geochim. Cosmochim. Acta*, **38**, 1719 (1974).

52 E.R. Sholkovitz, *Estuar. Coastal Marine Sci.*, **8**, 523 (1979).

53 D.W. Evans and N.H. Cutshall, in *Radioactive Contamination of the Marine Environment*, IAEA, Vienna, 125 (1973).

54 J.C. Duinker and R.F. Nolting, *Marine Pollut. Bull.*, **8**, 65 (1977).

55 J.C. Duinker, R.F. Nolting and H.A. Van der Sloot, *Neth. J. Sea Res.*, **13**, 282 (1979).

56 R. Wollast, G. Billen and J.C. Duinker, *Estuar. Coastal Marine Sci.*, **9**, 161 (1979).

57 H.A. Van der Sloot, R. Massee and G.D. Wals, *Third Internat. Conf. on Nuclear Methods in Environmental and Energy Research (Columbia, Missouri)* (1978).

58 D.W. Evans, N.H. Curshall, F.A. Cross and D.A. Wolfe, *Estuar. Coastal Marine Sci.*, **5**, 71 (1977).

59 R.E. Cranston, *Estuar. Coastal Marine Sci.*, **4**, 695 (1976).

60 L.S. Conleyr, E.B. Baker and H.D. Holland., *Chem. Geol.*, **7**, 51 (1971).

61 C.I. Measures and J.D. Burton, *Nature*, **273**, 293 (1978).

62 Y. Sugimura, Y. Suzuki and Y. Miyake, *J. Oceanogr. Soc. Jpn*, **32**, 235 (1976).

63 E.A. Boyle, S.S. Huested and B. Grant, *Deep-Sea Res.*, **29**, 1355 (1982).

64 G. Klinkhammer, in *Trace Metals in Sea Water* (Eds. C.S. Wong, E. Boyle, K.W. Bruland, J.D. Burton and E.D. Goldberg) Plenum Press, New York and London, 317 (1983).

65 H. Windom, G. Wallace, R. Smith, N. Dudek, M. Maeda, R. Dulmage and F. Storti, *Marine Chem.*, **12**, 183 (1983).

66 H.N. Kerdijk and W. Salomons, Delft Hydraulics Report M1640/M1736, (1981).

67 J.M. Jouanneau, *Doctoral Thesis*, Univ. of Bordeaux, France (1982).

68 D. Boust, *Doctoral Thesis*, Univ. of Caen, France (1981).

69 G.B. Jones, *Estuar. Coastal Marine Sci.*, **2**, 185 (1974).

70 P.C. Head and J.D. Burton, *J. Marine Biol. Assoc. U.K.*, **50**, 439 (1970).

71 J.M. Martin, O. Høgdahl and J.C. Philippot, *J. Geophys. Res.*, **81**, 3119 (1976).

72 W.L. Bradford, Chesapeake Bay Institute, John Hopkins Univ., Baltimore, Tech. Report 76, 103 p. (1972).

73 D.P. Kharkar, K.K. Turekian and K.K. Bertine, *Geochim. Cosmochim. Acta*, **32**, 285 (1968).

74 C.N. Murray and L. Murray, in *Radioactive Contamination of the Marine Environment*, IAEA, Vienna, 105 (1973).

75 N. Rohatgi and K.Y. Chen, *J. Water Pollution Control Fed.*, **47**, 2298 (1975).

76 C.H. Van den Weijden, M.J.M.L. Arnoldus and C.J. Meurs, *Neth. J. Sea Res.*, **11**, 130 (1977).

77 W. Salomons, *Environ. Technol. Lett.*, **1**, 356 (1980).

78 W. Salomons and W.G. Mook, *Sci. Total Environ.*, **16**, 217 (1980).

79 G.E. Millward and R.M. Moore, *Water Res.*, **16**, 981 (1982).

80 M.M. Benjamin and J.O. Leckie, *Environ. Sci. Technol.*, **16**, 162 (1982).

81 J. Vuceta, *Ph.D. Thesis*, Calif. Inst. Technol., Pasadena (1978).

82 R. Kummert, *Doctoral Thesis*, Swiss Federal Inst. Technol., ETH Zürich (1979).

83 R.A. Neihof and G.I. Loeb, *Limmol. Oceanogr.*, **17**, 7 (1972).

84 R.A. Neihof and G.I. Loeb, *J. Marine Res.*, **32**, 5 (1974).

85 K.A. Hunter and P.S. Liss, *Nature*, **282**, 823 (1979).

86 V. Pravdic, *Limnol. Oceanog.*, **15**, 230 (1970).

87 J.M. Martin, J. Jednačak and V. Pravdic, *Thalassia Jugoslavica*, **7**, 619 (1971).

88 J.M. Martin, Ecole Normale Supérieure, Paris, personal communication (1980).

89 C. Mignot, *La Houille Blanche*, **7**, 591 (1968).

90 L. Balistrieri, P.G. Brewer and J.W. Murray, *Deep-Sea Res.*, **284**, 101 (1981).

91 A.C.M. Bourg, *Proc. Internat. Conf. Heavy Metals in Environment (Amsterdam)*, 690 (1981).

92 S.R. Aston and R. Chester, *Estuar. Coastal Marine Sci.*, **1**, 225 (1973).

93 A.J. de Groot, *Int. Soc. Soil Sci.*, Aberdeen, 267 (1966).

94 A.J. de Groot, in *North Sea Science* (Ed. E.D. Goldberg) MIT Press, Cambridge, 308 (1973).

95 G. Müller and U. Förstner, *Environ. Geol.*, **1**, 33 (1975).

Acknowledgements

This work would not have been possible without support from the Swiss National Foundation for Scientific Research. Many thanks are expressed to Paul Schindler (Bern) and to the Scientific Advisory Board (Mission Scientifique) of the B.R.G.M. for their interest in this review and to Anna Kay Bourg for editing the English and typing the manuscript.

8.9   References

1   E.A. Jenne, in Trace Inorganics in Water, *Am. Chem. Soc. Adv. Chem. Ser.*, **73**, ACS, Washington, D.C., 337 (1968).

2   J.D. Hem, *Geochim. Cosmochim. Acta*, **40**, 599 (1976).

3   K.K. Turekian, *Geochim. Cosmochim. Acta*, **41**, 1139 (1977).

4   C. Mouvet, P. Cordebar and A.C.M. Bourg, *J. Français Hydrol.*, **13**, 299 (1982).

5   C. Mouvet and A.C.M. Bourg, *Water Res.*, **17**, 641 (1983).

6   A.C.M. Bourg, *Document B.R.G.M. No. 62*, National Geological Survey, Orléans, France, p. 171 (1983).

7   L. Sigg, W. Stumm and B. Zinder, in *Complexation of Trace Metals in Natural Waters*, (Eds. C.J.M. Kramer and J.C. Duinker) Martinus Nijhoff and Dr. W. Junk Publ., The Hague, 251 (1984).

8   A.C.M. Bourg and C. Mouvet, in *Complexation of Trace Metals in Natural Waters*, (Eds. C.J.M. Kramer and J.C. Duinker) Martinus Nijhoff and Dr. W. Junk Publ., The Hague, 267 (1984).

9   W. Salomons and U. Förstner, *Metals in the Hydrosphere*, Springer Verlag, Berlin, p. 349 (1984).

10   D.K. Nordstrom, L.N. Plummer, T.M.L. Wigley, T.J. Wolery, J.W. Ball, E.A. Jenne, R.L. Bassett, D.A. Crerar, T.M. Florence, B. Fritz, M. Hoffman, G.R. Holdren Jr., G.M. Lafon, S.V. Mattigod, R.E. McDuff, F. Morel, M.M. Reddy, G. Sposito and J. Thrailkill, in Chemical Modelling in Aqueous Systems, (Ed. E.A. Jenne) *Am. Chem. Soc. Symposium Ser.*, **93**, ACS, Washington, D.C., 857, (1979).

11   J.C. Westall, J.L. Zachary and F.M.M. Morel, Tech. Note 18, Dept. Civil Engineering, Mass. Inst. Techhnol., Cambridge, Mass., p. 91, (1976).

12   H.C. Helgeson, *Geochim. Cosmochim. Acta*, **32**, 853 (1968).

13   H.C. Helgeson, *Am. J. Sci.*, **267**, 724 (1969).

14   H.C. Helgeson, R.M. Garrels and F.T. Mackenzie, *Geochim. Cosmochim. Acta*, **33**, 455 (1969).

15   H.C. Helgeson, T.H. Brown, A. Nigrini and T.A. Jones, *Geochim. Cosmochim. Acta*, **34**, 569 (1970).

16   M.H. Reed, *Geochim. Cosmochim. Acta*, **46**, 513 (1982).

17   A.C.M. Bourg, S. Joss and P.W. Schindler, *Chimia*, **33**, 19 (1979).

18   W. Stumm, H. Hohl and F. Dalang, *Croat. Chem. Acta*, **48**, 491 (1976).

19   S.A. Greenberg, *J. Phys., Chem.*, **60**, 325 (1956).

20   S. Ahrland, I. Grenthe and B. Norèn, *Acta Chem. Scand.*, **14**, 1059 (1960).

21   D.L., Dugger, J.H. Stanton, B.N. Irby, B.L. McConell, W.W. Cummings and R.W. Maatman, *J. Phys. Chem.*, **68**, 757 (1964).

22   H. Hohl and W. Stumm, *J. Colloid Interface Sci.*, **55**, 281 (1976).

23   P.W. Schindler, B. Fürst, R. Dick and P.U. Wolf, *J. Colloid Interface Sci.*, **55**, 469 (1976).

24   W. Stumm, R. Kummert and L. Sigg, *Croat. Chem. Acta*, **53**, 291 (1980).

25   W. Stumm and J.J. Morgan, *Aquatic Chemistry*, 2nd edn. Wiley-Interscience, N.Y., p. 780 (1981).

26   R.O. James and T.W. Healy, *J. Colloid Interface Sci.*, **40**, 42 (1972).

27   R.O. James and T.W. Healy, *J. Colloid Interface Sci.*, **40**, 53 (1972).

28   R.O. James and T.W. Healy, *J. Colloid Interface Sci.*, **40**, 65 (1972).

29   J. Vuceta, *Ph.D. Thesis*, Calif. Inst. Technol., Pasadena (1978).

30   J. Vuceta and J.J. Morgan, *Environ. Sci. Technol.*, **12**, 1302 (1978).

96   F.J. Millero, The Sea. In *Marine Chemistry*, Vol. 5, (Ed. E.D. Goldberg) Wiley Interscience, New York, 3 (1974).
97   M. Whitfield, *Geochim. Cosmochim. Acta*, **39,** 1545 (1975).
98   F.J. Millero and D.R. Schreiber, *Am. J. Sci.*, **282,** 1508 (1982).
99   C.W. Davies, *Ion Association*, Internat. Assoc. for Great Lakes Res., Butterworths, London (1962).
100   J. Kielland, *J. Am. Chem. Soc.*, **59,** 1675 (1937).
101   J.J. Morgan, in *Principles and Application in Water Chemistry*, (Eds. S.D. Faust and J.V. Hunter) Wiley, New York (1966).
102   D.A. Crerar and H.L. Barnes, *Geochim. Cosmochim. Acta*, **38,** 279 (1974).
103   W.G. Breck, *J. Marine Res.*, **30,** 121 (1972).
104   J. Lyman and R.H. Fleming, *J. Marine Res.*, **3,** 134 (1940).
105   A. Ringbom, *Complexation in Analytical Chemistry*, Interscience, New York (1963).
106   J. Buffle, *Anal. Chim. Acta*, **118,** 29 (1980).
107   A.C.M. Bourg, *Environ. Technol. Lett.*, **3,** 305 (1982).
108   O. Donard and A.C.M. Bourg, in preparation.
109   E.J. Reardon and D. Langmuir, *Geochim. Cosmochim. Acta*, **40,** 549 (1976).

Section 10   Chemical Mechanisms Operating in Sea Water

By

Commission V.6   J. W. MURRAY

IUPAC   *School of Oceanography*
University of Washington
Seattle, WA 98195
U.S.A.

10.1 Introduction

The field of marine chemistry has grown enormously during the past ten years. This is especially evident in the scope of the research. Early chemists were often only acting as support for biological and physical studies and much of the research was directed toward establishing the first order distributions of the elements. The ultimate study of this nature was the GEOSECS program which established detailed sections of chemical properties through the Atlantic, Indian and Pacific Oceans.

During this early period, interest in analytical methods and data dominated the field. A few volumes in the 1960s such as those edited by Faust and Hunter [1] and Stumm [2] sparked interest in chemical mechanisms. A classic volume on aquatic chemistry by Stumm and Morgan [3] initiated a period of considerable interest in chemical mechanisms. Now that the basic distribution of many elements in sea water is known, it becomes increasingly important to conduct laboratory or field experiments on the controlling mechanisms. This type of research is still expanding and has an important future as a subdiscipline of marine chemistry. In the foreseeable future one fundamental objective of marine chemistry will be to understand the processes which control the composition of the contemporary ocean. With improved understanding of these chemical processes we will be in a better position to translate the information contained in the geological record into unravelling the evolution of sea water chemistry with time, as well as predicting its course in future.

With regard to modelling the controlling mechanisms, important decisions have to be made regarding the suitability of using equilibrium versus kinetic models. In view of the complexity of sea water systems, equilibrium calculations offer much in the way of simplicity, and we clearly want to apply them wherever possible. However, for most oceanographic processes kinetic considerations are the factors that shape the actual distributions.

In this chapter several areas of research that have received considerable attention during the past few years are reviewed. These include oxidation–reduction chemistry and biologically controlled mechanisms at work in the photic zone, scavenging by particles in the water column and at the sea floor and solubility controls in oxidizing and reducing marine environments. Both thermodynamic and kinetic examples are available.

10.2 Oxidation–Reduction Reactions

The redox state of sea water has been of special interest because several trace metals can exist in different oxidation states in sea water. Furthermore, oxidation state changes can control the chemical reactivity of a trace metal. Reduction can turn a particulate-bound metal into soluble form (e.g. Mn) or a soluble metal into a reactive and adsorbable species (e.g. Cr). In addition to studying these natural distributions there are many studies that involve the kinetics of these reactions. These studies are important because, whereas Sillen's [4, 5] classic thermodynamic models of the composition of sea water set important boundary limits, the kinetic considerations are nowhere more striking than for the oxidation–reduction reactions. Redox reactions have notoriously slow kinetics. Furthermore, they are continuously perturbed by the biological cycle [6] (Fig. 10.1).

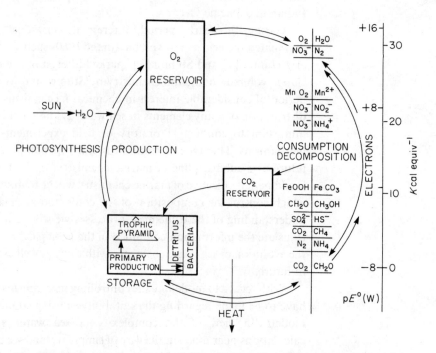

Fig. 10.1. Photosynthesis continuously perturbs the oxidation–reduction chemistry of the oceans. Photosynthesis traps light energy and fixes it in the form of high electron energy chemical bonds. Non-photosynthetic organisms then try to restore equilibrium by catalytically decomposing the unstable products of photosynthesis through energy-yielding redox reactions. The energy scales shown on the right in $pE°$ (at pH 7) and kcals define the electron free energy levels of the various oxidation–reduction couples [from Stumm and Morgan [3]).

Pankow and Morgan [7, 8] have recently reviewed chemical kinetics and natural water chemistry. They provide a good summary of the required mathematical (both analytical and numerical) methods available as tools for kinetic modelling. What is needed for these problems is the kinetic data for the chemical reactions that are rate limiting. The rates of the limiting chemical reactions are frequently expressed in the form of a rate law, i.e. the expression that describes the time-dependent velocity at which the reaction proceeds. The terms 'zero order', 'first order', and 'second order' refer to the sum of the powers of the concentration terms that appear in the rate law. For example:

Reaction	Rate Law
(a) $A \rightarrow B$	$\dfrac{d[A]}{dt} = -k_1[A]$
(b) $2A \rightarrow B$	$\dfrac{d[A]}{dt} = -k_2[A]^2$
(c) $A + B \rightarrow C$	$\dfrac{d[A]}{dt} = -k_3[A][B].$

In these examples, k_1 is an example of a first order rate constant while k_2 and k_3 are examples of second-order rate constants. If we assume there are no back reactions, the integration of the rate laws for reactions (a) and (b) for the initial conditions $[A] = [A]_0$ at $t = 0$ give the expressions:

$$\ln\frac{[A]_0}{[A]} = k_1 t \quad \text{and} \quad \frac{1}{[A]} - \frac{1}{[A]_0} = k_2 t.$$

These equations are used to derive the reaction half life. For first order reactions, $t_{1/2}$ is dependent only on the rate constant:

$$t_{1/2} = \ln 2 / k_1 = 0.69 / k_1.$$

For second-order reactions (b) and (c) the initial conditions must be considered; for example when $[A] = [A]_0 (= [B]_0)$, when $t = 0$, $t_{1/2} = 1/[A]_0 k_2 (= 1/[B]_0 k_3)$.

An overview of how the magnitude of the second order rate constant (k_2) affects the half-life for a number of second order reactions of interest to aquatic chemists is shown in Fig. 10.2. The effect of concentration is also shown. It is also worth noting that the slow reactions (half-lives of seconds or longer) are frequently oxidation reactions. Oxidation reactions are often slow unless kinetically mediated by bacteria. Hydrolysis reactions have half reaction rates of the order of microseconds, while complexation reactions are in the seconds to milliseconds range.

The kinetics of oxidation of Mn(II) have always been of great interest because of the mobility of Mn(II) and the scavenging ability of the surface of $MnO_2(s)$ [9]. Goldberg [10] first suggested that the oxidation of manganese, possibly catalysed by iron oxide surfaces, was an important mechanism for ferromanganese nodule formation. Subsequent laboratory studies [11–13] showed that Mn(II) oxidation is extremely slow at pH < 9 and the oxidation kinetics are autocatalytic. Several subsequent geochemical studies have suggested that the natural rates are much faster (10^3–10^5 times) than those observed in the laboratory.

Iron and manganese frequently occur together in nature. The kinetics of Fe(II) oxidation are much faster than those of Mn(II) and the solid phase formed is γ-FeOOH (lepidocrocite) [14, 15]. Sung and Morgan [16] undertook a laboratory study of the effects of γ-FeOOH on the Mn(II) oxidation removal rates. In this system the rate of Mn(II) removal was interpreted to be due to surface catalysis.

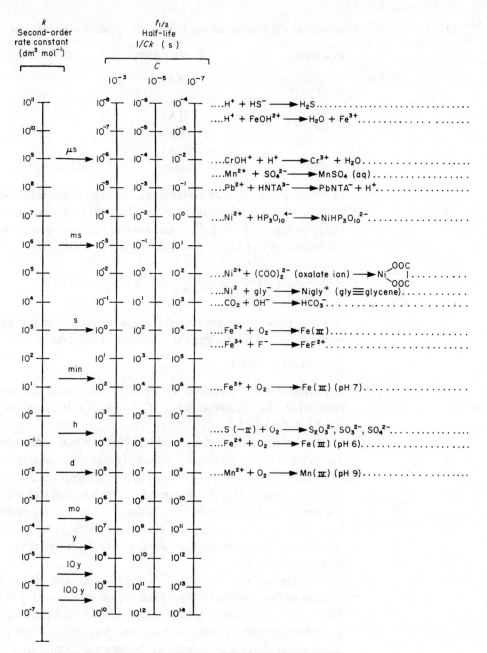

Fig. 10.2. Selected second-order rate constants and their corresponding $t_{1/2}$ values plotted as a function of the initial concentration C. The value of C is assumed to be equal for both reactant species (from Pankow and Morgan [8]).

Starting with a mass balance for manganese:

$$[\text{Mn}]_{\text{total}} = [\text{Mn(II)}] + [\text{Mn(II)}]_{\text{ads}} + [\text{MnO}_x]_{\text{surface}} \tag{10.1}$$

and the rate of change for a closed system:

$$\frac{d[\text{Mn}]_{\text{total}}}{dt} = 0 = \frac{d[\text{Mn(II)}]}{dt} + \frac{d[\text{Mn(II)}]_{\text{ads}}}{dt} + \frac{d[\text{MnO}_x]_{\text{surface}}}{dt} \tag{10.2}$$

and assuming that adsorbed manganese can be related to dissolved manganese by

$$[Mn(\text{II})]_{ads} \simeq K'[Mn(\text{II})] \tag{10.3}$$

where K' is a dimensionless adsorption constant, then

$$0 = \frac{dMn(\text{II})}{dt} + \frac{K'd[Mn(\text{II})]}{dt} + \frac{d[MnO_x]_{surface}}{dt} \tag{10.4}$$

If the rate of oxidation of $[Mn(\text{II})]$ in the surface is assumed to be first order with respect to adsorbed $Mn(\text{II})$ then

$$0 = \frac{d[Mn(\text{II})]}{dt} + \frac{K'd[Mn(\text{II})]}{dt} + \frac{k_s K'd[Mn(\text{II})]}{dt} \tag{10.5}$$

where k_s is the surface oxidation rate constant. When rearranged

$$\frac{-d[Mn(\text{II})]}{dt} = \frac{kK'}{1+K'}[Mn(\text{II})] \tag{10.6}$$

This first order equation for the surface oxidation rate holds well for the initial rates of $Mn(\text{II})$ removal from 0.7 mol dm^{-3} NaCl (Fig. 10.3). The initial rates were also found to be first order with respect to the total iron(III) concentration and second order with respect to pH. Presenting the rate law in an

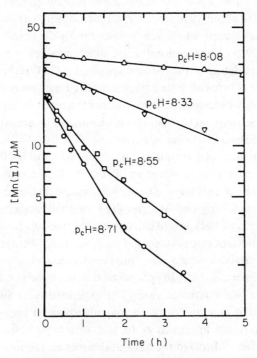

Fig. 10.3. Effect of pH on the oxygenation removal kinetics of $Mn(\text{II})$ in the presence of 10^{-3} mol dm^{-3} Fe(III) as γ-FeOOH. $P_{O_2} = 0.21$ atm, 0.7 mol dm^{-3} NaCl at 25°C; p_cH $= -\log$ [H$^+$] (from Sung and Morgan [16]).

oceanographically useful form gives:

$$-\frac{d[Mn(II)]}{dt} = k^*[OH]^2 (Fe(III))_{total}[O_2][Mn(II)] \tag{10.7}$$

where

$$k^* = 2.0 \times 10^8$$

For surface sea water of pH 8.1, with $O_2 = 250 \ \mu mol \ dm^{-3}$, and $[Fe(III)]_{total} = 10^{-8} \ mol \ dm^{-3}$, the rate is

$$-\frac{d[Mn(II)]}{dt} = 7.9 \times 10^{-6} d^{-1}[Mn(II)] \tag{10.8}$$

with a half life of 347 years. Recent experiments by S. Davies and J. J. Morgan (personal communication, 1984) have shown that in addition to γ-FeOOH, α-FeOOH will also greatly accelerate the rate. SiO_2 and γ-Al_2O_3 increase the rate slightly.

A recent field study that demonstrates an approach for studying chemical mechanisms in nature is that of Emerson et al. [17] of the redox interface in the water column of Saanich Inlet (a seasonally stratified anoxic fjord on Vancouver Island). Ten redox-sensitive chemical species were determined across this interface (Fig. 10.4). The O_2–H_2S boundary on the data concerned was found to be at 130 m. Nitrate and ammonia appear to respond rapidly to this redox front, but, several of the reduced species do exist in detectable concentrations above this boundary. Iodide decreases linearly above the interface to about 90 m while Cr(III) extends up to about 115 m. Reduced iron terminates sharply at the redox boundary while reduced manganese extends up to 120 m. These concentration differences result from differences in the rates of oxidation of the reduced species, I^-, $Cr(OH)_2^+$, and Mn^{2+}. Using the I^- profile to help calibrate a mixing model, Emerson et al. [17] derived a value for the eddy diffusion coefficient across the redox boundary. Using this value they calculated that the removal of chromium by adsorption and oxidation is controlled by a first order rate constant of 0.15–0.04 d^{-1}. This corresponds to a residence time for Cr(III) in the oxidizing water of 6–20 days. For Mn(II) the first order removal constant was 0.5 d^{-1} which results in a Mn^{2+} residence time of about two days. The half life calculated from the surface catalysis equations of Sung and Morgan [16] for similar conditions gives a value of about 35 days. Such in situ calculations of rate constants are very instructive. In this case, laboratory studies of the kinetics of Mn(II) oxidation without any type of catalysis would have predicted a residence time that is about 10^5 longer. Emerson et al. [18] proposed that bacterial catalysis could explain the enhanced in situ removal rates. The experiments by Sung and Morgan [16] suggest that inorganic surfaces may also play an important role.

An obvious approach is to test this hypothesis by adding spikes of radioactive [54]Mn(II) to the natural samples and to follow the removal of Mn from solution with and without bacterial presence. The results of these experiments as reported by Emerson et al. [18] clearly demonstrate a rapid removal of [54]Mn(II) with time (Fig. 10.5). When samples were poisoned with

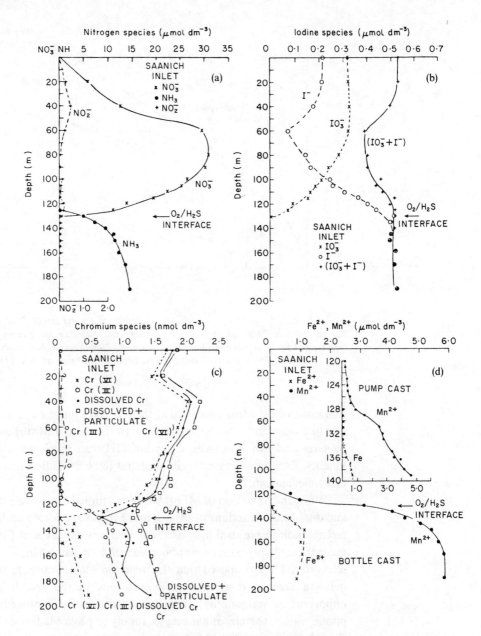

Fig. 10.4. Distribution of chemical species about the O_2/H_2S interface in Saanich Inlet in July 1977 (from Emerson *et al.* [17]). (a) Nitrogen species (NO_3^-, NO_2^-, and NH_3); (b) iodine species (I^-, IO_3^-); (c) chromium species [Cr(III), Cr(VI)]; (d) iron and manganese species.

formaldehyde, regardless of the time of addition, the subsequent rate of Mn(II) binding was always much slower than in the untreated control. Emerson *et al.* [18] were careful to consider the possibility of Mn(II) adsorption as a removal mechanism or that the addition of bacterial poisons slowed inorganic adsorption or oxidation by complexing the Mn(II). But, by careful experimental design and choice of controls, they were able to conclude satisfactorily

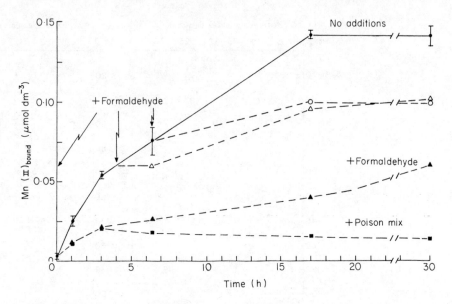

Fig. 10.5. Removal of Mn(II) as a function of time from a water sample from 100 m in Saanich Inlet. A ^{54}Mn spike was added increasing the total manganese concentration by 1.0 μmol dm^{-3} over the *in situ* concentration of 2.7 μmol dm^{-3}. Arrows show where formaldehyde or bacterial poisons were added (from Emerson *et al.* [18]).

that most of the Mn(II) removal at the O_2–H_2O interface in Saanich Inlet is due to oxidation by bacterial catalysis. In the case of Mn(II) oxidation both bacteria and other surfaces, e.g. γ-FeOOH, can accelerate otherwise slow kinetics. Each environmental situation has to be carefully studied to elicit the exact mechanism.

While the oxidation of Mn(II) has been studied extensively, the reduction and dissolution reactions of Mn(III) and Mn(IV) are poorly understood. Two recent studies have shed light on these reactions. Sunda *et al.* [19] found that manganese (III, IV) oxides were photoreduced by dissolved humic substances in sea water. They proposed that this reaction may contribute to the surface maxima seen in manganese concentrations in the sea. It may also be important in maintaining manganese in a dissolved and reduced form in photic waters, thereby enhancing its supply to phytoplankton. The details of the reduction of oxidized Mn with organic matter were explored by Stone and Morgan [20, 21]. They studied the rates of reaction of a mixed Mn(III, IV) oxide phase with 15 aromatic and 12 non-aromatic compounds chosen to represent the variety of structures and functional groups found in natural organic matter. Saturated alcohols, aldehydes, ketones and carboxylic acids showed no reactivity, except for pyruvic and oxalic acids. Catechols, hydroquinones, methoxyphenols, resorcinols and ascorbate reduced and dissolved manganese oxide at appreciable rates.

10.3 Biological Mechanisms

There are many ways that the activities of phytoplankton, zooplankton, and bacteria influence the distribution of minor elements in sea water. Most of these mechanisms are not well understood and are areas that deserve much greater research emphasis. The internal cycling of the ocean is very slow on the time scale of man. Thus, except for rare cases, we are not in a position to assign rates, much less understand detailed mechanisms. We therefore tend to approach the problem by inversion. From maps or profiles of the distribution of the various elements, we attempt to deduce the controlling mechanisms and their rates. Much of what we know about controlling mechanisms is based on inference as a result of correlations with other elements. In this regard, most of what we know about the distribution of trace metals in sea water has been learned since 1975. A paper by Boyle and Edmond [22] marked a new era in trace metal geochemistry. They stressed that acceptable data for dissolved trace elements in sea water must be oceanographically consistent. That is to say vertical profiles should be smooth with the absence of random spikes, and correlations should exist with other elements that share the same controlling mechanism.

10.3.1 Biological cycling

It has been discovered that a large number of elements are depleted in the surface water and enriched in the deep ocean. Examples include Sr, I, Ba, ^{226}Ra, Cr, Ni, Cu, Cd, Zn and Ge. The nutrient elements N, P, Ca, Si and C also show the same general pattern. Of the metals listed above some (e.g. Cd; Fig. 10.6) tend to correlate more closely with N and P while others like Ba, Ra, Zn and Ge tend to resemble silica or alkalinity (Fig. 10.7). Based on these correlations, the hypothesis has been advanced that the elements that correlate with N and P are associated with the organic tissue and are regenerated via oxidative processes, in a shallow regeneration cycle. The elements that correlate with Si and alkalinity are said to be involved in a deep regeneration cycle possibly due to association with skeletal materials. Neither of these mechanisms is understood well. Detailed chemical studies on the particulate carrier phases will be necessary to understand them in any detail.

Redfield [23] first noted that the nutrient ratios in sea water were similar to the ratios in marine plankton. He concluded that photosynthetic and oxidative decomposition processes were responsible for the bulk of the variations of the nutrient elements in the ocean. If certain trace metals have constant proportions to the nutrients, this might imply that the 'Redfield' model is applicable to them. The biological control may perhaps be similar for these trace metals and nutrients even though some of the trace metals have no known biological function. Laboratory data on trace metal uptake by phytoplankton confirms the great affinity of algae for trace metals. These data also demonstrate that much of the metal uptake is by binding to high affinity

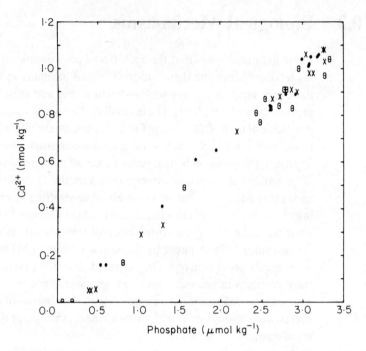

Fig. 10.6. Cadmium (nmol kg^{-1}) versus phosphate (μmol kg^{-1}) (from Bruland [55]).

surface ligands. Such binding is effectively 'passive' and phytoplankton surface ligands, on live or dead organisms or incorporated into fecal pellets, may be the site of the metal incorporation into this material.

10.3.2   Photic zone and reduced species

One of the major problems to be resolved in minor and trace element marine chemistry is to understand the mechanisms and rates of formation of reduced chemical species in the photic zone. In a region that has uniformly high dissolved oxygen concentrations, there are anomalous concentrations of reduced forms of elements such as Cr(III), As(III), I($-$1), Se(IV), and reduced gases such as CH_4, CO, and N_2O. Methylated forms of many metals have also been detected. Example profiles for Cr and I are shown in Fig. 10.8. Frequently the high concentrations of these reduced chemical species are found to be associated with high values of NO_2^- and/or NH_4^+. These associations have been used to infer a biological origin, but depending on the element, both algae and bacteria have been involved. Zooplankton have not been as frequently implicated but must be seriously considered from the point of view of excretion as well as acting as a site of residence for bacteria [24]. An understanding of the origin of the NO_2^- and NH_4^+ maxima may give us clues regarding the mechanisms producing the reduced elements, but at present the biological elements are not well understood. For example, there are at least three different explanations for the primary NO_2^- maxima [25–27].

Conventional wisdom is that bacteria must play a dominant role [28, 29], yet when specific elements have been studied in detail, phytoplankton have

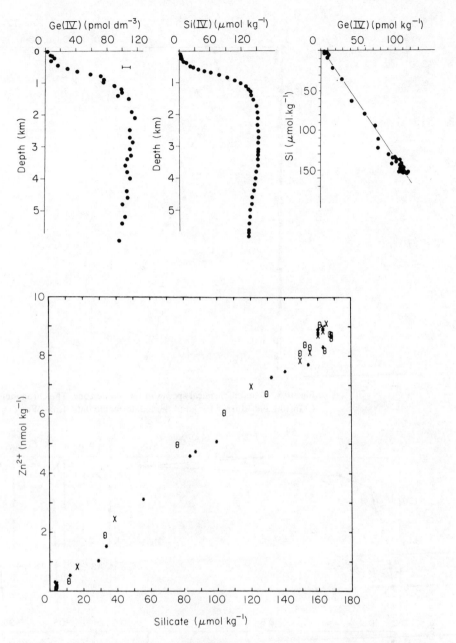

Fig. 10.7. Metal–silica correlations. The Ge–Si data from Froelich and Andrae [91] and the Zn–Si data (from Bruland [55]).

been found to be important. For example, arsenite [As(III)] profiles show a maximum in the photic zone (e.g. Fig. 10.9). Andreae [30] suggested that these high arsenite concentrations are a consequence of high arsenate stress on the phytoplankton population which responds by attempting to 'detoxify' arsenate by transforming it to arsenite. It has been shown that arsenate toxicity for algae is increased drastically at very low phosphate concentrations. The chemical similarity between phosphate and arsenate suggests that

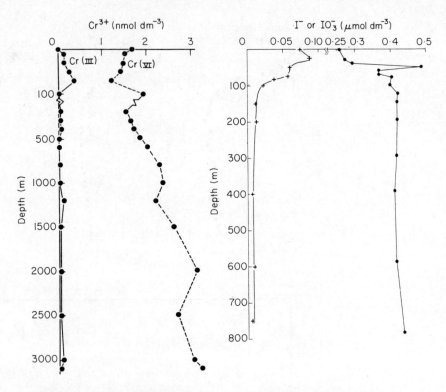

Fig. 10.8. Reduced chemical species in the photic zone. The chromium(III) and (VI) data from Cranston and Murray [92] and the iodide–iodate data (from Wong and Brewer [93]).

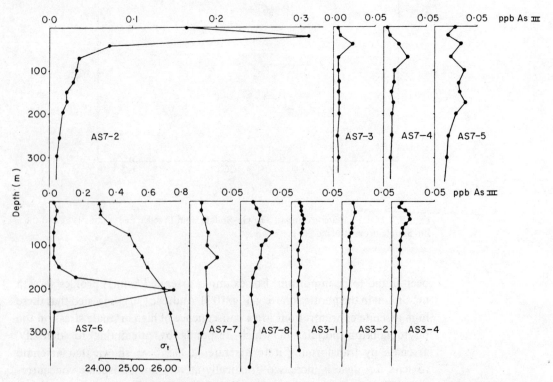

Fig. 10.9. Vertical profiles of arsenite through the photic zone (from Andreae [30]).

these ions should show competitive behaviour with regard to the phosphate uptake system in marine algae. Taking this idea one step further Andreae and Klumpp [31] studied the uptake of arsenate from bacteria-free pure cultures of marine phytoplankton species. All species studied released significant amounts of arsenite and methyl and dimethyl arsenic into their environment.

Another, more indirect example, concerns the ocean as a source of mercury to the atmosphere. Recent analyses by Fitzgerald et al. [32] indicate that biologically productive regions in the equatorial Pacific Ocean may be a source of gaseous Hg species to the marine troposphere. Total gaseous mercury in the marine atmosphere increased and the dissolved concentrations in the surface ocean decreased in the region between $4°N$ and $10°S$. The suggestion of these field and other laboratory studies is that dissolved inorganic Hg is converted to volatile forms of Hg such as dimethyl mercury by algae and bacteria in the photic zone. The process of biomethylation has been shown to be potentially important for a wide range of metals. The mechanism is probably evolved to give certain micro-organisms selective advantages for the elimination of heavy metals normally toxic to the cell [29].

10.3.3 Effect of trace metals on primary productivity

Copper concentrations in seawater at natural levels have long been known to be very poisonous for algal photosynthesis [33]. This toxic effect of copper can be an especially critical problem for biologists who are attempting to measure primary productivity by the ^{14}C method. Fitzwater et al. [34] have shown that by preparing ^{14}C-labelled bicarbonate, as recommended by Steeman-Nielsen [35] and Strickland and Parsons [36], the copper concentration in a productivity experiment can increase by about 1 nmol dm^{-3}. This can more than double the ambient Cu level if the sample is surface sea water. Large quantities of Mn, Zn, Pb, Ni, and Fe are also added. When all reagents are cleaned of contaminating trace metals and the productivity measurements are conducted in a 'clean' manner, the productivity results are 25–35% higher than when the precautions are not taken. The implications are that the primary productivity of phytoplankton is extremely sensitive to trace metal contamination.

Algae are sensitive to the cupric ion activity not the total copper concentration. This conclusion was first shown by Sunda and Guillard [37]. They conducted culture experiments with the estuarine diatom *Thalassiosira pseudonana* in which the cupric ion activity was altered independently of total copper concentration by varying the chelator concentration and the pH. Cupric ion [expressed as $pCu = -\log(Cu^{2+})$] inhibits the growth of *Thalassiosira* at pCu values of about 10.6 and completely inhibits growth when pCu $\leqslant 8.3$ (Fig. 10.10). When growth rate is plotted versus copper concentration the data do not fall on a single curve (Fig. 10.10). Not only is growth inhibited, but the cells are elongated and morphologically distorted. Similar results were found for an estuarine green alga *Nannochloris atomus*. In this case growth was totally inhibited for pCu $\leqslant 8.7$ and initial effects were noted at pCu $= 10.4$. Similar results have since been reported for the dinoflagellate

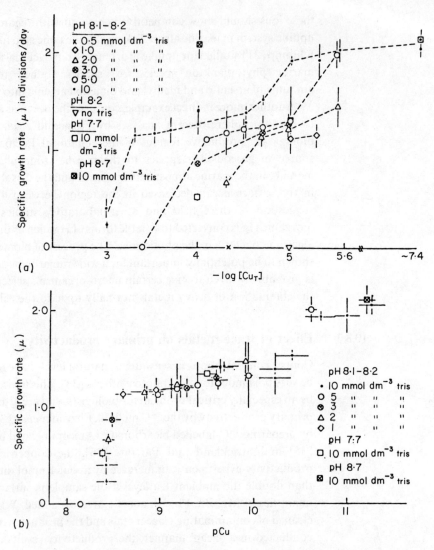

Fig. 10.10. (a) Growth rate of clone 3H versus the negative log of the total copper concentration. (b) Growth rate of clone 3H versus the negative log of the activity of free copper ion (pCu) (from Sunda and Guillard [37]).

Gonyaulax tomarensis by Anderson and Morel [38]. Both motility and photosynthesis were reduced with increasing cupric ion activity. The cells were 100% nonmotile at pCu = 9.7.

The importance of these experimental toxicity studies is emphasized by comparison with the cupric ion activity in sea water. Assuming that Cu^{2+} is complexed only by inorganic ligands, at pH 8.2 and 25°C, Sunda and Guillard [37] calculated that the ratio of $p(Cu^{2+}/Cu_T) = 1.8$. For a total Cu concentration of 2 nmol dm^{-3} for coastal sea water [39, 90] the resulting pCu = 9.0. This value is in the correct range at least to inflict partial growth inhibition of all species studied. From these studies it is easy to see why primary productivity measurements are so sensitive to copper contamination.

The mechanism of this inhibition is still unclear. Sunda and Guillard argued [37] that the two-stepped nature of the inhibition curve suggested that Cu^{2+} inhibition involves at least two separate inhibition sites. The dependency on Cu^{2+} has also led to speculation regarding the role of organic compounds released by phytoplankton that might complex trace metals and reduce their toxic effects [40] or make essential metals 'available' [41, 42]. Swallow *et al.* [43] studied eight species known for their copious release of organic matter and found that only one (*Gloeocystis gigas*) appeared to produce extracellular organic compounds that can reduce the activity of 10^{-6} mol dm^{-3} copper. Thus the question of the release of strong copper chelators by algae under conditions of copper toxicity remains uncertain.

Competition of copper with other metals such as manganese can also be important [44]. This may occur through a physiological interaction between copper and manganese in which copper competes for manganese nutritional sites, thereby interfering with manganese metabolism. Stimulation of phytoplankton growth by EDTA or NTA appears to result directly from the ability of these chelators to complex copper strongly without appreciably binding manganese.

10.4 Adsorption and Scavenging

The mechanism of scavenging, or adsorption onto solid surfaces has been determined to be an important control on the distribution of chemical elements in sea water and between sediment particles. Early workers [45–47] recognized that the low dissolved trace metal content of sea water could not be attributed to lack of supply through geologic time but must be caused by efficient and rapid removal processes. In general, metal concentrations in sea water are too low for solubility equilibrium and instead adsorption of dissolved species onto marine particles is believed to be the dominant control [48, 49].

10.4.1 Evidence based on dissolved elemental or isotopic distributions

Field evidence of the role of scavenging has been obtained from the deficiency of metals in the water column with respect to conservative mixing models, or, for radioactive species, with respect to a parent nuclide. For example, several workers [50–52] have examined ^{226}Ra–^{210}Pb disequilibria in the deep sea. ^{210}Pb is approximately 50% deficient in the deep sea relative to its parent ^{226}Ra. Craig *et al.* [50] attributed this to *in situ* scavenging by falling particles and calculated the first order scavenging rate constant from the equation below for a one-dimensional vertical advection–diffusion model.

$$\omega \frac{\partial C}{z} - K_z \frac{\partial^2 C}{\partial z^2} - \lambda C - \psi C = 0 \tag{10.9}$$

where ω = vertical advection velocity, K = vertical eddy diffusion coefficient,

λ = radioactive decay constant, ψ = chemical removed rate constant and C = dissolved concentration. The 'best fit' value of the scavenging residence time ($\tau = 1/\psi$) in the deep Pacific was found to be about 54 years (Fig. 10.11). As pointed out by Nozaki and Tsunogai [52] such residence times are actually net values, probably reflecting simultaneous regeneration and scavenging. Another limitation, pointed out by Bacon *et al.* [51], is that vertical models do not give an unambiguous explanation for where the scavenging takes place. The sediment boundaries (sides and bottom of the ocean) may also be important sites of scavenging.

Deep ocean scavenging has also been required to explain the vertical profiles of ^{230}Th [53, 54], and dissolved copper [39, 55] nickel, and cadmium [56]. The activity of ^{230}Th increases about almost linearly with depth (Fig. 10.12). The simple scavenging model described by eqn (10.9) could not adequately describe the data so Nozaki *et al.* [53] modified the model to include *in situ* production or desorption as well as *in situ* removal or adsorption. They calculated that the residence times with respect to simultaneous scavenging and regeneration were 235 and 57 days, respectively.

Surface water and coastal zone scavenging have been studied using ^{234}Th

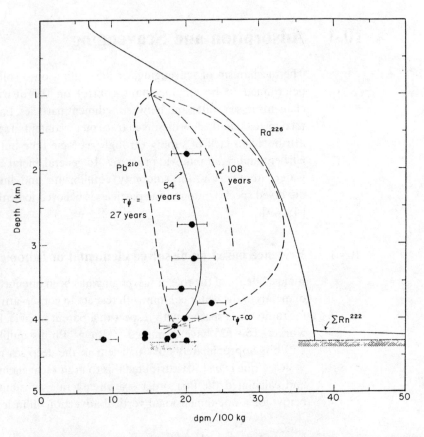

Fig. 10.11. Profiles of ^{210}Pb and ^{226}Ra at a station in the North Pacific at 28.5°N, 121.5°W (dpm = decays per minute). Model profiles for ^{210}Pb are shown for scavenging residence times of 27 years, 54 years (best fit), 108 years and infinity (from Craig *et al.* [50]).

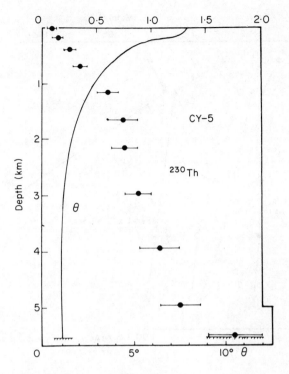

Fig. 10.12. Vertical profile of ^{230}Th from the north western Pacific (from Nozaki *et al.* [53]).

and ^{228}Th as tracer isotopes reflecting the properties of reactive pollutants. Broecker *et al.* [57] found that ^{228}Th was deficient relative to its parent ^{228}Ra in the surface open ocean and that the average ratio of daughter to parent was ^{228}Th/^{228}Ra $= 0.21$. Using a simple box model, in which the production of ^{228}Th is balanced by its own decay and chemical scavenging, they calculated that the mean life-time of thorium with respect to chemical removal was about 0.7 year. Similarly in the coastal waters of the New York Bight ^{234}Th and ^{228}Th have been used to model non-radioactive removal from the water column [58]. There the scavenging residence time varies from 10 days nearshore to 70 days at the shelf break.

10.4.2 Evidence from sediment trap samples

The *in situ* scavenging inferred from the distribution of dissolved isotopes in the water column has been confirmed by study of samples from sediment traps moored in the deep sea. In particular, results by Brewer *et al.* [59] from a site in the Atlantic near Barbados (Parflux E) [60], show that Mn, Cu, Fe, Sc, and ^{230}Th are characterized by an increase in the metal/Al ratio with depth as shown in Fig. 10.13. All of these elements are known to be highly surface active in sea water and are expected to be scavenged by particles settling through the water column.

The scavenging rate, ψC, was determined from the observed profiles by

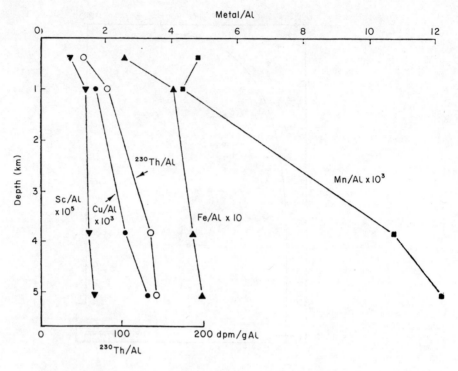

Fig. 10.13. Depth profiles of metal to Al ratios for Sc, Cu, ^{230}Th, Fe, and Mn in sediment trap samples from the Parflux E site in the Atlantic (13°N, 54°W) (from Brewer *et al.* [59]).

means of a steady state vertical scavenging model described by:

$$S\frac{\partial X}{\partial z} + \psi C + \lambda_p X_p - \lambda X = 0 \qquad (10.10)$$

where S = mean settling velocity of the particles, ψ = first order scavenging rate constant, λ = decay constant, C = dissolved concentration, X = particulate concentration, X_p = parent concentration. The ratios of the thorium isotopes ^{228}Th/^{234}Th were used to calibrate the model and resulted in a settling rate of 21 m d^{-1}. From this the scavenging residence times of Th, Pa, Po, Pb, Mn, Fe, Cu, and Sc were determined. In most cases the values agree well with scavenging rate constants determined by independent means.

10.4.3 Surface chemistry model for marine particulate matter

In an attempt to quantify the field observations of trace metal scavenging, Balistrieri *et al.* [48] developed a surface chemistry model for marine particulate matter. This approach involved combining the field observations of trace metal scavenging (e.g. Table 10.1) with recent theoretical models of the surface chemistry of oxide particles [61, 62]. It was assumed that the surface chemistry concepts that have been derived from study of well-defined substances can be applied to the heterogeneous surfaces of marine particulate matter. In addition, the surface sites on marine particulate matter were

Table 10.1. Residence of particulate matter and scavenging residence times for metals (from Balistrieri et al. [48])

Element	τ_ψ(years)
Mn	20
Fe	77
Cu	500
Sc	230
Th	22
Pb	47
Po	27
Pa	31
Particles	0.365

considered to be similar to either the –OH groups of metal oxides or the –COOH groups of organic ligands. Adsorption equilibrium reactions of the following types were considered

$$SOH + Me^{z+} \xrightarrow[K_{Me}]{} SOMe^{(z-1)^+} + H^+ \qquad (10.11)$$

$$SOH + Me^{z+} + H_2O \xrightarrow[K_{MeOH}]{} SOMeOH^{(z-2)^+} + 2H^+. \qquad (10.12)$$

With apparent adsorption equilibrium constants defined as:

$$K_{Me} = \frac{[SOMe^{(z-1)^+}]\,(H^+)}{[SOH]\,[Me^{z+}]} \qquad (10.13)$$

and

$$K_{MeOH} = \frac{[SOMeOH^{(z-2)^+}]\,(H^+)^2}{[SOH]\,[Me^{z+}]} \qquad (10.14)$$

where [ ] is the concentration of the surface complexes or dissolved species in mol dm^{-3} and ( ) is the activity. SOH stands for a surface –OH group. These adsorption equilibrium constants have been determined for the adsorption of a number of elements on pure metal oxides such as SiO_2, α-FeOOH, $Fe_2O_3 \cdot H_2O$ and γ-Al_2O_3, but they have not been determined for adsorption on natural marine particulate matter. For the solid phases, where stability constants have been determined, they have been found to correlate strongly with the hydrolysis constants of the elements in solution. This concept was developed by Balistrieri et al. [48] to estimate adsorption constants on marine particles from scavenging residence times. The greatest uncertainty in these calculations was the choice of the concentration of free surface sites on marine particulate matter. A range of 0.1–10 mol kg^{-1} was chosen based on the literature values for various oxides and natural marine organic matter. A comparison of the equilibrium constants estimated for natural marine particulate matter with some pure reference metal oxide compounds is shown in Fig. 10.14. In general, metals interact more strongly with the natural material than with the model oxides. Pure oxides do not appear to be suitable model substances for marine particulate matter [63]. Similar comparison with model organic compounds suggests that the surface chemistry of marine particles is controlled by organic compounds. This is not a surprising result and agrees with the electrophoretic mobility measurements of Neihof and Loeb [64], Loeb and Neihof [65], and Hunter and Liss [66].

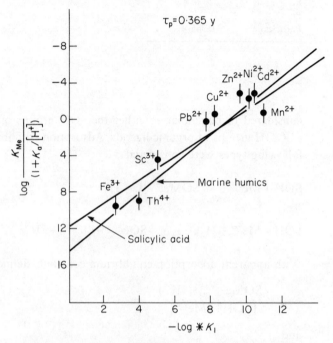

Fig. 10.14. Comparison of the adsorption equilibrium constants calculated for natural marine particulate material with literature values for model metal oxide compounds (from Balistrieri et al. [48]).

10.4.4 **Surface chemistry of model metal oxide compounds in sea water**

Even though it appears that the surface chemistry of natural marine particles is controlled by organic coatings, it is still very beneficial to study the surface chemistry of model oxide compounds, e.g. α-FeOOH and δ-MnO$_2$ in organic-free sea water. Sea water is a very complex medium and an excellent test of the various surface chemistry models is to see if equilibrium constants determined under non-competitive conditions can be applied to complex ionic mixtures. This type of problem is essentially an extension of the Garrels and Thompson [67] sea water speciation calculation to include the metal oxide surface as an additional ligand. The electrical double layer at the metal oxide interface makes this a more complicated calculation. But the effort is worth while because understanding the electrical interface of model compounds will help us understand the more complex natural surfaces. In addition, understanding the organic-free system provides a starting point for evaluating the effect of organic coatings.

Balistrieri and Murray [68–70] have determined the distribution of species on the surface of goethite (α-FeOOH) and δ-MnO$_2$ in major ion sea water. Emphasis has been placed on these phases because of their importance as scavenging surfaces in marine sediments and ferromanganese nodules. The site binding model of Yates et al. [71] and Davis et al. [62] was used to describe the electrical aspects of the interface. The results for goethite were encouraging because adsorption equilibrium constants determined in dilute single salt solutions could be combined to predict the surface speciation correctly in the complex sea water medium. The pH dependence of the surface speciation of goethite in major ion sea water is shown in Table 10.2. The resulting predicted titratable surface charge agreed very well with that actually measured.

Table 10.2. Surface complexes of goethite in major ion sea water as a function of pH (from Balistrieri and Murray [68])

	% of total surface sites				
	pH 5	pH 6	pH 7	pH 8	pH 9
FeOH	29.8	37.1	40.2	36.7	24.5
FeOH$_2^+$	2.0	1.7	0.4	0.2	0.3
FeO$^-$	0.1	0.2	0.8	1.8	0.4
FeO$^-$Na$^+$	0.3	0.4	0.9	1.4	0.5
FeO$^-$K$^+$	<0.01	<0.01	<0.01	<0.01	<0.01
FeO$^-$Mg^{2+}	11.9	15.2	17.4	16.5	10.6
FeO$^-$MgOH$^+$	<0.01	<0.01	0.6	9.3	35.7
FeO$^-$Ca^{2+}	5.8	7.5	8.5	8.1	5.3
FeO$^-$CaOH$^+$	<0.01	<0.01	0.6	0.9	3.5
FeOH$_2^+$SO$_4^{2-}$	19.6	24.0	24.6	21.5	14.9
FeOH$_2^+$HSO$_4^-$	18.8	1.9	1.0	<0.01	<0.01
FeOH$_2^+$Cl$^-$	11.7	11.9	6.2	3.6	4.1

When the experimental system was enlarged to include trace metals, the effect of the major sea water ions on the adsorption of trace metals on goethite could be properly evaluated [70]. Magnesium and sulfate are the major ions that influence the adsorption of Cu, Pb, Zn, and Cd on α-FeOOH in sea water. Sulfate adsorption acts to enhance the adsorption of trace metal ions at low pH, while magnesium adsorption tends to suppress the adsorption of trace metals at high pH. At the pH of sea water, only the competitive effects of Mg are significant.

The surface chemistry of δ-MnO$_2$ in major ion sea water is more complicated because ion exchange reactions other than with H$^+$ are important (Fig. 10.15).

1 The number of surface sites bound by Mg and Ca increases with increasing pH.

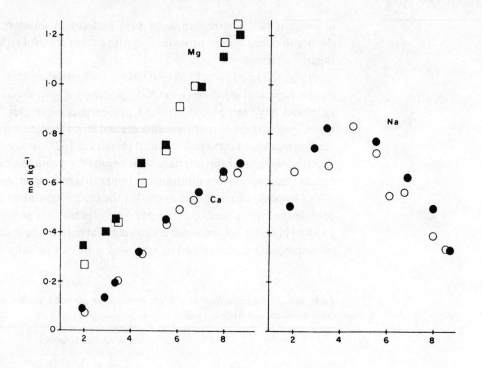

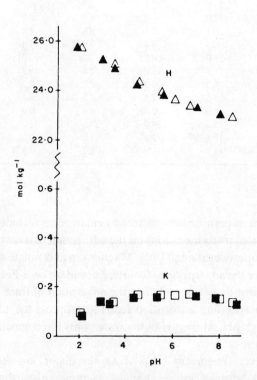

Fig. 10.15. Experimental data for the simultaneous adsorption of Na, Mg, Ca, K and H on δ-MnO$_2$ from synthetic sea water. The adsorption data are plotted as moles of adsorbed ion per kilogram of δ-MnO$_2$ as a function of pH. The open and filled symbols indicate replicate experiments (from Balistrieri and Murray [69, 70]).

2 The number of sites bound by Na increases to a maximum and then decreases with pH.

3 The number of sites bound by K^+ increases slightly and then levels off or decreases slightly with pH.

4 The number of sites blocked by H decreases as pH is increased. These results suggest that Mg and Ca ions adsorb on δ-MnO_2 by exchange with Na and K ions as well as protons. In all cases equivalent charge is conserved. This is analogous to the ion exchange process on clay surfaces.

10.4.5 Surface chemistry of natural marine particles

The deep ocean scavenging model calibrated using scavenging residence times based on sediment-trap data suggests that the scavenging component of marine particles has an organic nature. This result is consistent with early studies by Neihof and Loeb [64] and Loeb and Neihof [65] using micro-electrophoresis. In their experiments, a wide variety of model particles all acquired a uniform surface charge when exposed to natural sea water. In organic-free sea water the surface charge characteristics were different. These results suggest that adsorbed organic constituents play an important role in influencing the surface charge of natural particles. These results were further developed, using model particles in lake water, by Davis [72, 73] Davis and Gloor [74] and Tipping [75]; and in coastal sea water by Hunter [76]. Hunter exposed ion exchange resins, Al_2O_3 and SiO_2 to successive volumes of natural sea water. All particles, regardless of their initial surface charge, became uniformly negatively charged after exposure to sea water (Fig. 10.16). The studies cited above for lake water showed similar results. The variation in electrophoretic mobility with pH and metal ion concentration suggests that carboxyl and phenol groups are the major functional groups.

Work on natural particles is scarce although much more will require to be done in this direction in the future. Tipping *et al.* [77] studied ferric oxide particles that formed *in situ* in a stratified lake. The natural particles were negatively charged probably due to adsorbed humic substances. Figure 10.16 compares the electrophoretic mobility of these natural particles with synthetic iron gel added to the same lake water and to organic free NaCl. Hunter and Liss [66] have measured the electrophoretic mobility on river and estuarine particles. The mobility as a function of salinity for several estuaries is shown in Fig. 10.17. Again, natural particles appear to be uniformly negatively charged probably due to surface-active organic matter.

The most recent work has focussed on evaluating the effects of organic coatings on metal uptake as well as determining apparent binding constants with natural material [78, 79]. These experiments have shown that the surface chemistry concepts that have been developed for simple metal oxide surfaces can be applied to describe adsorption on complex mixtures of marine particles. Of the phases present in marine sediments, the presence or absence of MnO_2 seems to have the greatest influence on the magnitude of the adsorption binding constants.

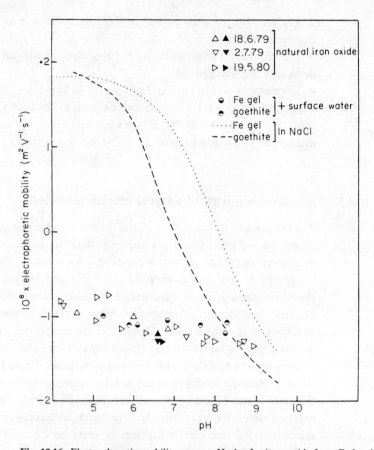

Fig. 10.16. Electrophoretic mobility versus pH plot for iron oxide from Esthwaite Water, an eutrophic lake in the UK. The solid symbols are natural iron oxides from the lake. The open triangles show the mobility of the natural particles after pH adjustment. The circles show the results using synthetic iron oxides added to the lake water. The dotted lines indicate the mobility of the synthetic iron oxides in organic clean NaCl (from Tipping *et al.* [77]).

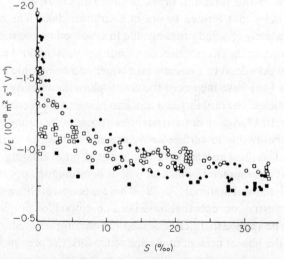

Fig. 10.17. Electrophoretic mobility (U_E) of natural particles as a function of salinity for four river estuaries in Great Britain (from Hunter and Liss [66]).

10.5 Solubility Control

There are few elements for which solubility is a controlling factor in open ocean sea water. Even when solid phases do form, in many cases, it is due to biological processes overcoming a state of undersaturation. Opal (SiO_2) and celestite ($SrSO_4$) shells are typical examples. Only for calcite and aragonite (both polymorphs of $CaCO_3$) are surface sea waters saturated. Even in these cases, without living organisms, the solids would not spontaneously precipitate because of specific inhibitors (e.g. Mg^{2+}). Many unusual minerals have been identified in particulate sea water samples that suggest that in very local conditions biologically created supersaturation may exist. The mineral barite ($BaSO_4$) is a specific example [80]. Discrete micron-sized barite particles are present in suspended matter everywhere in the world ocean. A wide variety of exotic compounds like malachite, tenorite and spalerite have also been reported [81]. Their origin is unknown.

10.5.1 Solubility and kinetics: the constant composition approach

The equilibrium solubility and rates of precipitation of sparingly soluble salts, e.g., $CaCO_3$ and $MgCO_3$, calcium phosphates (e.g. apatites), and Ba and Sr sulfates are of interest in marine chemistry. Laboratory studies have been difficult in the past because of kinetic problems. Tomson and Nancollas [82] described a new and highly reproducible method for studying rates of mineralization even at very low supersaturation. This is essentially a pH-stat method in which the goal is to add the reagents at the same rate as which precipitation is occurring. The chemical potential of the solution species remains constant and the rate of precipitation can be readily calculated. Tomson and Nancollas [82], Nancollas *et al.* [83], and Koutsoukos *et al.* [84] have described the application of this approach for the study of calcium phosphates.

Jahnke [85] recently applied this technique to determine the solubility of synthetic carbonate fluorapatite. The precipitation of authigenic phosphate phases is one of the important mechanisms for removing phosphate from the oceans and the most abundant authigenic phosphate phase in marine sediments is francolite. Francolite is a highly substituted (mostly by carbonate) form of fluorapatite. Jahnke was able to synthesize fluorapatite with varying degrees of carbonate substitution. The stoichiometry was expressed as:

$$Ca_{10}(PO_4)_{5.83-0.57(x)}(CO_3)_x F_{2.52-0.3(x)}.$$

The free energy of formation of the carbonate fluorapatite increased with increasing carbonate content, indicating a decrease in the stability of fluorapatite with increased substitution.

The importance of including the effect of carbonate substitution on the thermodynamic stability of fluorapatite in sea water is demonstrated in Fig. 10.18. The curves show the equilibrium phosphate concentration as a function of carbonate substitution. There is a dramatic increase in dissolved PO_4^{3-}

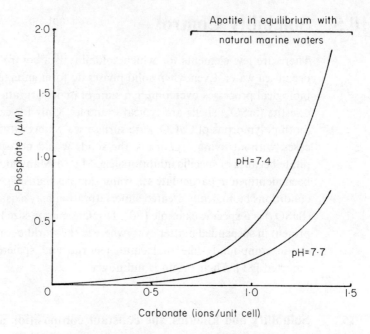

Fig. 10.18. Phosphate concentration in sea water in equilibrium with carbonate fluorapatite as a function of carbonate content in the solid. The range predicted to be in equilibrium for naturally occurring carbonate activities is also shown (from Jahnke [85]).

concentration with CO_3^{2-} substitution in the solid phase. For a carbonate content of 1.4 ions/unit cell, the equilibrium PO_4^{3-} concentration is about 100 times greater than for pure fluorapatite.

10.5.2   **Solid solution formation**

Even though trace constituents may not form pure solid phases in sea water, solid solution formation with a major host phase is an intriguing possibility. Equilibrium solid solution formation is known to reduce the solubility of the trace component [3]. In addition, geochemists find that the amounts of minor and trace metals coprecipitated into calcite and aragonite are a record of the chemical environment during crystallization. A recent application of this feature was demonstrated by Boyle and Keigwin [86]. The cadmium content of the skeletons of foraminifera depends on the cadmium concentration in sea water [87]. In sea water, cadmium and phosphate are closely correlated. Thus the careful study of the cadmium content of buried foraminifera can give insight into ancient ocean circulation patterns. There are many factors other than equilibrium solid solution involved in trace metal uptake by forams, but solid solution is a good starting point for understanding the chemistry.

There have been few careful studies of solid solution formation in sea water but one fine example is the laboratory study by Lorens [88]. He evaluated the effect of the rate of precipitation on the distribution coefficients of Sr^{2+}, Co^{2+}, Mn^{2+} and Cd^{2+} in calcite. A pH-stat was used to maintain a constant degree of saturation and as a result a constant precipitation rate in 0.69 mol dm^{-3} NaCl. The precipitation rate was proportional to the degree of supersaturation and the mass of seed crystal introduced.

At equilibrium, the distribution of a trace metal, Me, in calcium carbonate is described by:

$$\left(\frac{Me}{Ca}\right)_{CaCO_3} = D\left(\frac{Me}{Ca}\right)_{solution} \tag{10.15}$$

where D is the distribution coefficient. This relation assumes that the solid phase is homogeneous and contains no concentration gradients. When changes in solution composition occur during precipitation, a heterogeneous distribution of trace metals occurs within the precipitated calcite and the Doerner–Hoskins heterogeneous distribution law applies:

$$\log\left(\frac{Me_0}{Me_f}\right) = \log\left(\frac{Ca_0}{Ca_f}\right) \tag{10.16}$$

where 0 and f indicate initial and final concentrations in solution.

Lorens [88] found that substantial changes in the value of the distribution coefficient occur with variations in calcite precipitation rate. The Sr distribution coefficients increase with increasing precipitation rate, while Co, Mn, and Cd values decrease (Fig. 10.19). The following rate expressions (at 25°C) were derived:

$\log \lambda_{Sr} = 0.249 \log R - 1.57$
$\log \lambda_{Mn} = -0.266 \log R + 1.35$
$\log \lambda_{Co} = -0.173 \log R + 0.68$
$\log \lambda_{Cd} = -0.194 \log R + 1.46$

where R is the precipitation rate in nmoles $CaCO_3$ per mg seed crystal per minute. These experiments indicate that both solution composition and growth rate are significant factors influencing the trace metal uptake by calcite.

10.5.3 Solubility in reducing environments

Although solubility control of trace metals in aerobic sea water is probably unimportant, dramatic changes in the solubility occur across the redox front located at the oxygen–hydrogen sulfide boundary found in environments with restricted circulation. Locations where these changes occur are within sediment pore waters, in marine hydrothermal systems, and in closed anoxic basins. The solubility of metals in sulfide-containing waters is controlled by the degree to which they form solid sulfides and metal sulfide complexes. The detailed behaviour across the oxygen–hydrogen sulfide boundary is made complicated by the complex-dissolved sulfide speciation and metal-sulfide complexes.

Emerson *et al.* [89] have recently presented a summary of these complicated solubility calculations together with comparisons with published analyses from marine sulfide containing waters. The thermodynamic approach provides a valuable frame of reference and Emerson *et al.* [89] also

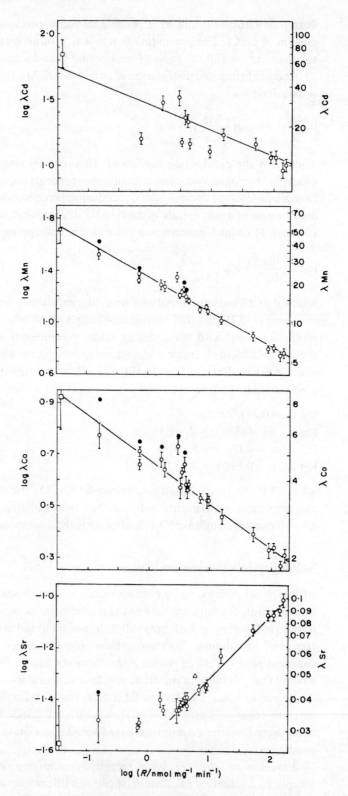

Fig. 10.19. Adsorption corrected distribution coefficients for Sr, Co, Mn and Cd in calcium carbonate as a function of the log growth rate (from Lorens [88]).

stress that the relation between the kinetics of the specific chemical reactions and the rates of mixing greatly influence the distributions across an anoxic interface.

A major factor influencing the solubility trend across a sulfide interface is whether a metal ion is classified as a class B metal or a transition metal [3]. Transition metal cations [e.g. Mn(II), Fe(II), Ni(II), Co(II) and Cu(II)] should decrease in concentration with increasing total dissolved sulfide because they form solid sulfides but do not form strong dissolved sulfide complexes. Class B metal cations [e.g. Cu(I), Cd(II), Pb(II) and Zn(II)] should show solubility increases with increasing sulfide concentration because of the strong complexation by reduced sulfide ligands. The solubility of metals thus is dependent on the oxidation state of the metal as well as the concentration of dissolved sulfide ligands and the magnitude of the equilibrium solubility and complexation constants. Emerson *et al.* [89] present detailed solubility calculations for Fe, Mn, Cu, Cd and Ni using the ratio of $(HS^-)/(SO_4^{2-})$ as a master variable. Examples for iron and cadmium are shown in Fig. 10.20. Solid lines represent individual species activity and the dashed (pH = 7.4) and dotted (pH = 7.8) lines represent the concentration of total soluble metal in sea water.

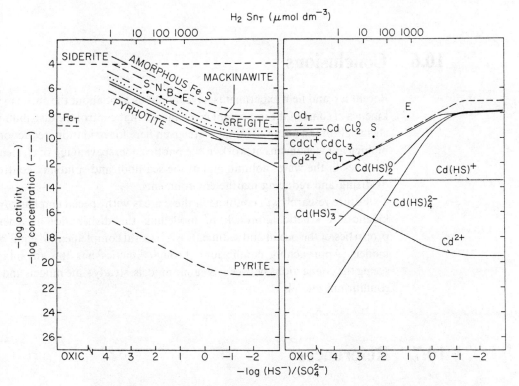

Fig. 10.20. Activity and concentration of iron and cadmium in sea water as a function of $-\log (HS^-)/(SO_4^{2-})$. Solid lines represent calculated individual species activities. The dashed lines on the left are estimates of the upper limit of the dissolved Fe and Cd concentration in open ocean oxic sea water. The calculations are for pH 7.4. The symbols represent measured values from Saanich Inlet (S), Black Sea (B), Lake Nitinat (N), and Enghien-les-Bains (E) (from Emerson *et al.* [89]).

Horizontal lines at the far left represent observed metal concentrations in oxic sea water. The redox front is seen to be a major discontinuity between observed oxic concentrations and predicted anoxic concentrations. The dissolved iron responds to increasing sulfide levels with a decrease in concentration as predicted for transition metal cations. Several kinetic intermediate iron sulfide solid phases were considered and iron measurements from anoxic waters tend to fall on the greigite solubility line. The manganese and nickel solubility diagrams resemble iron. Cadmium on the other hand is representative of class B metals. Cadmium complexes strongly with bisulfide resulting in an enhanced cadmium solubility of 3–10 orders of magnitude over the predicted free ion activity. Starting with aerobic sea water, cadmium concentrations should go through a minimum due to the low solubility at low total sulfide. Analyses of anoxic water samples are in close agreement with the predicted Cd solubility trends. The Cu diagram is similar to that of Cd with the complication that Cu changes oxidation states (II to I) in the anoxic zone.

Much research remains to be done in this field. The first requirement is accurate sets of field data from anoxic environments with a range of sulfide values. Laboratory experiments are needed to establish equilibrium constants for bisulfide and polysulfide complex formation and to provide more accurate metal sulfide solubility products.

10.6   Conclusions

Recent lab and field experiments have taught us a lot about the chemistry and kinetics of some of the important mechanisms that control the distributions of chemicals in the ocean. Examples are given here for oxidation–reduction and biological mechanisms at work in the photic zone, scavenging of elements by particles in the water column and at the sea floor and solubility controls in oxidizing and reducing marine environments.

Future research will continue in these areas with special emphasis on the kinetics. A kinetic approach to modelling the distribution of chemical properties of the ocean and sediments is a natural complement to equilibrium models. Approaching equilibrium through kinetics has the advantage of seeing the connections between dynamic models, steady state models and local equilibrium models.

10.7   References

1   S.D. Faust and J.V. Hunter, *Principles and Applications of Water Chemistry*, John Wiley, NY, Pp. 643 (1967).
2   W. Stumm, *Advances in Chemistry Series No. 67*, ACS, Washington DC, Pp. 344 (1967).
3   W. Stumm and J.J. Morgan, *Aquatic Chemistry*, Wiley-Interscience, New York, Pp. 780 (1981).
4   L.G. Sillen, *Am. Assoc. Adv. Sci. Publ.*, **67**, 549 (1961).

5   L.G. Sillen, *Science*, **156**, 1189 (1967).

6   W. Stumm, *Thalassia Jugoslavia*, **14**, 197 (1978).

7   J.F. Pankow and J.J. Morgan, *Environ. Sci. Technol.*, **15**, 1155 (1981).

8   J.F. Pankow and J.J. Morgan, *Environ. Sci. Technol.*, **15**, 1306 (1981).

9   J.W. Murray and P.G. Brewer, In *Marine Manganese Deposits*, (Ed. by G.P. Glasby), Elsevier, 291 (1977).

10 E.D. Goldberg, In *Oceanography*, (Ed. by M.N. Hill) AAAS Pub. No. 67 (1961).

11 J.D. Hem, *U.S. Geol. Surv. Water – Supply Pap.*, 1667-A, Pp. 64 (1963).

12 J.J. Morgan, *Chemistry of Aqueous Manganese(II) and (IV)*. PhD Thesis, Harvard Univ., (1964).

13 J.J. Morgan, In *Principles and Applications of Water Chemistry*, (Eds. S.D. Faust and J.V. Hunter), J. Wiley and Sons, 561 (1967).

14 J.W. Murray, In *Marine Minerals*, (Ed. by R.G. Burns) (Min. Soc., Am. Short Course Notes), Chap. 2, 47 (1979).

15 W. Sung and J.J. Morgan, *Environ. Sci. Technol.*, **14**, 561 (1980).

16 W. Sung and J.J. Morgan, *Geochim. Cosmochim. Acta*, **45**, 2377 (1981).

17 S. Emerson, R.E. Cranston and P.S. Liss, *Deep-Sea Res.*, **26**, 859 (1979).

18 S. Emerson, S. Kalhorn, L. Jacobs, B.M. Tebo, K.H. Nealson and R.A. Rosson, *Geochim. Cosmochim. Acta*, **46**, 1073 (1982).

19 W.G. Sunda, S.A. Huntsman and G.R. Harvey, *Nature*, **301**, 234 (1983).

20 A.T. Stone and J.J. Morgan, *Environ. Sci. Technol.*, **18**, 450 (1984).

21 A.T. Stone and J.J. Morgan, *Environ. Sci. Technol.*, **18**, 617 (1984).

22 E.A. Boyle and J.M. Edmond, *Nature*, **253**, 107 (1975).

23 A.C. Redfield, In *James Johnstone Memorial Volume*, Liverpool University Press, Liverpool, 176 (1934),

24 E.D. Traganza, J.W. Swinnerton and C.H. Cheek, *Deep-Sea Res.*, **26**, 1237 (1979).

25 J.H. Steele and C.S. Yentsch, *J. Marine Biol. Assoc.*, *U.K.*, **39**, 217 (1960).

26 R.W. Eppley, E.H. Renger, E.L. Venrick and M.M. Mullin, *Limnol. Oceanogr.*, **18**, 534 (1973).

27 D.A. Kiefer, R.J. Olson and O. Holm-Hansen, *Deep-Sea Res.*, **23**, 1199 (1976).

28 R.S. Braman and C.C. Foreback, *Science*, **182**, 1247 (1973).

29 J.M. Wood and H.-K. Wang, *Environ. Sci. Technol.*, **17**, 582 (1983).

30 M.O. Andreae, *Limnol. Oceanogr.*, **24**, 440 (1979).

31 M.O. Andreae and D. Klumpp, *Environ. Sci. Technol.*, **13**, 738 (1979).

32 W.F. Fitzgerald, G.A. Gill and J.P. Kim, *Science*, **224**, 597 (1984).

33 E. Steeman-Nielsen and S. Wium-Anderson, *Marine Biol.*, **6**, 93 (1970).

34 S.E. Fitzwater, G.A. Knauer and J.H. Martin, *Limnol. Oceanogr.*, **27**, 544 (1982).

35 E. Steeman-Nielsen, *J. Cons. Cons. Int. Explor. Mer.*, **18**, 117 (1952).

36 J.D.H. Stickland and T.R. Parsons, *A Practical Handbook of Seawater Analysis*, Fisheries Res. Board of Canada, Pp. 311 (1968).

37 W. Sunda and R.R.L. Guillard, *J. Marine Res.*, **34**, 511 (1976).

38 D.M. Anderson and F.M.M. Morel, *Limnol. Oceanogr.*, **23**, 283 (1978).

39 E.A. Boyle, F.R. Sclater and J.M. Edmond, *Earth Planet. Sci. Lett.*, **37**, 38 (1977).

40 J.J. Jackson and J.J. Morgan, *Limnol. Oceanogr.*, **23**, 268 (1978).

41 R. Johnston, *J. Marine Biol. Assoc.*, *UK*, **44**, 87 (1964).

42 R.T. Barber, In *Trace Metals and Metal Organic Interactions in Natural Waters*, (Ed. by P.C. Singer), Ann Arbor Science Publ., 321, (1973).

43 K.C. Swallow, J.C. Westall, D.M. McKnight, N.M.L. Morel and F.M.M. Morel, *Limnol. Oceanogr.*, **23**, 538 (1978).

44 W.G. Sunda, R.T. Barber and S.A. Huntsman, *J. Marine Res.*, **89**, 567 (1981).

45 V.M. Goldschmidt, *J. Chem. Soc.*, 655 (1937).

46 E.D. Goldberg, *J. Geol.*, **62**, 249 (1954).

47 K.B. Krauskopf, *Geochim. Cosmochim. Acta*, **12**, 331 (1956).

48 L. Balistrieri, P.G. Brewer and J.W. Murray, *Deep-Sea Res.*, **28**, 101 (1981).

49 Y.-H. Li, *Geochim. Cosmochim. Acta*, **45**, 1659 (1981).

50 H. Craig, S. Krishnaswami and B.L.K. Somayajulu, *Earth Planet. Sci. Lett.*, **17**, 295 (1973).

51 M.P. Bacon, D.W. Spencer and P.G. Brewer, *Earth Planet. Sci. Lett.*, **32**, 277 (1976).

52 Y. Nozaki and S. Tsunogai, *Earth Planet. Sci. Lett.*, **32**, 313 (1976).

53 Y. Nozaki, Y. Horibe and H. Tsubuta, *Earth Planet. Sci. Lett.*, **54**, 203 (1981).

54 M.P. Bacon and R.F. Anderson, *Earth Planet. Sci. Lett.*, **87**, 1045 (1982).

55 K. Bruland, *Earth Planet. Sci. Lett.*, **37**, 38 (1980).

56 P.G. Brewer and W.M. Hao, In *Chemical Modelling in Aqueous Systems – Speciation*,

Sorption, and Kinetics, (Ed. by E.A. Jenne), *Am. Chem. Soc. Symp. Ser.*, 93, 261 (1979).

57 W.S. Broecker, A. Kaufman and R. Trier, *Earth Planet. Sci. Lett.*, **20**, 35 (1973).

58 A. Kaufman, Y.-H. Li and K.K. Turekian, *Earth Planet. Sci. Lett.*, **54**, 385 (1981).

59 P.G. Brewer, Y. Nozaki, D.W. Spencer and A.P. Fleer, *J. Marine Res.*, **38**, 703 (1980).

60 S. Honjo, *J. Marine Res.*, **38**, 53 (1986).

61 H. Hohl and W. Stumm, *J. Colloid Interface Sci.*, **55**, 281 (1976).

62 J.A. Davis, R.O. James and J.O. Leckie, *J. Colloid Interface Sci.*, **63**, 480 (1978).

63 P.W. Schindler, *Thalassia Jugoslavia*, **11**, 101 (1975).

64 R.A. Neihof and G. Loeb, *J. Marine Res.*, **32**, 5 (1974).

65 G.I. Loeb and R.A. Neihof, *J. Marine Res.*, **35**(2), 283 (1977).

66 K.A. Hunter and P.S. Liss, *Nature*, **282**, 823 (1979).

67 R.M. Garrels and M.E. Thompson, *Amer. J. Sci.*, **260**, 57 (1962).

68 L.S. Balistrieri and J.W. Murray, *Amer. J. Sci.*, **281**, 788 (1981).

69 L.S. Balistrieri and J.W. Murray, *Geochim. Cosmochim. Acta*, **46**, 1041 (1982).

70 L.S. Balistrieri and J.W. Murray, *Geochim. Cosmochim. Acta*, **46**, 1253 (1982).

71 D.E. Yates, S. Levine and T.W. Healy, *J. Chem. Soc. Faraday Trans. 1*, **70**, 1807 (1974).

72 J.A. Davis, In *Contaminants and Sediments*, Vol. 2 (Ed. by R.A. Baker) Ann Arbor Science Publ., 279 (1980).

73 J.A. Davis, *Geochim Cosmochim. Acta*, **46**, 2381 (1982).

74 J.A. Davis and R. Gloor, *Environ. Sci. Technol.*, **15**, 1223 (1981).

75 E. Tipping, *Geochim. Cosmochim. Acta*, **45**, 191 (1981).

76 K.A. Hunter, *Limnol. Oceanogr.*, **25**, 807 (1980).

77 E. Tipping, C. Woof and D. Cooke, *Geochim. Cosmochim. Acta*, **45**, 1411 (1981).

78 L.S. Balistrieri and J.W. Murray, *Geochim. Cosmochim. Acta*, **47**, 1091 (1983).

79 L.S. Balistrieri and J.W. Murray, *Geochim. Cosmochim. Acta*, **48**, 921 (1984).

80 F. Dehairs, R. Chesselet and J. Jedwab, *Earth Planet. Sci. Lett.*, **49**, 528 (1980).

81 J. Jedwab, *Geochim. Cosmochim. Acta*, **43**, 101 (1979).

82 M.B. Tomson and G.H. Nancollas, *Science*, **200**, 1059 (1978).

83 G.H. Nancollas, Z. Amjad and P. Koutsoukos, In *Chemical Modelling in Aqueous Systems*, (Ed. by E.A. Jenne), Am. Chem. Soc. Symp. Ser., #93, 475 (1979).

84 P. Koutsoukos, Z. Amjad, M.B. Tomson and G.H. Nancollas, *J. Am. Chem. Soc.*, **102**, 1553 (1980).

85 R.A. Jahnke, PhD Thesis, *Current Phosphorite Formation and the Solubility of Synthetic Carbonate-Fluorapatite*, University of Washington, Pp. 210 (1981).

86 E.A. Boyle and L.D. Keigwin, *Science*, **218**, 784 (1982).

87 K. Hester and E. Boyle, *Nature*, **298**, 260 (1982).

88 R.B. Lorens, *Geochim. Cosmochim. Acta*, **45**, 553 (1981).

89 S. Emerson, L. Jacobs and B. Tebo, In *Trace Metals in Seawater*, (Eds. Wong, Boyle, Bruland, Burton and Edberg) Plenum Press, 579 (1983).

90 E.A. Boyle, S.S. Husted and S.P. Jones, *J. Geophys. Res.*, **86**, 8048 (1981).

91 P.N. Froelich and M.O. Andreae, *Science*, **213**, 205 (1981).

92 R.E. Cranston and J.W. Murray, *Anal. Chim. Acta*, **78**, 269 (1978).

93 G.T.F. Wong and P.G. Brewer, *Anal. Chim. Acta*, **81**, 81 (1976).

Subject Index